Y0-EGF-717

Design of Industrial Catalysts

CHEMICAL ENGINEERING MONOGRAPHS

Edited by Professor S.W. CHURCHILL, Department of Chemical Engineering, University of Pennsylvania, Philadelphia, Pa. 19104, U.S.A.

Vol. 1 Polymer Engineering (Williams)

Vol. 2 Filtration Post-Treatment Processes (Wakeman)

Vol. 3 Multicomponent Diffusion (Cussler)

Vol. 4 Transport in Porous Catalysts (Jackson)

Vol. 5 Calculation of Properties Using Corresponding-State Methods (Štěrbáček et al.)

Vol. 6 Industrial Separators for Gas Cleaning (Štorch et al.)

Vol. 7 Twin Screw Extrusion (Janssen)

Vol. 8 Fault Detection and Diagnosis in Chemical and Petrochemical Processes (Himmelbau)

Vol. 9 Electrochemical Reactor Design (Pickett)

Vol. 10 Large Chemical Plants. Efficient Energy Utilisation – Plant Design and Analysis – Processes – Feedstocks. Proceedings of the 4th International Symposium held in Antwerp, October 17–19, 1979 (Froment, editor)

Vol. 11 Design of Industrial Catalysts (Trimm)

Design of Industrial Catalysts

DAVID L. TRIMM

The University of New South Wales
Kensington, N.S.W., Australia

ELSEVIER SCIENTIFIC PUBLISHING COMPANY
Amsterdam – Oxford – New York 1980

ELSEVIER SCIENTIFIC PUBLISHING COMPANY
335 Jan van Galenstraat
P.O. Box 211, 1000 AE Amsterdam, The Netherlands

Distributors for the United States and Canada:

ELSEVIER/NORTH-HOLLAND INC.
52, Vanderbilt Avenue
New York, N.Y. 10017

Library of Congress Cataloging in Publication Data

Trimm, David L 1937-
Design of industrial catalysts.

(Chemical engineering monographs ; v. 11)
Includes bibliographical references and index.
1. Catalysts. I. Title.
TP159.C3T74 660.2'995 80-17002
ISBN 0-444-41906-3 (Elsevier/North-Holland)

ISBN 0-444-41906-3 (Vol. 11)
ISBN 0-444-41296-6 (Series)

Printed in The Netherlands

ACKNOWLEDGEMENTS

The author is grateful for the permission granted by publishers and authors of several books and journals to reproduce material in this text. These include:

Academic Press Inc.	Figs. 3.12, 7.5
D.A. Dowden	Figs. 3.13, 3.16, Table 7.3
Elsevier Scientific Publishing Company	Figs. 2.2, 2.3, 2.4
Elsevier Sequoia	Fig. 7.2
Institution of Chemical Engineers, U.K.	Fig. 6.4
Marcel Dekker Inc.	Fig. 3.13.

CONTENTS

To D.A. Dowden - without whom this book would never have been written, and to P.B. - with whom this book would never have been written.

CHAPTER 1

INTRODUCTION

Although catalysis is a subject of tremendous industrial importance, it has long been regarded by the general scientific population as being the last stronghold of alchemy. The reasons for this seem to originate with the complexity of the subject. Most catalysts are inorganic materials that catalyse organic reactions by accelerating the rate of a thermodynamically favourable process. As a result, some knowledge of inorganic, organic and physical chemistry is required. In addition, since many catalysts are porous solids, familiarity with solid state chemistry and with chemical engineering principles of mass and heat transfer are also desirable.

These attitudes are intensified in the context of catalyst design. In dealing even with other catalytic scientists, comments range from the polite statement that it is very interesting - but will never be of any use, through the recommendation as to where to buy a crystal ball, to the extreme statement relating monkeys, typewriters and Shakespeare. Nonetheless, it is clear that the basis of a method of design does exist, and that - although the design methods are far from perfect - they can lead to the development of improved catalysts in shorter times.

The purpose of this book is to describe the methods that form the basis of the scientific design of catalysts. It must be emphasized that this subject is in the course of development, and that this text must be regarded as a signpost rather than as recording a final position. The arrival of new or improved theories or the increase in published information can only result in improving and refining methods of catalyst design.

Catalyst design can be regarded, in many ways, as a logical application of available information to the selection of a catalyst for a given reaction. Over the years it has been possible to formalise a procedure for doing this and this is described below. It will be seen that the degree of complexity put into the design is largely a matter for the individual, and alternative approaches are described. However, design introduces no new concepts nor requires new theories in itself. What it does require is a new way of looking at a problem in the light of established ideas and available experimental data. It requires that the literature should be studied from different angles, and that information as to the mode of action of some of the components of a catalyst should be available. This is not strictly necessary, but the chances of success improve with the volume of published information.

It is regrettable that design methods are not sufficiently accurate to pinpoint exactly a single catalyst for a given reaction. Instead, a design will normally lead to the suggestion of several catalysts which may be suitable. Choice between these is a matter of experimental testing and it will be seen that a successful design involves interaction between the design and the experimental tests.

What has become obvious in the course of writing this book is that the amount of information relevant to a catalyst design is very large indeed. The process of design, as a result, could be regarded as a structuring of this information into a logical arrangement. The same information may be useful in more than one context, but the distinction between various stages of design is valuable. As a result, this book has been split into two. In the first part, the overall design procedure is reviewed, and information pertinent to the execution of each stage of the design is presented. Where it is helpful, individual steps are illustrated by reference to different reactions.

Because the experience of the author lies mainly in the field of heterogeneous catalysis, attention has been focused on the design of solid catalysts. However, it should be pointed out that a similar approach could also pay dividends in the context of homogeneous catalysis. Of course, some of the basic approaches must be adjusted but there is now sufficient evidence on factors such as the dependence of catalytic activity on the strength of the metal-ligand bond to suggest that homogeneous catalysts could also be designed. Indeed, as will be seen below, many of the arguments pertinent to the design of solid catalysts are based firmly on concepts originating in the context of metal ion catalysts in solution.

Experience has shown that catalyst design is best appreciated in carrying out an actual design. The second half of this book presents some examples that have been completed by groups of students from all over the world, but mainly at Imperial College: to these people I record my gratitude. It seemed possible either to report examples of near-perfect designs, or to report actual designs which differ widely in the amount of effort devoted to them. A deliberate choice was made in favour of the latter since, to my continuing surprise, catalyst design can be successful even with a limited effort, or even if based on incorrect assumptions. As a result it seemed useful to record the designs as they were done, and to comment on their successes and failures separately.

It is fitting at this point to acknowledge the inspiration and help given by D.A. Dowden in the preparation of this text. The process of catalyst design as described in this book was initiated by Professor Dowden, and my interest in the subject was stimulated by discussions with him. My sincere thanks are due for all the advice and assistance that has been freely offered over the years.

To echo many of my colleagues, let me, then, pick up my crystal ball, put it by my typewriter and try to reproduce one of the plays of Shakespeare! Whether it should be "Much Ado about Nothing" or a "Midsummer Night's Dream" must be left to the reader to decide.

CHAPTER 2

THE OVERALL DESIGN OF CATALYSTS

I. INTRODUCTION

The design of catalysts has been described as the application of established concepts to the problem of choosing a catalyst for a given reaction. In a sense, it is a reversal of the normal approach to catalysis, in which a series of experimental observations are explained (as best one can) in terms of existing or new theoretical postulates. Instead, what is done is to take established theoretical ideas and to predict experimental behaviour from these. Because our background knowledge is far from complete, it is impossible to predict with complete accuracy and, as a result, predictions are made on the basis of several theoretical concepts. The accuracy of these predictions can only be checked by experimental testing, but the numbers of catalysts to be tested can be substantially reduced by proper application of the design procedure.

As a result, catalyst design is better described as a new application of existing ideas rather than as a new theory of catalysis. It is, however, a complicated process, since predictions are made (and compared) on the basis of several different theoretical approaches. It is the purpose of this chapter to provide an overview of the process of catalyst design, in which various approaches to design are related one to another within the overall framework. Particularly important aspects are dealt with in detail in subsequent chapters of part 1, while several case studies (which illustrate their application) are presented in part 2.

This attempt to present a logical framework of design begins by describing the place of catalyst design in the development of an industrial catalyst. Subsequent to this, various "chemical" and "chemical engineering" aspects are considered, and the information that can be obtained from each approach is evaluated. Where useful, illustrative examples are presented, but the detailed case studies presented in part 2 give a more balanced picture of the overall process.

II. THE OVERALL DEVELOPMENT OF AN INDUSTRIAL CATALYST

The development of a new industrial catalyst can be an expensive and time-consuming operation, depending on the chemical reaction to be catalysed, on the knowledge and equipment already available and on the proposed scale of the process. The three main motivations result from a change in marketing conditions, from a change in government legislation or from the desire to optimise the performance of a plant already in existence. Because of the high and increasing cost of new plant, it is rare to find a new catalyst developed which involves construction of a large new plant. Instead, it is more common to develop a catalyst that can optimise the performance of existing equipment. In considering catalyst design, however, it is possible to ignore economic arguments of this type, on the grounds that understanding the basis of design allows application of the principles to any chosen situation.

The process of bringing a new catalyst on line involves several stages of differing cost and complexity, which are summarised in Fig. 2.1. The exploratory stage, although costing comparatively little, is unpopular among industrial management since it is regarded as too basic or too speculative. By the time the catalyst has reached the more expensive larger scale testing procedures, most of the speculation has been removed. None-the-less, it is rare to find a catalyst which has been taken through the entire flow sheet.

Considering the various steps in this flow sheet, the most difficult step is often the decision as to which catalyst should be designed for which reaction. It is theoretically possible to predict market requirements, on the basis of economic or legislative pressures, but such predictions are notoriously dangerous. In essence, it is necessary either to obtain maximum information for a calculated gamble or it is necessary to attack a problem which exists in the present and the foreseeable future. It is not surprising that the latter approach is favoured! However, throughout this book, attention will be focused on new catalysts for new reactions, on the grounds that understanding of the most complex problem will allow application of the principles to any case of interest.

Once the reaction to be catalysed can be defined, it is necessary to describe the reaction in chemical terms, so as to allow some check of the thermodynamic and economic feasibility of the reaction. A reaction may be thermodynamically feasible, but if economics show that a yield of over 90% is necessary to make a profit then some rapid redefinition of the problem is recommended! Factors such as future feed supply, product demand and possible competition should be examined at this stage.

If this appraisal is satisfactory, catalysts may be designed, prepared and tested. At the moment this is done largely on the basis of experience and of trial and error, and a logical basis of selection of catalysts to be tested is the subject of the text below. If some of the catalysts are found to be satisfactory, then the preparation and testing must be repeated and the catalysts re-evaluated.

The next stage is to test exhaustively the activity, selectivity, life and strength of catalysts under industrial conditions and to ascertain the performance of the best catalysts prepared by available industrial techniques. Although the science of catalyst preparation is the subject of much attention (ref. 1), it still involves a considerable amount of trial and error. Assessment of the

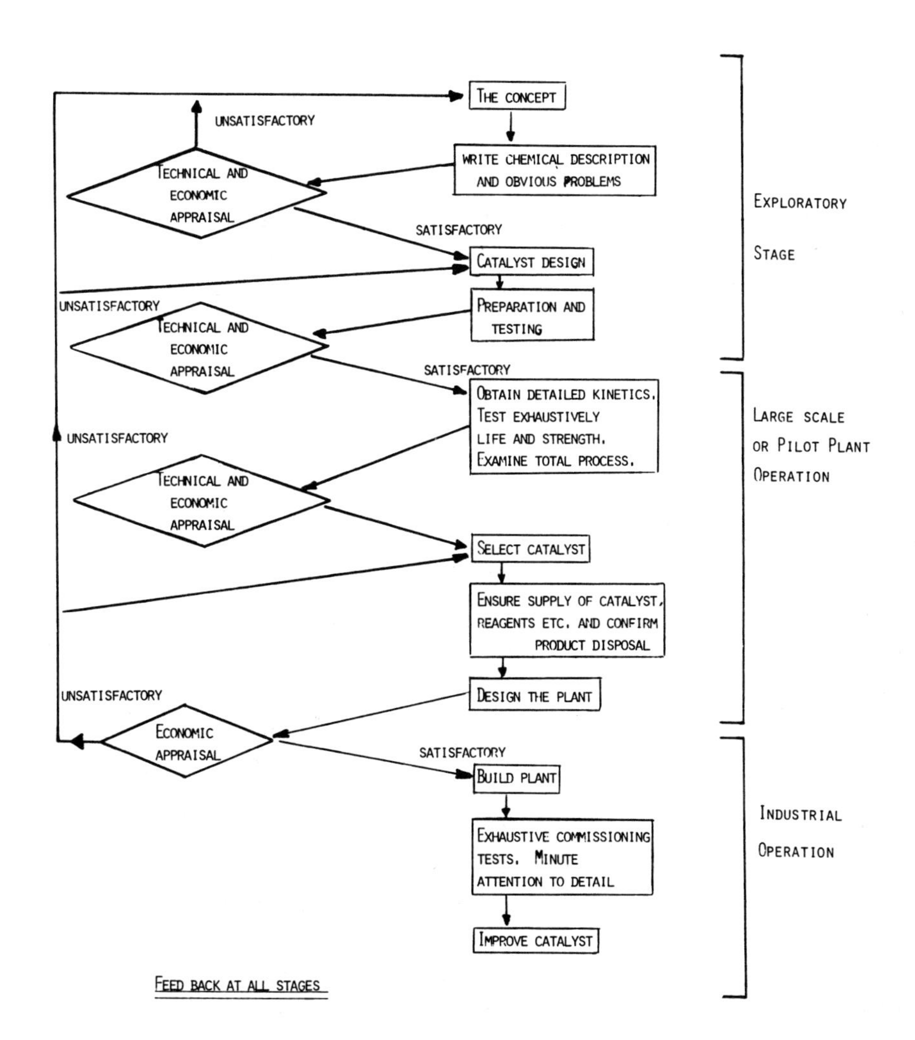

FIGURE 2.1. THE DEVELOPMENT OF AN INDUSTRIAL CATALYTIC REACTION.

overall process involves consideration of the effect of such factors as feed composition, variation, possible poisons, inhibitors or abrasives, chemical changes during pre-heating and so on, alongside such factors as reagent/product transport and storage or effluent treatment (including spent catalyst). A continuing supply of catalyst must be ensured, either from a specialist firm or by building a small plant to manufacture the material.

At this stage, the costs of the operations start to escalate. Large scale laboratory testing may well involve the use of pressure vessels containing about 100 cm^3 of catalyst, at a cost of ca. 5 times as much as exploratory testing. This is, however, much less than the costs of pilot plant testing, to which an acceptable catalyst should next be subjected. A pilot plant is operated by process workers on a continuous basis for long periods, and the costs reflect this. If the pilot plant already exists, the costs will be reduced, but if it is necessary to construct or to modify equipment, then costs will be higher.

Constructing and commissioning of the full scale plant must involve minute attention to detail. Once the plant has been handed over to operators, the plant manager will be most reluctant to allow anyone from the research or development team near the plant, except when things go wrong. The only data available for giving the answers then required are plant records and the commissioning tests. Under these circumstances, construction of the plant must be followed by exhaustive testing of the products and the catalyst, both new, spent and poisoned. In addition, the catalyst should be examined in the laboratory and, as far as possible, in situ in the reactor. It cannot be emphasized enough that minute attention to all teething troubles can save large amounts of money during the life of the plant.

To summarise, then, the development of a new industrial catalyst is an expensive operation, in which the costs escalate as a successful conclusion is approached. Nonetheless, costs can be minimised if the initial stages of the process are carried out carefully and accurately. How this may be done forms the basis of the bulk of this book, and we may begin by considering the framework of the design of catalysts.

III. THE SCIENTIFIC BASIS OF DESIGN

One of the features of research in catalysis in recent years has been an attempt to provide a logical scientific basis for many of the processes that have been developed arbitrarily. Thus, for example, considerable effort is now being extended to this end in the field of catalyst preparation (ref. 1), and this is resulting in improved catalysts. Similarly, some attention has also been directed at developing a logical process for selection of catalysts for particular reactions (refs. 2-6), with particular reference to the initial choice of catalyst. The degree of effort involved in these so-called catalyst design exercises varies widely between authors, but a general sequence of operations is apparent. This is summarised and discussed below, as are the simplifications and complications of the basic procedure (Chapters 3-7).

As stated above, the object of the design exercise is to predict the best catalyst for a given reaction. Development of an original concept can be unrewarding, but it is often desired to improve catalysts for an existing process rather than to develop an entirely new system. Since both approaches depend on essentially the same line of reasoning, the flow sheet below is a convenient summary on which to base the discussion. For clarity, this discussion uses the oxidative dehydroaromatisation of olefins as an illustrative example.

(a) The idea

A catalytic process resulting in the formation of aromatics may be required because of the necessity to raise the octane number of gasoline as a result of the legislative pressure to reduce pollution and remove lead additives. Octane ratings can either be increased by increasing the percentage of highly branched aliphatic hydrocarbons, which have a high octane number and a high volatility, or by increasing alkyl aromatics, which have a high octane number but are less volatile. The production of aromatics seems more feasible, on thermodynamic grounds.

Another reason for production of aromatics is associated with the production of terephthalic acid from p-xylene. Existing routes to p-xylene also result in the production of the o- and m-isomers. Any process which resulted in the selective production of p-xylene would be attractive.

The choice of the starting reagent is illustrative of the gamble in developing a new catalyst. At the start of the illustrative design example, light olefins were freely available and the choice of a process to convert olefins to aromatics made economic sense. Since then, the increased importance of hydrotreating processes has changed the picture, to the extent that a process to convert aromatics to olefins may be almost as attractive as the reverse reaction!

However, at the start of the project, a catalyst for the general reaction

$$R_1C = CR_2 \longrightarrow R_1\text{-}C_6H_4\text{-}R_2$$

was an attractive proposition and the conversion of propylene to benzene seemed a good example to consider.

(b) Preliminary checking

The first essential is to check whether or not the reaction is possible. Simple

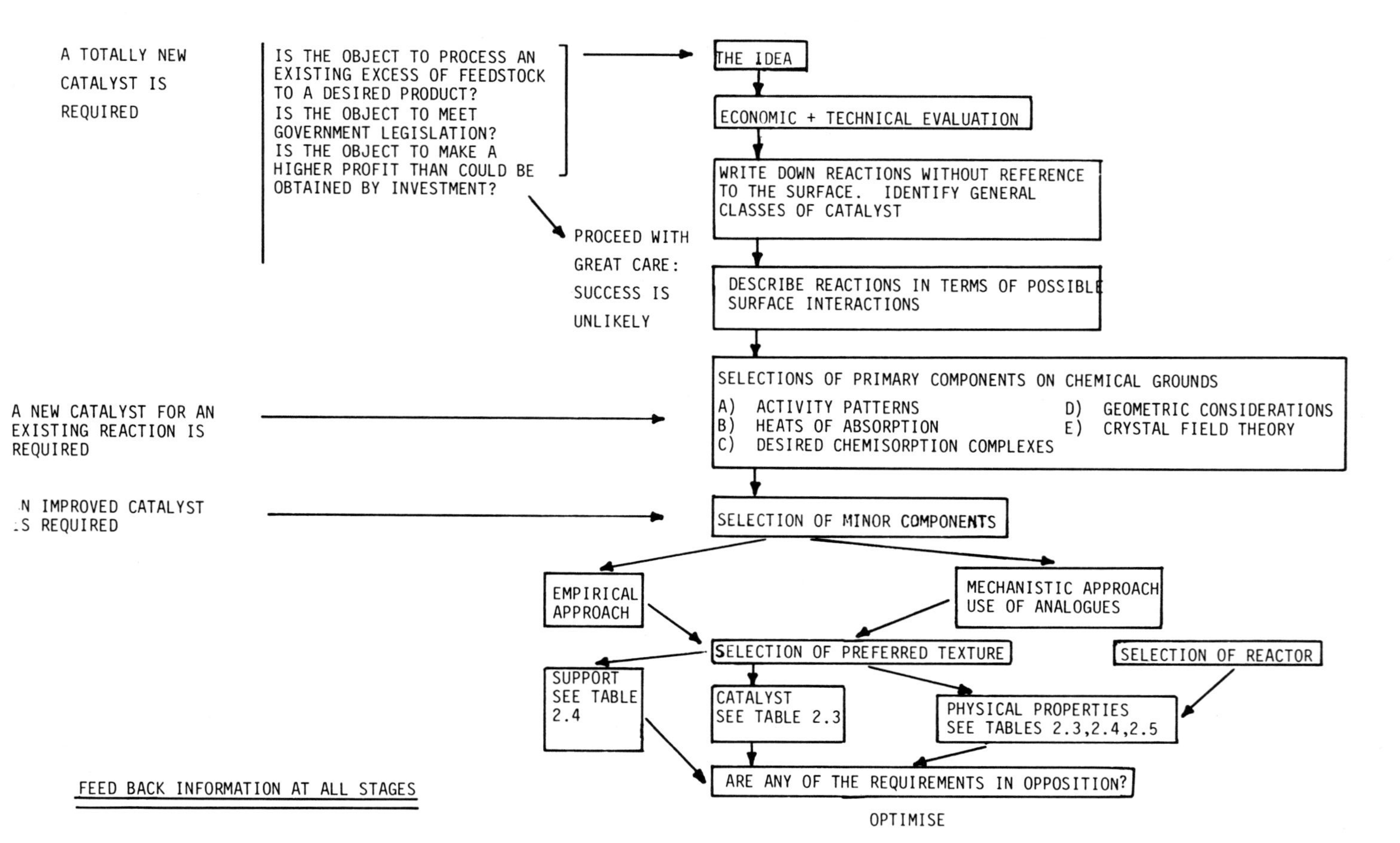

Fig. 2.2. Overall catalyst design procedure.

thermodynamic calculations show that it is.

Secondly, a literature search is carried out with the aim of locating information on the specific reaction, on analogous reactions and on activity patterns for the processes. Experience has shown that one difficulty can arise at this point. Because of our scientific training, we tend to think linearly and, as a result, some novel approaches may not be considered. Thus, for example, the oxidative dimerisation of olefins was (at the time that the propylene to benzene design was completed) a novel reaction route, which would not normally be considered. Instead, the obvious reaction was found to be the polymerisation/cyclisation of olefins, which is known to occur over acidic or basic catalysts (refs. 3,7). As shown below, this reaction does not give the desired products, while the oxidation route does.

Unless the literature survey produces a catalyst that meets the exact requirements of the design, the results of the survey are noted but should not be allowed to influence unduly the next stage of the procedure.

(c) The description of the idea

The next step in the design is to write down stoichiometric equations that describe the overall process, without reference to reactions that could occur on the surface. This may be carried out in two degrees of complexity, and there are advantages to both.

The simple approach is to write down only the reaction in which one is interested and obvious unwanted side reactions. Thus, for example, the conversion of propylene to benzene may be written as

$$2CH_2 = CH - CH_3 \xrightarrow{1} CH_2 = CH - CH_2 - CH_2 - CH = CH_2 + H_2 \qquad (1)$$

$$\downarrow 2$$

$$\text{cyclohexadiene} + H_2 \qquad (2)$$

$$\downarrow 3$$

$$\text{benzene} + H_2 \qquad (3)$$

while obvious unwanted side reactions could include

$$2CH_2 = CH - CH_3 \longrightarrow CH_3.CH_2 - CH_2 - C(CH_3) = CH_2 \qquad (4)$$

$$CH_2 = CH - CH_3 \longrightarrow 3C + 3H_2 \qquad (5)$$

For the desired reaction, dehydrogenation is seen to be common to all steps, and linear dimerisation (step 1) and cyclisation (step 2) also occur. Branched chain dimerisation (step 4) and carbon formation (step 5) are unwanted.

Description of theidea in these terms leads, as we shall see, to a situation in which experimental testing becomes much more important. Feedback to the design from experimental results is essential in all catalyst designs, but it is particularly important if the idea is described in these simple terms.

The alternative approach is strongly favoured by Dowden (ref. 6) and should (but not necessarily does) minimise experimental testing. Essentially, he advocates that the idea should be described in such a way that every possible reaction should be recognised. Adopting a logical approach to the problem, all possible reactions are written down in detail before elimination of unlikely reaction sequences. He then recognises desired and undesired classes of reaction, and seeks catalysts that favour or inhibit these. The basis of the description is given below, and a fairly complete example is presented in the case study of the hydrogenation of benzaldehyde to benzyl alcohol (part 2). The somewhat unusual nomenclature adopted by Dowden is given below.

The approach is best illustrated for the reaction

$$A + B = C + D$$

This desired reaction (target transformation) is written down and other possible reactions are listed. Atoms, radicals and ions are not considered at this stage, and molecularities greater than two are ignored.

This 'characteristic chemistry' then includes the target transformation together with the following

a) Primitive processes

This involves reactions of one reactant molecule, such as isomerisation or cracking

$A \rightarrow A'$ $\qquad B \rightarrow B'$

$A \rightarrow A_1 + A_2$ $\qquad B \rightarrow B_1 + B_2$

b) Self-interactions

These involve reactions between two molecules of the same reactant

$A + A \rightarrow A_3$ $\qquad B + B \rightarrow B_3$

c) Cross-interactions

This term includes reactions between different reactant molecules

$A + B \rightarrow C + D$

$A + B \rightarrow E$

d) Derived primitive processes

Single product molecules may also break down

$C \rightarrow C^{l}$ $D \rightarrow D^{l}$ $E \rightarrow E^{l}$

$C \rightarrow C_1 + C_2$ $D \rightarrow D_1 + D_2$ $E \rightarrow E_1 + E_2$

e) Derived self-interactions

Involving two of the same product molecules, the reactions may be written

$2C \rightarrow F$ $2D \rightarrow G$

f) Derived cross-interactions

These involve reactions between two product molecules

$C + E \rightarrow H$

$D + E \rightarrow I$

At this stage, the number of reactions to be written down begins to increase, and Dowden recommends the use of a matrix description to ensure all reactions are considered.

g) Sequential reactions

These involve reactions between reactants and products obtained from themselves (alone or in combination)

$A + A^1 \rightarrow J$ $B + B_3 = K$

h) Sequential cross-interactions

Reactions can also occur between reactants and products of primitive processes, self interactions or cross interactions

$A + C \rightarrow L$ $B + H \rightarrow M$

i) Interjacent primitive processes

Product molecules can isomerise or crack

$L \rightarrow L^1$ $L \rightarrow L_1 + L_2$

j) Interjacent self-interactions

Pairs of product molecules may also react

$2L \rightarrow N$ $2M \rightarrow O$

k) Interjacent cross-interactions

This is a general term involving reactions between dissimilar molecules produced from any of the above sequences.

It is obvious that, if the original molecules have any degree of complexity, then the description of the idea can become very complex indeed. Thus, for example, the first few steps in the description of the steam reforming process are given in Table 2.1, and a more or less complete description of the hydrogenation of benzaldehyde to benzyl alcohol is presented in a case study in part 2.

There is, however, one saving grace. Although the number of reactions may be large, the general types of reaction are much smaller. It will be seen later that a subsequent stage in the design is based upon the recognition of activity patterns for desired or undesired types of reaction (refs. 4,5,6). As a result, it is often unnecessary to do other than to recognise the class of reaction to which a particular reaction belongs.

The choice between the simple and the complex description of the idea is essentially subjective, but depends on a number of factors. Using the simple description, the design may lead to a catalyst which is unselective to some degree, but this may be corrected using the feedback from experimental testing. The more complex description can well produce a more selective catalyst but experimental testing (and feedback) is still essential. In effect, the simple description saves time earlier in the design, but demands more experimental testing. The complex description costs time early in the design, but may minimise experimentation. In both cases experimental testing is necessary, even though this is usually straightforward. The choice then, seems to rest on the ease with which experimental assessment of the different catalysts can be carried out, although it should be pointed out that Dowden would insist that a full description is necessary in all cases.

In both cases, subsequent treatment of the design is similar and, for clarity, a design framework based on the simple description has been adopted. As stated above, the example chosen involves the reaction sequence

$$2CH_3\text{-}CH = CH_2 \xrightarrow{1} CH_2 = CH - CH_2 - CH_2 - CH = CH_2 + H_2 \quad (1)$$

$$\xrightarrow{2} \text{cyclohexadiene} + H_2 \quad (2)$$

$$\xrightarrow{3} \text{benzene} + H_2 \quad (3)$$

In handling this, the first thing to look for is activity patterns for the general classes of reaction.

For the general dehydrogenation reaction, metals and metal oxides are known to be active catalysts. Alternatively, oxidative dehydrogenation is also possible.

TABLE 2.1
Steam reforming: stoichiometric statement (ref. 4)

Reaction	Type	ΔG, 900 K (kcal mole^{-1})
Target transformation		
$C_4H_{10} + 4H_2O \rightarrow 4CO + 9H_2$		-48.2
$C_4H_{10} + 3H_2O \rightarrow 4CO_2 + 13H_2$		-53.9
Primitive processes		
$C_4H_{10} \rightarrow C_4H_8 + H_2$	dehydrogenation	+ 1.9
$C_4H_8 \rightarrow C_4H_6 + H_2$	dehydrogenation	+ 2.1
$C_4H_{10} \rightarrow C_3H_6 + CH_4$	demethanation	-13.2
$C_4H_8 \rightarrow 2C_2H_4$	cracking	- 3.5
Cross-interactions		
$C_4H_8 + H_2O \rightarrow C_4H_9OH$	hydration	+ 21.5
$CH_3\text{-}CH{=}C{=}CH_2 + H_2O \rightarrow C_2H_5\text{-}CO\text{-}CH_3$	hydration	- 4.3
$CH_3\text{-}C{=}CH + H_2O \rightarrow CH_3\text{-}CO\text{-}CH_3$	hydration	- 4.8
$CH_3\text{-}CH_2\text{-}CH{=}CH_2 + H_2O \rightarrow C_3H_8 - CH_2O$	steam cracking	+ 7.75
$CH_3\text{-}CH_2\text{-}CH{=}CH_2 + H_2O \rightarrow C_2H_5.CHO + CH_4$	steam cracking	- 2.9
$CH_3\text{-}CH{=}C{=}CH_2 + H_2O \rightarrow CH_2O + CH_2{=}CH\text{-}CH_2$	steam cracking	- 3.0
Primitive processes of intermediates		
$C_4H_9OH \rightarrow C_3H_7\text{-}CHO + H_2$	dehydrogenation	- 9.7
$C_3H_7\text{-}CHO \rightarrow C_3H_8 + CO$	decarbonylation	-28.8
$CH_3\text{-}CO\text{-}CH_3 \rightarrow CH_4 + CH_2{=}C{=}O$	cracking	- 8.0
$CH_2{=}C{=}O + H_2O \rightarrow CH_3\text{-}COOH(SQ)$	hydration	- 4.9
$CH_3\text{-}COOH \rightarrow CH_4 + CO_2$	decarboxylation	-29.7
"Equilibration" reactions		
$CH_4 + H_2O{=}CO + 3H_2$	methane reforming	- 0.5
$CO + H_2O{=}CO_2 + H_2$	water gas shift	- 1.4

Supported metal oxides such as chromia/alumina or molybdena/alumina can catalyse both dehydrogenation and polymerisation. Cyclisation can be carried out over acidic and basic catalysts, over metal oxides and over free radical initiating catalysts. As a result, it is possible to assign, from the literature, a class of catalyst for each general reaction. Preliminary testing can be carried out on the basis of this assignment.

Dehydrogenation over metallic catalysts can result in heavy carbon deposits but, based on the activity pattern reported for similar reactions (ref. 4), the order of activity is likely to be

Precious metals $>$ Ni $>$ Co $>$ W $\sim$ Cr $>>$ Fe

Oxidative dehydrogenation is favoured over bismuth molybdate and over tin/antimony oxide catalysts (ref. 8). At the time of this design, there was no published evidence of polymerisation or cyclisation activity, although Seiyama (ref. 9) has since reported the activity patterns

$$ZnO > Bi_2O_3 > In_2O_3 > SnO_2 > CdO$$

and

$$2Bi_2O_3.P_2O_5 > BiAsO_4 > BiPO_4 > Bi_2O_3.2TiO_2 > (BiO)_2.SO_4 > Bi(PO_4)$$

Free radical processes were thought to be too unselective for the linear dimerisation required, and ionic polymerisation catalysts were considered. Anionic catalysts, based on supported alkaline metals (ref. 7) favour dimerisation rather than polymerisation, but cationic polymerisation/cyclisation reactions may be carried out over a Ziegler-type catalyst (ref. 10), over a solid acidic catalyst (ref. 3) or over a dual-function supported metal catalyst (ref. 11). The thermodynamics of dehydrogenation/dimerisation favours high yields at temperatures less than 300°C and at high pressure (ref. 12), branched isomers being the major products. Formation of branched isomers can be expected to decrease the yield of the desired products.

As a result of these arguments, it was possible to select a few catalysts for preliminary testing. It should be emphasized that, unless one is very fortunate, it is unlikely that the exact catalyst will be selected from the simple application of activity patterns. It may be possible, however, to check that a class of catalysts may be suitable, as a result of the fact that at least some of the desired products are obtained over the example selected.

In the case of the conversion of propylene to benzene, this was not possible. Catalysts selected and tested included platinum, chromia, molybdena, thoria and cobalt salts, all supported on alumina. These catalysts gave no significant production of benzene, although branched chain dimers and polymers of propylene were obtained. Oxidative dehydrogenation catalysts such as bismuth molybdate or tin/antimony oxide catalysed the oxidation of propylene to acrolein and to carbon dioxide: no significant production of hexadiene or of benzene was observed under the conditions tested. As a result, it was necessary to proceed to the next stage of the design without any general pointers.

(d) Theoretical design: primary components of the catalyst

Since the various theories of catalysis are unsatisfactory in at least some respects, it is not surprising that their application to catalyst design can be inadequate. It is true to say, however, that their application has been shown, in general, to be very revealing and to lead to the successful identification of novel catalysts.

The principle of the theoretical design is simple. The stoichiometric reactions described above are first considered in the light of possible surface reactions. Using the literature as a general guide, the reactions are translated into possible sequences on the surface, having due regard to the possibilities of adsorption on different types of catalyst. In addition, as discussed in more detail below, valency changes and the electronic and geometric properties of different catalysts must be matched with the reaction sequences proposed. The different approaches are found to suggest possible catalysts, which may be tested experimentally.

It is perhaps most useful to give an example of this process before attempting a general classification. Although the example chosen, the conversion of propylene to benzene, does not use all of the approaches that are possible, it does illustrate and clarify the general method.

Starting at the description of the idea, it is seen that the initial reaction involves linear dimerisation, with loss of hydrogen and retention of the double bond

$$2CH_3\text{-}C{=}CH_2 \longrightarrow CH_2{=}CH\text{-}CH_2\text{-}CH_2\text{-}CH{=}CH_2 + H_2 \qquad (6)$$

Inspection of the literature showed that an analogous reaction existed, involving the oxidation of olefins to unsaturated aldehydes or to dienes. The reaction involves π-allylic intermediates produced by the oxidative removal of hydrogen to the double bond (8)

$$Me\text{-}CH{=}CH_2 \xrightarrow{O_2} [CH_2\text{-}CH\text{-}CH_2]_{ads} \xrightarrow{O_2} CH_2{=}CH.CHO \qquad (7)$$

A reaction sequence for the linear dimerisation can also be postulated in terms of the same intermediate:

$$2Me\text{-}CH{=}CH_2 \xrightarrow{O_2} \underset{*}{\underline{CH_2\text{-}CH\text{-}CH_2}} + \underset{*}{\underline{CH_2\text{-}CH\text{-}CH_2}} + 2OH^- \rightarrow CH_2{=}CH\text{-}CH_2\text{-}CH_2\text{-}CH{=}CH_2 \qquad (8)$$

The essential difference between the two reactions lies in the number and position of the allylic intermediates (dimerisation would obviously be favoured by adjacent adsorption) and by the availability of oxygen at the active site. This should be limited, both to reduce the possibility of producing acrolein and to avoid over-oxidation to carbon oxides. It may well be desirable to design the catalyst such that hexadiene is desorbed and readsorbed at a fresh active site.

The mechanism of formation of the π-allylic intermediate is known to involve redistribution of charge within the adsorbed complex (ref. 13):

$$Me\text{-}CH{=}CH_2 + OMO \rightarrow \begin{array}{c} \ominus \\ \underline{CH_2\text{-}CH\text{-}CH_2} \\ \downarrow \\ O^{2-}\text{-}M^{n+}\text{-}(OH)^- \\ \downarrow \\ \delta+ \\ \underline{CH_2\text{-}CH\text{-}CH_2} \\ O^{2-}\text{-}M^{(n-1)+}\text{-}(OH)^- \end{array} \qquad (9)$$

where π-adsorption of olefins is possible on metal ions with the electronic structure d^0, d^1, d^2, d^3, d^8, d^9 or d^{10}. The production of hexadiene via the π-allylic route should be carried out on such an ion, but since hexadiene itself can π-adsorb, desorption would be favoured if the metal ion, after reaction, did not have any of these electronic structures.

As a working assumption, production of the dimer from two allylic species on the same ion centre was considered, inferring that the metal ion must be capable of accepting two electrons. If the ion has only two valency states separated by two units, the reduction of the ion demands two electrons in quick succession. Abstraction of a second electron from one allylic intermediate is probably more difficult than abstraction of one electron from two allylic species, and the easier path should lead to two allylic species π-adsorbed next to each other in a position which should favour dimerisation.

As a result of these arguments, the periodic table was inspected for metal oxides with valency states separated by two and which could π-adsorb olefins in the oxidised but not the reduced state. Examples include Sn^{2+}/Sn^{4+}, Tl^+/Tl^{3+}, Pb^{4+}/Pb^{2+}, Bi^{5+}/Bi^{3+}, In^+/In^{3+}, Cd^{2+}/Cd^0 and Hg^{2+}/Hg^0. Since the lower valency states tend to become more stable as Groups IIIb, IVb and Vb are descended (ref. 14) and the electron attraction increases with the stability of the lower valency state, oxides of thallium, lead and bismuth would be expected to be good catalysts, neglecting mercury because of the difficulty in handling Hg^0. Bismuth and lead oxides are not very stable thermally, and preliminary attention was focused on the oxides of thallium, indium, cadmium, tin and antimony.

Similar reasoning can be applied to reactions 2 and 3, on rewriting these in terms of desired chemisorption states:

$$CH_2{=}CH{-}CH_2{-}CH_2{-}CH{=}CH_2 \longrightarrow CH\,(M^{n+})\,CH + 2\,O^{2-}$$

$$\downarrow$$

$$C_6\text{–}M^{(n-2)+} + 2(OH)^-$$

$$\downarrow$$

$$C_6 + M^{(n-2)+} \qquad (2)$$

$$\downarrow \text{ OMO} \qquad (3)$$

$$C_6H_6$$

from whence it can be seen that the same type of catalyst would appear to be useful for all stages.

Now, although this example illustrates the approach, it does not illustrate all of the avenues which may be used. These are considered briefly below, and are discussed in more detail in subsequent chapters.

The reason for starting this section with an example is clear when the interactive nature of the different approaches is considered. The general plan of action is shown in Fig. 2.3, and this should be borne in mind when considering individual factors.

The first stage is to write down reactions which are possible on different surfaces: these will, of course, vary as the nature of the general catalytic route proposed. In particular, the forms of adsorption and the mechanisms of reactions will vary between types of catalyst. With these surface reactions in mind, it is

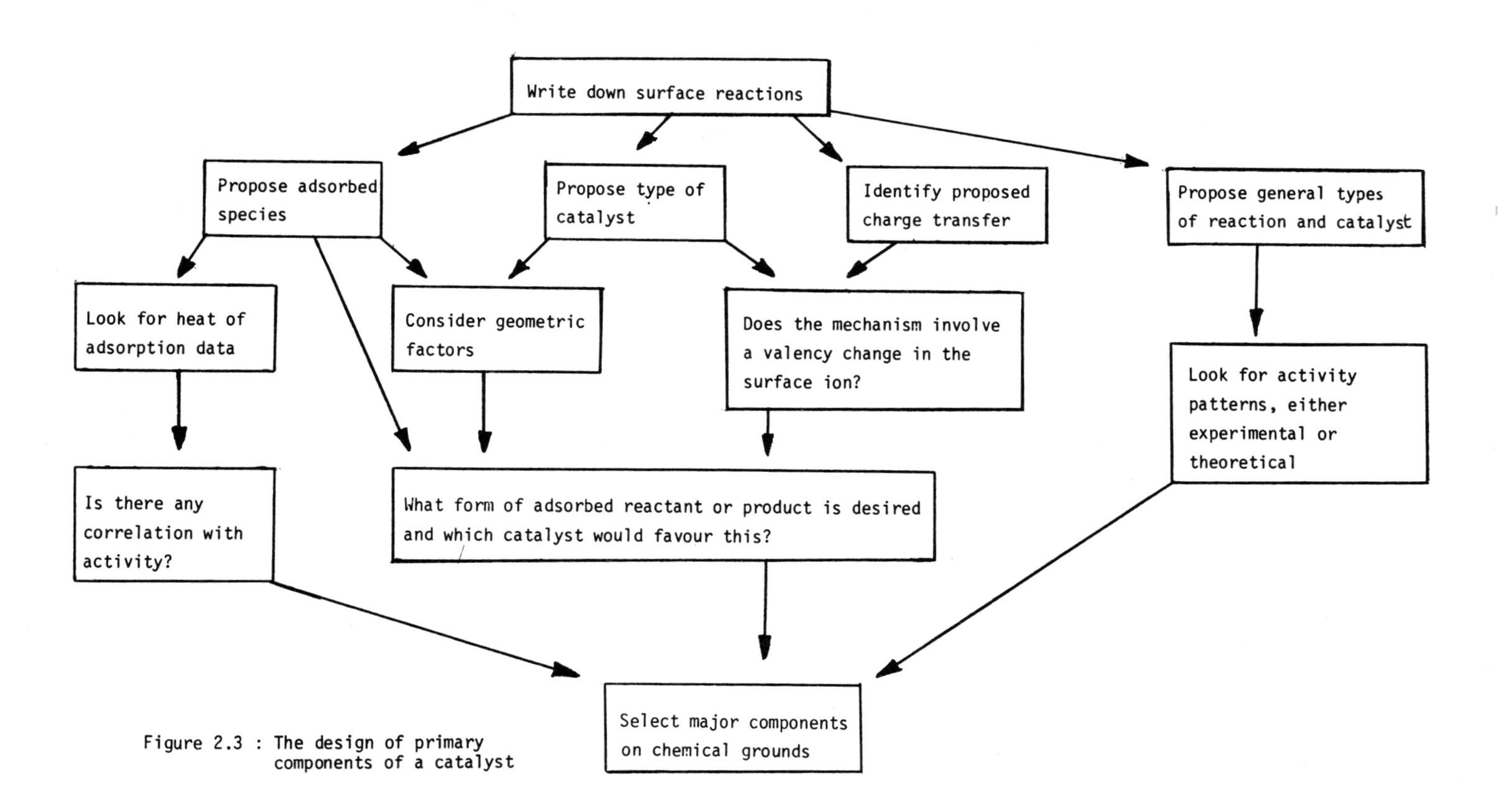

Figure 2.3 : The design of primary components of a catalyst

now possible to consider individual approaches.

(i) Activity patterns

Catalytic activity patterns are, without question the most useful tool available to the catalyst designer. Experience has shown that the relative activity of different solids is often the same for the same class of reactions, for reasons which may or may not be theoretically understood. Activity patterns may be simple or complex, but they represent a rich source of prediction.

In view of their importance, activity patterns are dealt with in more detail in later chapters, and they form a continuing pattern in the text. It should be emphasized that any pattern of activity is useful, no matter the soundness of the theoretical argument behind the pattern. In addition, even where an activity pattern does not exist for a particular class of reaction, it is possible to use a pattern available for an analogous reaction. Thus, for example, an activity pattern for hydrogen-deuterium exchange could provide information of the relative activities of solids for hydrogenation, on the grounds that both reactions may involve H-H fission as the first step.

(ii) Correlations of activity with bulk properties of the catalyst

In the past there have been several attempts to relate catalytic activity with the bulk properties of solids. Although these have fallen into disrepute with the realisation that the surface may have very different properties to the bulk, it is possible to obtain valuable pointers for catalyst design from the approach. Thus, for example, the concept of percentage d bonding (ref. 11) is open to severe criticism, and yet there are some systems in which it is possible to relate catalytic activity with this factor. Similarly, attempts to relate adsorption and catalysts with bulk semiconductor properties of oxides have been reasonably successful in explaining some patterns of catalytic activity (ref. 3), and should certainly not be ignored at this stage of the design. However, more accurate predictions should be obtained by considering the gas-solid interface.

(iii) Predictions on the basis of geometric considerations

One of the more useful methods of predicting catalytic activity arises from consideration of geometric factors. The concept that the geometry of the catalyst can affect activity has been recognised for many years, and formed the basis of the multiplet theory of catalysis (ref. 15). It is particularly useful in prediction, partly because data is readily available and can be easily applied, and partly because the predictions turn out to be reasonably accurate. We shall see, however, that the basis of the prediction may be open to question.

Predictions are made on the basis of matching bond lengths of adsorbed species with crystal parameters of catalyst. Thus, for example, alkylation reactions to produce a compound such as limonene could favour the products

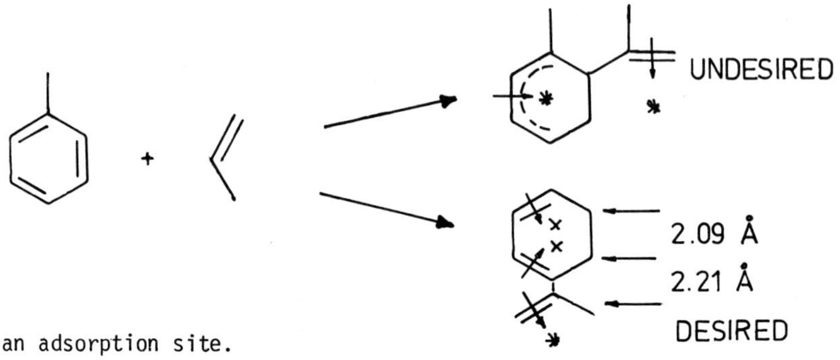

UNDESIRED

2.09 Å
2.21 Å
DESIRED

where x is an adsorption site.

Assuming that the products come from an adsorbed methyl cyclo-olefin and from adsorbed propylene, it is possible to generate a table of distances between adsorption centres which are desired or undesired, depending on the modes of adsorption and the reaction path:

Mode of adsorption		Distance between adsorption centres (Å)	
Propylene	Cyclodiene	Desired*	Undesired*
pi-allyl	pi	1.98	2.51
pi	pi	2.15	2.51
pi-allyl	pi-allyl	2.92	1.46, 3.84
pi	pi-allyl	3.26	1.91, 2.13, 3.84, 4.33

* In terms of a reaction leading to limonene.

Comparison with the lattice parameters of metals and of metal oxides gives a strong indication of which metals or metal oxides can be considered and which should be avoided.

Although the argument is both simple and obvious, it is, in fact, open to question. The problem is very simple. Lattice parameters of metals or metal salts are based on those available for perfect crystals. Almost by definition, catalysis occurs on imperfect crystals which contain many defects. Although the arguments can be extended to consider such imperfections (ref. 16), further problems arise in the fact that the surface of a catalyst often reorganises under reaction conditions (ref. 17) and, as a result, predictions may be based on a surface geometry which is simply not present under operating conditions.

Under these circumstances, it is surprising that geometric arguments offer any assistance to the catalyst designer, but the fact of the matter is that they can often be very useful indeed. As with other approaches, however, geometric arguments must be regarded as indicative rather than definitive. A more detailed description of the approach is given in the following chapters.

(iv) Chemisorption on the catalysts

Even from the reactions on the surface listed above, it is obvious that the direction of reaction must be very dependent upon the nature of the adsorbed complex. Since more than one form of adsorption is always possible, a catalyst should be selected that can be expected to favour the desired adsorption state. In addition, where this is possible, the optimal heat of adsorption should be chosen.

This latter case is the basis of a well known means of selecting catalysts which can be applied, regrettably, in only a few cases. The basis of the approach is that, if adsorption is too strong, then a gas will not be displaced from the surface or will not react. If adsorption is too weak, the residence time of the adsorbed gas on the surface will be too short to favour reaction. Using arguments of this type, it has been possible to identify the most active catalysts for the hydrogenation of nitrogen and for ethylene (ref. 18), and similar arguments can be used for analogous reactions. A plot of activity vs heats of adsorption on different catalysts passes through a maximum where the heat of adsorption is optimised with respect to the above. Prediction of optimal catalysts from this plot does depend, however, on the availability of heat of adsorption data for different gases on a range of catalysts. This is available for only a few systems.

A somewhat similar approach has been used for oxidation catalysis, where various properties that indicate the strength of adsorption of oxygen have been correlated with catalytic activity. Boreskov (ref. 19) has measured the heat of adsorption of oxygen on some oxides and has also measured isotopic oxygen exchange between gas and oxides (refs. 20,21). Moro-oka (ref. 22), on the other hand, relates catalytic activity to the heat of formation of the bulk oxide. It will be seen that all of these measurements may have a satisfactory theoretical explanation, even when the authors were not aware of it. Certainly, the plots offer a guide to the selection of the most active catalysts, although it must be remembered that activity and selectivity in catalysis are often inversely related.

A more quantitative approach to the problem can be obtained by considering the nature of the chemisorbed complex. This may be done in several ways.

From general inorganic chemistry (ref. 23) it is known, for example, that the formation of a pi bond requires the overlap of a filled, bonding pi orbital from an olefin with an empty sigma (dz^2) orbital of a metal if it is to be strong. Back donation from occupied d_{xy}, d_{yz} orbitals of the metal to the empty pi antibonding orbitals of the olefin is also desirable. Considering, for example, square pyramidal coordination on the surface, such adsorbed species can only be formed on d^1, d^2, d^3 metals if only the D orbitals are involved in bonding (V(2), (3),

(4), Ti(1), (2), (3), Cr(3), (4), (5), etc.) and on d^8, d^9, d^{10} metals when both D and P orbitals are involved (Fe(0), Co(0), (1), Ni(0), (1), (2), Cu(1), (2) and Zn(2)).

Similar arguments can be applied to other adsorbed species to produce a table of solids that can adsorb different reactants or products in the desired form.

Thus, for example, in considering the oxidation of olefins, possibilities of adsorption of reactants can be summarised as in Table 2.2 (ref. 5).

This type of approach is very useful in limiting the number of catalysts that should be considered in the design. It does not take too much time since, once available, such a table is widely applicable.

Similar arguments can be applied to one other, more complex, method of assessing the importance of different chemisorbed complexes. This involves the application of molecular orbital calculations to chemisorption and catalysis. The calculations are complex, and are not to be undertaken lightly. However, several papers have appeared in recent years which assess the probability of finding a given adsorption form on a catalyst (refs. 24, 25). One particular useful review of molecular orbital calculations of chemisorbed molecules and intermediates in heterogeneous catalysis has been published by Beran and Zagradnik (ref. 26), covering mainly Russian work up to the second half of 1975. For the normal catalyst design, individual calculations of this type are probably unrewarding: where such information is available, it can be used to good effect.

The dependence of the formation of chemisorbed species on the directional properties of bonds emerging from a surface is an approach which combines the present concept with geometric effects. In an elegant paper covering the possibility of adsorption on different crystal faces, Bond (ref. 16) has considered the fact that e_g and t_{2g} orbitals are spatially directed. As a result, location of particular chemisorbed species in positions favourable to reaction can be envisaged, and predictions based on this were found to be accurate. Regrettably, for the catalyst designer, this approach is of limited value, in that we know little about how we should prepare a catalyst with a desired structure, and even less about how we should stabilise that structure during reaction.

Relatively modern theories advanced to explain the behaviour of inorganic complexes have been found to offer a good description of chemisorption and catalysis: these include the crystal field and the ligand field theories (refs. 23,27). The basis of the theories lies in the fact that d orbitals are known to have directional properties and, if a transition metal ion is associated with ligands, the energies associated with these orbitals can vary. The nature of the ligand and the nature of the complex (high spin or low spin) can obviously affect the energies, but the geometry of the complex (as dictated by the coordination) is very important. Now the chemisorption of a reactant on a metal ion centre can also

TABLE 2.2

No. of d electrons	0	1	2	3	4	5	6	7	8	9	10	s^1	s^2
		Ti(3) V(4) Cr(5) Mo(5) W(5)	V(3) Cr(4) Mo(4) W(4)	V(2) Cr(3)	Cr(2) Mn(3)	Fe(3) Mn(2)	Fe(2) Co(3)	Co(2) Ni(3)	Pd(2) Pt(3) Ni(2)	Cu(2)	Sn(4) Sb(5) Cu(1) Te(6)	Zn(1)	Sn(2) Sb(3)
Adsorption of olefins													
(pi bonded)	+	+	+	+					+	+	+		
(sigma bonded) d_{z^2}					+	+	+	+	+			+	+
Oxygen (radical)				+	+	+	+	+	+				
(pi bonded)		+	+	+					+	+	+	+	+
(donor lone pair) (atoms)	+	+	+	+	+	+	+	+	+	+	+	+	+

be described as the formation of a complex, whether the ion is isolated (say in solution) or is located on the surface of a crystal matrix. Obviously, the geometry will be more constrained by the general form of the matrix, but the principle is the same.

The energy changes that occur on formation of such "complexes" depend on many factors, of which the crystal field stabilisation energy is one. The five degenerate d orbitals of the free ion are split by crystal fields of different symmetry, and the amount of the splitting is measured in terms of an energy parameter, 10 D_q, which is usually obtained from optical data. Chemisorption, the addition of a ligand to a complex, results in a change in geometry of the complex, for example from square pyramid to octahedron or from tetrahedron to square pyramid to octahedron. This, in turn, alters the crystal field stabilisation energy, and calculations show that a characteristic twin peak pattern is obtained. What is interesting is that this pattern is similar to the patterns of chemisorption and catalytic activity for many metals and oxides. Since catalysis involves reaction of the chemisorbed complex (either to a new chemisorbed molecule or to a desorbed product), it is not surprising that the same effect can be seen in both cases. Indeed, if catalysis involves transfer of an electron, rearrangement of the coordination to a different complex (which has the electrons distributed to give overall lower energy) could well be an important driving force for the reaction.

Arguments on this basis provide a useful diagnostic tool for prediction and provide a sound theoretical basis for many of the observed activity patterns. A full description of the approach is given below and, for the purpose of design, the information obtained can be very useful indeed.

There is no doubt that consideration of desired and undesired chemisorbed complexes can be of great importance to a catalyst design. Success depends on how accurately the reaction has been transcribed to the surface (both desired and undesired reactions), and on how feasible the proposed surface reactions are. This can only be determined by advanced surface analysis or, by analogy, from experimental testing of proposed catalysts.

Obviously, as shown in Fig. 2.3, all of these individual approaches are very interactive, and the same catalysts can be suggested by several methods. Application of the methods can be expected, as shown in the illustrative example, to lead to the identification of several potential catalysts, which may be distinguished only on the basis of experimental testing. As will be seen from the second part of this book, most designs suggest about 10-20 potential "primary" catalysts, which may be reduced to 2-3 possible catalysts by testing. The procedures used for testing are also discussed in some detail at a later stage.

Given that this initial testing does identify potential catalysts, one problem will not be apparent as a result of preliminary tests. This is the problem of catalyst deactivation important to the design of both the primary and secondary components of the catalyst, it is convenient to discuss it, in general terms, at this stage.

(e) Catalyst deactivation

Although the definition of a catalyst includes the fact that it shall itself be recovered unchanged after the reaction, this is usually not the case. What is much more common is that a catalyst will lose activity slowly during operation: the object of the designer is to minimise this loss of activity.

Catalyst deactivation may be permanent or temporary. Permanent deactivation requires replacement of the catalyst, and may be caused by factors such as loss of surface area (sintering) or poisoning by compounds hard to remove (such as arsenic). Temporary deactivation may be reversed, but usually only by stopping the reaction. Thus, for example, deposition of carbon on the surface will decrease activity, and this may be restored (usually only partially) by gasification by air/steam mixtures. Permanent deactivation is dealt with primarily in Chapter 5: high temperatures, steam and the presence of unwanted compounds in the feedstock should be avoided to minimise permanent deactivation.

Temporary deactivation can originate from a number of causes. In some cases, effects that would be thought to cause permanent deactivation can be reversed. Thus, for example, the redistribution of precious metal in reforming catalysts can be effected by adding chlorine to the gases used to burn off carbon (ref. 28). Usually, however, temporary deactivation is caused by chemical reasons such as the deposition of carbon or of sulphur on the surface. Minimisation of deactivation and regeneration of the catalyst are the two problems facing the designer.

Regeneration is a specific problem. Thus, for example, sulphides can be rendered non-toxic by oxidation but, upon exposure to hydrogen, reduction to toxic sulphide occurs. Instead it is better to pass hydrogen over the catalysts, thereby generating hydrogen sulphide which may (or may not) elute from the catalyst. Carbon formation, on the other hand, can be reversed by gasification in either oxidising or reducing conditions. Usually, carbon is gasified in a mixture of oxygen and steam, and the major problem lies in the control of temperature. Since the oxidation is exothermic, a large rise in temperature could cause sintering and permanent deactivation.

Minimisation of deactivation is an important object for the designer. Stability can be induced in the catalyst by the correct choice of materials (Chapter 5), but chemical poisoning may be harder to avoid. As a general rule, it is desirable to avoid acidity in a catalyst (unless it is necessary), since this will favour carbon formation. As a result, catalysts for reactions such as steam reforming are alkalised (ref. 29) with the object both of minimising carbon deposition and of catalysing carbon gasification. In general, however, it is difficult to give rules of wide applicability. Instead it is necessary to consider problems that could arise in individual systems. It is particularly desirable, in this context, to consider the possibilities of impure feed stocks, since they are often a source of poisons. The choice of primary (or secondary) components of catalysts is then made in the context of both desired and undesired reactions.

(f) Secondary components of a catalyst

The objective of this part of the design is easily stated, in that the primary catalyst is performing less well in some respect(s) than is desired. The question is how to improve the performance.

Two approaches are possible. The simplistic approach is easy to apply and often produces results. Thus, for example, if the reaction produces chemicals via a reaction path involving an excess of one reagent (say oxygen), addition of a minor component designed to decrease the amount of oxygen adsorbed is easily carried out (ref. 2) and often improves selectivity. Although such an approach is pragmatic, it often works.

The second approach is more intellectually stimulating and almost certainly will work: the disadvantage is that it usually takes considerable time and effort. The basis of the method is to delve deeply into the mechanism of the reaction, on the grounds that understanding the mechanism allows optimal fine tuning of the catalyst. This is generally correct, but there is a closed circle in the sense that a catalyst must be of considerable interest to warrant the necessary attention and yet the catalyst may only be of sufficient interest once it has been improved. As a result, detailed studies of this kind are usually carried out only for catalysts which are in current use, but which could be improved.

There are, in fact, two ways of studying the mechanism in order to fine-tune the catalyst. The most widely used way is to study reactions on the surface, using recently developed analytical techniques. Of these, electron spin resonance and electron paramagnetic resonance spectroscopy (ref. 30), together with infra-red spectroscopy (ref. 32) have proved particularly useful, while electron spectroscopy (ref. 32) holds out much hope for the future. It should also be noted that isotopic labelling experiments, while not fashionable, can be very revealing (ref. 33). The object of the study is to try and identify the active site or the desired intermediate, with the aim of adding components or changing the catalyst in order to optimise the occurrence of the preferred reaction route.

The second method is less direct, but appears to be very interesting. It involves studies of analogues of the catalyst, in which it is possible to control, for example, the location or valency of one of the components of the original catalyst. Several such systems have now been identified, varying from solid solutions to compounds such as scheelites (ref. 34), perovskites (ref. 35), palmierites (36) and tungstates (ref. 37).

Again, the object of the exercise is to identify the role of different additives or of given intermediates in order to optimise the catalyst. Thus, for example, in an elegant study of the oxidation of olefins on bismuth molybdate catalysts and their scheelite analogues (ref. 34), Sleight and Linn found that defects promoted the formation of allyl radicals, while the role of bismuth appeared to be mainly to replenish the active site with oxygen.

There is no doubt that detailed studies are needed to understand the mode of action of additives, and that such studies can also be used to predict which additives could be useful. However, the time needed to be invested is high, and one must be sure that the catalyst will be useful before undertaking the effort. Where this is not certain, it is probably best to use a blend of empericism (based, where possible, on similar catalysts) and as deep a level of thought as is possible, based on the proposed mechanism. This latter process has been illustrated for different catalysts designs (refs. 2,3,4,5,6). Where a large effort is justified, a more detailed description of the approach is given below.

(g) Selection of the preferred form of the catalyst

The morphology of a catalyst is of interest to catalyst design for two reasons, which may be called micro and macro effects. "Micro-effects" is used as a general term covering the desired crystallinity, surface area, porosity, etc. of the catalyst, while "macro-effects" cover such factors as pellet size and strength. With a few notable exceptions, macro-effects have not received the attention in the literature that they deserve, since mechanical breakdown is a very common cause of catalyst replacement. However, micro- and macro-effects, although showing individual characteristics, are closely tied together. This is clearly shown, for example, in the case of alumina, where variation in crystallite size, localised variation in phase change or heating, and homogeneity (in crystallite dimensions) were found to have a major effect on attrition and crush strength (ref. 38).

Andrew has presented a good, but too short, review on macro-effects in catalysts (ref. 29), in which the desired properties are related by the diagram shown in Fig. 2.4. While correctly identifying physical strength characteristics as being very important, the relation with other factors important in catalysis does not emerge from the review. Thus, for example, the strength of catalysts is related to surface area and porosity, but these factors also have a large influence on activity and selectivity, in that they influence mass and heat transfer in the system.

Considering the problem in terms of catalyst preparation, we would expect that a high surface area should give highest activity. However, high surface areas are difficult to prepare, are difficult to maintain (because of the possibility of sintering) and are associated with high porosity. This may introduce mass transfer limitations and will certainly give rise to a weaker catalyst.

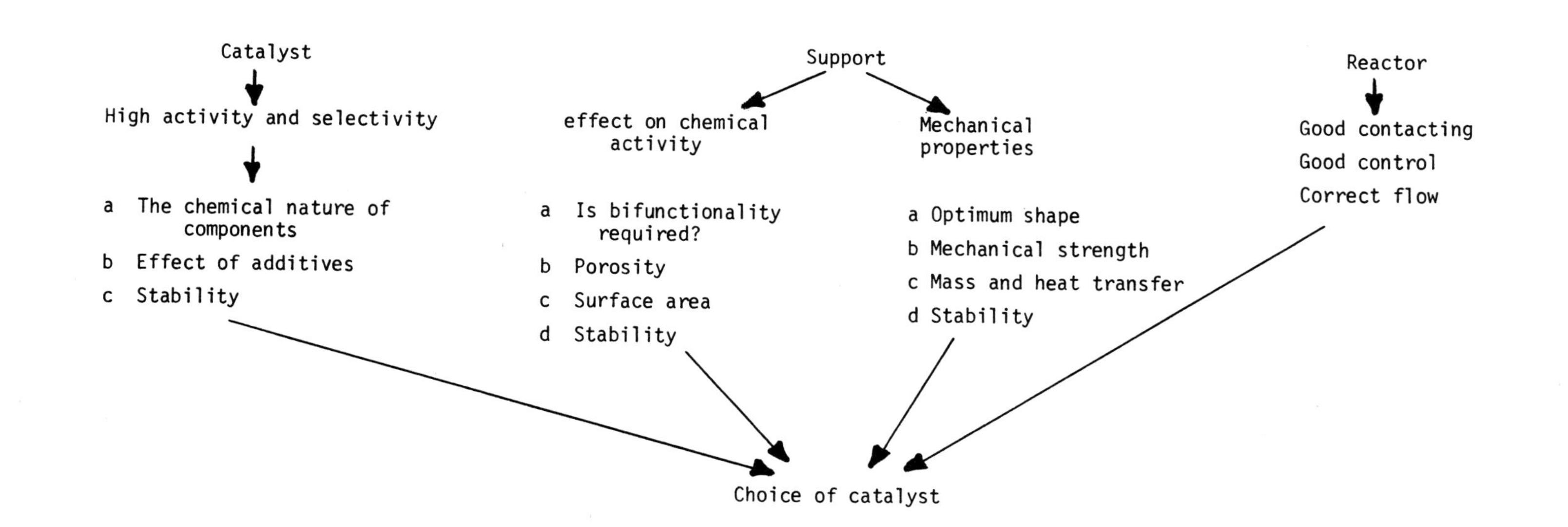

Fig. 2.4. Factors influencing catalyst selection.

TABLE 2.3

The reaction produces	Temperature control* is		Diffusional effects are		Surface area **	Porosity**	Thermal conductivity
	Desirable	Unnecessary	Desirable	Undesirable			
Terminal products such as CO_2, CH_4 etc.	✓		✓		Medium	Medium : maxima in PSD is 50-100 A	High
		✓		✓	High	High, provided temperature is not too high Low if temperature rise is very large	Any value
Two products concurrently, one of which is desired	✓		✓		Medium	Medium : maxima in PSD at 50-100 A	High
	✓			✓	Medium	Low porosity or very wide pores	High
Two products consecutively, the first of which is desired	✓		✓		Medium	Medium : maxima in PSD at 50-100 A	High
	✓			✓	Medium	Low porosity or wide pores	High
A product, but there is a potential poison in the feed or produced	✓		✓		Medium	Medium. Pores must be such either not to allow the poison to enter the pores or to avoid pore blocking by poison accumulation	
				✓	Medium		High
	✓						
A very high temperature rise	✓				Low	Non porous	

*For any reason: it is assumed that the catalyst is stable at the highest temperature liable to be observed

**Consistent with desired strength

The problem can be neatly illustrated by the oxidation of methane to formaldehyde and to carbon dioxide. Formaldehyde is unstable and, at the temperatures needed to oxidise methane, will further oxidise to carbon dioxide very easily. Since both reactions are exothermic, the catalyst temperature will tend to rise and this will favour over-oxidation and catalyst sintering. As a result, it is necessary to remove formaldehyde from the active catalyst rapidly, and to equalise temperature in the catalyst bed at as low a value as possible. This requires a low porosity catalyst with good thermal conductivity. For similar reasons, platinum-rhodium gauzes are widely used for the oxidation of ammonia (ref. 18) and novel geometries have been developed for methanation catalysts, including the idea of plating them on the surface of a heat exchanger (ref. 39). A general guide to selection of the properties of the catalyst is given in Table 2.3. It should be emphasised that this table is general, since particular systems may show individual effects. Thus, for example, mass transfer limitations may be desirable, in that they improve selectivity (ref. 11) or even, in some cases, increase rate (ref. 40).

The variation in porosity and surface area that can be achieved using a pure catalyst is limited by the preparation methods available and by the fact that such materials tend to sinter rapidly. As a result, it is usual to introduce the desired characteristics primarily through the use of a suitable support. Here, too, there are several other factors which can influence the choice. These may be discussed with the aid of Table 2.4.

TABLE 2.4
Choice of support

Chemical Factors
Is the support required to show catalytic activity?
Are chemical interactions with the catalyst possible? If so, are these desired or undesired?
Can the support interact with reactants or products? Is this desired or undesired?
How resistant is the support to poisoning?
Can the catalyst be deposited on the support in the desired form?
Does the support induce a particular coordination geometry on the catalyst (refs. 5, 53, 54)
Is the support stable under operating conditions?
Physical Factors
What is the desired surface area and porosity?
What is the desired thermal conductivity?
Is the support mechanically strong (ref. 45)? Would this be affected by deposition of poisons, such as carbon?
Is the support stable under the operating conditions?
What is the desired form of the pellet?

Recent studies have revealed that chemical interactions between the catalyst and the support may be more important than was previously thought. Of course, it has been long established that bifunctional catalysis is of major industrial importance (ref. 11), and that chemical interaction between the support and the catalyst may be desirable (ref. 41) or undesirable (ref. 42). What has emerged comparatively recently is that the support may be able to induce a given geometry on a catalyst without formal chemical interaction, and that this will influence adsorption and catalysis. Such effects have been reported for supported silver (ref. 43) and for supported metal oxides (refs. 44,45), and may be widely spread. No general guidelines for catalyst design can, as yet, be listed.

However, it is probably the physical properties of the support which are primarily responsible for its selection, provided that these are consistent with desired chemical properties. Both mechanical strength and porosity/surface area are important, as well as stability with respect to temperature, etc. In addition, we have to consider the environment in which the catalyst will be used, both with respect to stability and with respect to the reactor. A general guide to the factors that influence the choice of the latter is given in Table 2.5.

Considerably more is known of micro-effects in catalysis, although these may be of more academic interest than applied. This results from the fact that although we begin to know how to prepare a catalyst with a given structure, we know little about how to retain this structure under reaction conditions. This may be illustrated by considering supported metal catalysts.

It is generally desirable to optimise the use of the metal by preparing it in as high a dispersion as possible, and, in practice, this usually means depositing the metal in small crystallites on a support. Now, theoretically, this can have other advantages, in that small crystallites are most likely to show geometrical effects (ref. 46) in their catalytic action. The concept of structure-sensitive and structure-insensitive reactions is well established (refs. 47,48), and some assessment can be made of the nature of a given reaction, using the criteria outlined in Table 2.6. Although it should be emphasised that this table has no theoretical justification, experimental observations indicate that the greater the probability that a reaction is demanding or is facile. Similarly, on a more theoretically sound base (ref. 46), it can be shown that the number of edge and corner sites on a catalyst will be much higher with smaller crystallites.

In practice, there are two difficulties with this approach. With the exception of catalytic reforming, in which the overall reaction involves both facile and demanding reactions (49), most reactions of industrial interest are, in fact, facile. Secondly, given that a reaction is demanding, it can be very difficult to achieve or to maintain a given geometry in a catalyst prepared on anything but the smallest scale, since reorganisation of the catalyst - either within the particle (ref. 50), or between particles (refs. 51,52) - is often rapid under reaction conditions - or even during preparation and activation.

TABLE 2.5

REACTOR	ADVANTAGES	DISADVANTAGES	FORM OF CATALYST PELLETS
GAS-SOLID REACTOR (TUBULAR)	WIDELY USED AND WELL UNDERSTOOD	TEMPERATURE CONTROL MAY BE DIFFICULT	PELLETS. EFFECTIVENESS FACTOR ~ 0.6; GOOD THERMAL STABILITY
FLUID BED	GOOD TEMPERATURE CONTROL. USEFUL WHERE CATALYST NEEDS FREQUENT REGENERATION	BAD MIXING. CATALYST ATTRITION. DIFFICULT TO OPERATE.	FLUIDISABLE PARTICLES (E.G. 40-70 MICRONS): GOOD ATTRITION STABILITY
TRICKLE BED	ALLOWS GOOD GAS-LIQUID-SOLID CONTACTING GOOD TEMPERATURE CONTROL.	DIFFICULT TO OPERATE. FOAMING. SPLASHING.	SMALL PARTICLES: OPEN POROSITY: HIGH SURFACE AREA.
HOMOGENEOUS CATALYST REACTOR	MAY GIVE GOOD SELECTIVITY AT LOW TEMPERATURES	MAY BE DIFFICULT TO SEPARATE PRODUCTS AND CATALYST, ALTHOUGH "ANCHORED" CATALYSTS ARE NOW FASHIONABLE (55, 56).	
SLURRY REACTOR	GOOD TEMPERATURE CONTROL	MAY INVOLVE DIFFICULTIES IN GAS-LIQUID-SOLID CONTACTING	

TABLE 2.6

Structure sensitive and insensitive reactions

Structure insensitive (facile) reactions may be
(a) Addition reactions or elimination reactions
(b) Reactions involving a large decrease in free energy
(c) Reactions involving reactants with lone pair electrons or pi bonds or strain energy
(d) Reactions that do not require a multifunctional catalyst
(e) Reactions occurring on an active catalyst whose lattice parameters do not change with dispersion
(f) According to Ponec (refs. 54,55), for a series of C_4 based hydrocarbons, the following reactions are structure insensitive
(i) Hydrogenation-dehydrogenation of multiple C-C bonds
(ii) Deuterium exchange
(iii) Hydrogenation of ketones
Structure sensitive (demanding) reactions
(a) May occur on certain sites, e.g. N_2 chemisorbs only on W(111) and NH_2 is produced only on W(111)
(b) Involve single C-C bond breakage
(c) May involve reactants with no lone pair electrons, pi bonds or strain energy
(d) Need a multifunctional catalyst
(e) Need a less active catalyst whose lattice parameters change with dispersion
(f) May involve reactants with unpaired electrons (e.g. NO)
(g) According to Ponec (refs. 54,55), for a series of C_4 hydrocarbon reactions, the following are structure sensitive
(i) Hydrogenolysis
(ii) The conversion of alcohols to ethers
(iii) $R-NH_2$ disproportionation
(iv) C-C bond breakage
and, in addition, the methanation reaction

N.B. It is unlikely that a reaction can be assigned on the basis of meeting only one criterion. The more criterion that can be met, the more likely the reaction is to fall into a classification.

Gross rearrangements (sintering) can be prevented to some extent by the use of spacers. Conventionally, these involve a high melting point non-deleterious metal salt which is co-precipitated on a support together with the catalyst, thereby acting as a physical barrier to agglomeration. However, there are other forms of spacer, although they are not considered in these terms. Thus, for example, a solid solution not only dictates the geometry of the solute ion (ref. 53) but also acts to separate solute ions. Similarly, an alloy distributes one metal in another, and the resulting effect can be due either to dilution or to chemical or electronic interaction (ref. 54). In both cases, one component could be regarded as a spacer for the other.

Again, a more detailed discussion of the selection of supports during catalyst design is given in later chapters.

(h) The overall design

A general programme for the design of a catalyst has been discussed above. Although brief, it does attempt to review important areas in the context of catalyst design. However, it is worthwhile emphasizing some points at this stage.

Regrettably, we have insufficient knowledge to ensure that the catalyst design is absolutely correct, and experimental testing must be carried out. However, it should be stated that, on the thirty-odd occasions that the author has carried out a catalyst design, the procedure has shown up a catalyst that has been found - either previously or subsequently - to be active for the reaction under consideration. It should be hastily added, however, that several inactive catalysts have also been suggested by the design. As a reasonable assessment, the design procedure offers a guide to experimentation that can often be successful and that requires the investment of only a little time. As the knowledge and experience of the designer improves, so the accuracy of the design can also be expected to improve.

Secondly, it is necessary to emphasize the feedback cycle in the design. Experimental testing is necessary at various points, and the results of these experiments can be used to modify the conceptional basis of the design. Thus, for example, if experiments show that reaction path A leads to a more desirable product spectrum than reaction path B, the design can be adjusted to put more weight on reactions of type A.

Thirdly, it must be remembered that there are always factors that have not been considered. Catalysis is a complex subject, involving interrelated phenomena from a wide variety of fields. As a result, a set of experiments, carried out for a given reason, may give the "wrong" results because of a second factor that has not been considered. Perhaps the most obvious case of this was the work carried out on the design of a catalyst to convert propylene to benzene (ref. 2): at the end of the design it was discovered that changing economics would make the reverse action more attractive! This, again, emphasizes the importance of feedback at all stages.

It is also essential to point out that, although the procedure above gives a summarised description of catalyst design, there is considerably more information that can be used than is presented above. At least some of this information is considered in the subsequent chapters.

REFERENCES

1 Proc. 1st Int. Symp. on the Scientific Bases for the Preparation of Heterogeneous Catalysts, Brussels, Elsevier, Amsterdam, 1975.
2 D.L. Trimm, Chemistry and Industry, p.1012, 1973.
3 O.V. Krylov, Catalysis by Non-metals: Rules for Catalyst Selection, Academic, New York, 1970.
4 D.A. Dowden, Proc. IV Int. Congr. on Catalysis, Moscow, p.201, 1968.
5 D.A. Dowden, Chem. Eng. Progr. Symp. 63, No. 73, 90 (1967).
6 D.A. Dowden, La Chimica e l'industria., 55, 639 (1973).
7 A.W. Shaw, C.W. Bittner, W.V. Bush and G. Hobzurah, J. Org. Chem. 30, 3286 (1965).
8 J.C. Germain, Intra-Science Reports, 6, 101 (1972).
9 T. Seiyama, M. Egashira, M. Iwamoto, Some Theoretical Problems in Catalyst Research, Report Soviet-Japanese Seminar on Catalysis, 1, 35 (1971).
10 K. Ziegler, Angew. Chem., 72, 829 (1952).
11 J.M. Thomas and W.J. Thomas, Introduction to the Principles of Heterogeneous Catalysis, Academic, London, 1967.
12 J.E. Kilpatrick, G.J. Prosen, K.S. Pitzer and F.D. Rossini, J. Res. Nat. Bur. Stand., 36, 559 (1946).
13 J.M. Peacock, A.J. Parker, P.G. Ashmore and J.A. Hockey, J. Catal., 15, 373, 379, 387, 398, (1969).
14 P.A. Batist, A.H.W.M. der Kinderis, Y. Leeuwenburgh, P.A.M.G. Metz and G.C.A. Schuit, J. Catal., 12, 45 (1968).
15 A.A. Balandin, Russ. Chem. Revs., 31, 589 (1962).
16 G.C. Bond, Discuss. Faraday Soc., 41, 200 (1966).
17 G.A. Somorjai, J. Catal., 27, 453 (1972).
18 G.C. Bond, Catalysis by Metals, Academic, London, 1962.
19 G.K. Boreskov, Adv. in Catal., 15 (1964).
20 G.K. Boreskov and V.V. Popovskii, Kinet. Catal., 2, 593 (1961).
21 G.K. Boreskov, A.P. Dzisyak and L.A. Kasatkina, Kinet. Catal., 4, 335 (1963).
22 Y. Moro-oka, Y. Morikawa and A. Ozaki, J. Catal., 7, 23 (1967).
23 J.A. Duffy, General Inorganic Chemistry, Longman, London, 1966.
24 V.D. Sutula and I.I. Zakharov, Kinet Catal., 12, 980 (1971).
25 G.F. Kventsel' and G.I. Golodets, Kinet. Catal., 13, 193 (1972).
26 S. Beran and R. Zagradnik, Kinet. Catal., 18, 299 (1977).
27 F. Basolo and R.G. Pearson, Mechanisms of Inorganic Reactions, John Wiley, London, 1960.
28 M. Primet, M. Dufaux and M.V. Mathieu, C.R. Hebd. Seances Acad. Sci. Ser. C., 280, 419 (1975).
29 Catalyst Handbook, Wolfe Scientific Texts, London, 1970.
30 J.H. Lunsford, Adv. in Catal., 22, 265 (1972).
31 R.P. Gischens and W.A. Pliskin, Adv. in Catal., 102 (1955).
32 L. Lee, Characterisation of Metal and Polymer Surfaces, Vols. 1 and 2, Academic, London 1977.
33 M. Ozaki, Isotopic Studies of Heterogeneous Catalysis, Academic, London, 1977.
34 A.W. Sleight and W.J. Linn, Annals. N.Y. Acad. Sci., 272, 22 (1976).
35 R.J.H. Voorhoeve, J.P. Remeikan and L.E. Trimble, Annals. N.Y. Acad. Sci., 272, 3 (1976).
36 J.M. Longo and L.R. Clavenna, Annals. N.Y. Acad. Sci., 272, 45 (1976).
37 S. De Rossi, E. Iguchi, M. Schiavello and R.J.D. Tilley, Z. Phys. Chem., 103, (1976).
38 R. Gauguin, M. Graulier and D. Papce, Catalysts for the Control of Automotive Pollutants, Adv. in Chem., 143, 147 (1975).
39 R.R. Schechl, H.W. Pennline, J.P. Strakey and W.P. Haynes, Amer. Chem. Soc. Div. Fuel Preprints, 21, 2 (1976).
40 J. Wei, Adv. in Chem., 148, 1 (1975).
41 P.B. Weisz, C.D. Prater and K.D. Rittenhouse, J. Chem. Phys., 21, 2236 (1953).
42 J.R.H. Ross, Surface and Defect Properties of Solids, 4, 34 (1975).
43 K.K. Kakati and H. Wilman, J. Phys. D., 6, 1307 (1973).
44 V.A. Shvets and V.B. Kazansky, Kinet. Catal., 25, 123 (1972).
45 M. Goldwasser and D.L. Trimm, Ind. Eng. Chem., Prod. Res. and Dev. (in press).

46 R. Van Hardeveld and F. Hartog, Surface Sci., 15, 189 (1969).
47 C. Bernardo and D.L. Trimm, Carbon, 14, 225 (1976).
48 M. Boudart, Proc. VI. Int. Congr. on Catal., London, 1 (1976).
49 M.J. Sterba and V. Haensel, Ind. Eng. Chem. Prod. Res. Dev., 15, 3 (1976).
50 B.J. Cooper, B. Harrison and G. Shutt, SAG paper 770367.
51 A.E.B. Presland, G.L. Price and D.L. Trimm, J. Catal., 26, 313 (1972).
52 R.M.S. Fiederow and S.E. Wanke, J. Catal., 43, 34 (1976).
53 F.S. Stone, Adv. in Catal., 13, 1 (1962).
54 V. Ponec, Cat. Rev. Sci. Eng., 11, 41 (1975).
55 A. Van der Burg, J. Doornbos, N.J. Kos, W.J. Ultee and V. Ponec, J. Catal., 54, 243 (1978).

CHAPTER 3

DESIGN OF THE PRIMARY CONSTITUENTS OF THE CATALYST

The process of catalyst design described in Chapter 2 is seen to be a sequence of actions, decisions and testing. Perhaps the most important part of the sequence, however, involves the selection of the primary components since, if no active component can be found, the design must fail.

As described briefly above, selection of the primary component involves several different approaches, all or none of which may be useful. These are summarised in Fig. 3.1. It is, in fact, possible to apply these without regard to their basis (as generations of students can testify!), but it is more useful to have some appreciation of the grounds of the approaches. Some attempt to give the necessary background is presented in this chapter, but a more complete account is given in many other texts (refs. 1,2,3,4).

I. THEORIES OF CHEMICAL BONDING

It is well known that catalysis usually involves the adsorption of reactants, the reaction of adsorbed reactants (or an adsorbed reactant with another in the gas phase), followed by the desorption of products. As such it is obvious that chemical bonds are being formed and broken in the process, and that one good place to start to look at theoretical concepts is with the nature of these bonds.

We shall see that much of catalysis can be related to coordination chemistry, involving the behaviour of a central atom or ion (usually a metal) surrounded by a cluster of molecules or ions (known as ligands). The number of attachments to the central atom (the coordination number) and the geometry of the complex may vary, but it is a feature of a single complex that it tends to retain identity, even when dissolved in solution. The bonding in the complex may be purely ionic or purely covalent or, more probably, may involve some qualitites of each.

It is useful to review briefly the three main theories that are used to describe the properties of these bonds: these are the valence bond theory, the molecular orbital theory and the electrostatic theory (which includes corrections due to the crystal field). In order to do this, it is necessary to have some knowledge of the atomic orbitals on the central atom that would exist in the gas phase in the absence of ligands. These are represented in Fig. 3.2 for s, p and d orbitals as contours which indicate roughly the region in space which will contain most of the electronic charge of an electron orbital. Each atomic orbital can hold one or

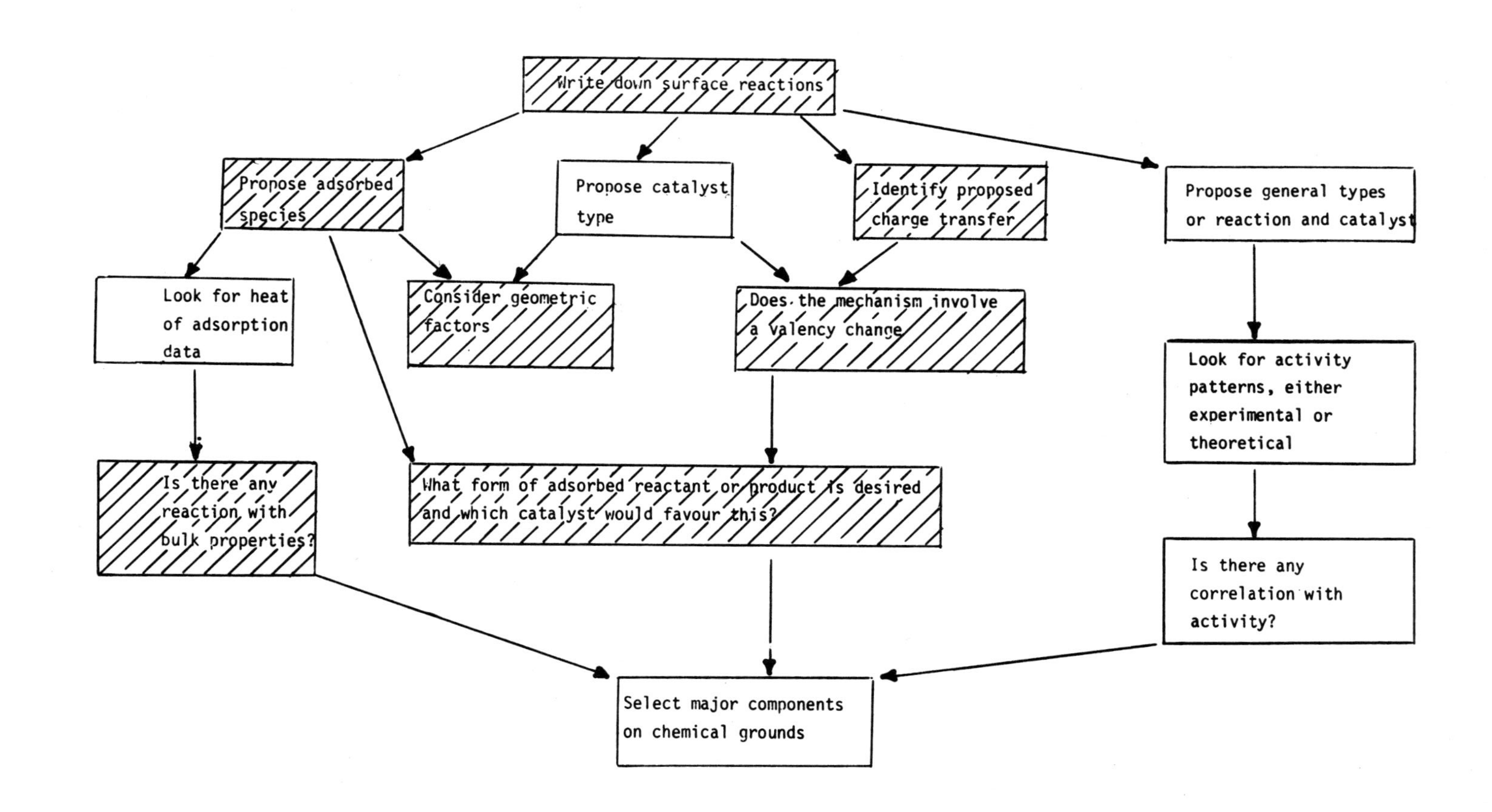

Figure 3.1. Summarised design of primary components.

two electrons and, in the free atom, two electrons avoid being in the same orbit if this is possible.

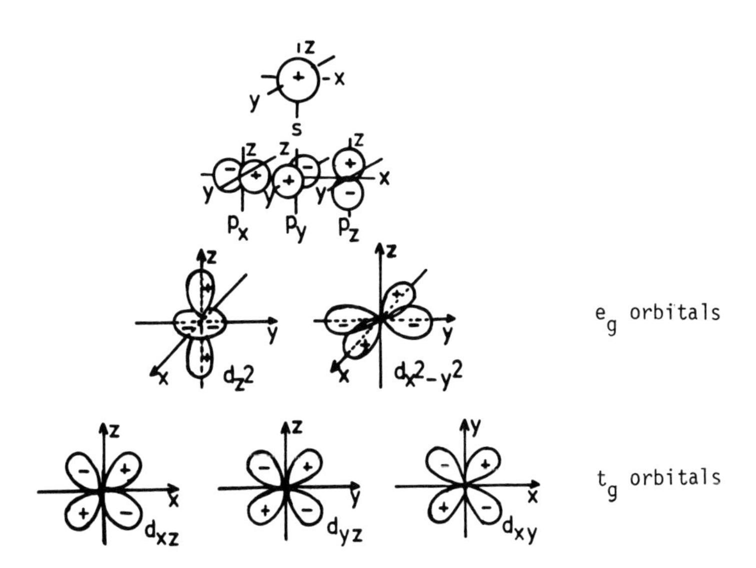

Fig. 3.2. Boundary surfaces for s, p and d orbitals, showing the sign of the wave function.

In the valence bond theory, developed by Pauling (refs. 4,5), bonding between atoms is envisaged as arising from the overlap of orbitals originating from each of the species involved in the bond: generally speaking, the greater the overlap, the stronger the bond. The theory suggests that a number of orbitals from the central atom equal to the number of ligands should be made available to form co-ordinate covalent bonds. This could be done, for example, in the case of carbon ($1s^2\ 2s^2\ 2p^2$) by uncoupling the pair of s electrons and promoting one to the p orbitals ($1s^2\ 2s^1\ 2p^3$): this would result in four orbitals, only three of which would be equivalent. Since it is a fact that all four bonds in CX_4 are equivalent, it is necessary to introduce the concept of hybridisation, in which the four electrons should be mixed (or hybridised) to produce four equivalent orbitals. Of course, other types of hybridisation can occur, involving s, p and d electrons. In view of the directional properties of the d electron orbitals (Fig. 3.2) hybridisation usually involves the $d_{x^2-y^2}$ and the d_{z^2} orbitals. However, for transition metal compounds, hybridisation can occur between s and p orbitals of one quantum level with d electrons of the same quantum level or of the previous quantum level.

Thus, for example, the 4s and 4d orbitals can be hybridised with either the 4d or the 3d orbitals.

The combination of different orbitals results in hybrid orbitals which have different directional properties, the most important of which are shown in Table 3.1.

TABLE 3.1
The direction and strength of hybrid orbitals

Coordination number	Orbital	Strength	Shape
1	s	1.00	spherical
1	p	1.73	dumb-bell
2	sp	1.93	linear
3	sp^2	1.99	trigonal
4	sp^3	2.00	tetrahedral
4	dsp^2	2.69	square planar
5	spd^3, dsp^3		trigonal bi-pyramid
5	spd^3, d^2sp^2		square pyramid
6	d^2sp^3	2.92	octahedral

Since the directional properties dictate the shape of the resulting complex, the geometric form is given in the right hand column. The bond strengths are calculated on the basis of the angular orbital overlaps, each hybrid orbital combining with a suitable ligand orbital to form a valence bond.

Such bonds are called σ bonds because the electron density of the bond is symmetrical about the bond axis. In addition to the hybrid orbitals there are, however, d_{xy} etc. atomic orbitals which, because of their spatial direction, are not usually used in hybrid bonding. They are, in fact, suitably placed to interact with either p or d orbitals on the ligand. If such interaction is possible (and cases will be discussed below) the resulting bond (π) will have a nodal plane (or a minimum of electron density) along the bond axis. Of course, both a σ and a π bond may be formed in the same molecule, as is the case with ethylene. π bonding is almost always synonymous with double bonding.

The valence bond theory has been decried in recent years, but it can explain the main features of complexes, such as structure, coordination number, magnetic properties etc. However, this is also true for the other theories of bonding, as can be seen from brief discussion of the molecular orbital theory (ref. 6).

If we consider the combination of two atoms into a molecule, the behaviour of an electron can be expected to have some of the properties of its behaviour in the isolated atoms. As a result, a molecular orbital could be considered as the linear combination of atomic orbitals (LCAO) divided by a factor. Thus the wave function of the molecular orbital could be expressed as $(\psi_A (1) + \psi_B (1))/ 2$ where $\psi_A (1)$ is the wave function of electron 1 on nucleus A, and 2 is a normalisation factor. If the electrons from the atoms combine to form a bond of high electron density, then the molecular orbital may be described as bonding. However, it is also possible that the atomic orbitals can be combined in such a way that there is minimal electron density between the atoms. As a result, the molecule will be destabilised, and an anti-bonding orbital will be produced. In general a bonding orbital is known as, for example, a σ 2s bond since it originates from 2s atomic orbitals and is responsible for a σ bond: the anti-bonding orbital is known as a σ^*2s orbital.

Possible combinations can be represented both in terms of orbital envelopes and in terms of energy as in Fig. 3.3 and, for more complex molecules such as nitrogen, as in Fig. 3.4.

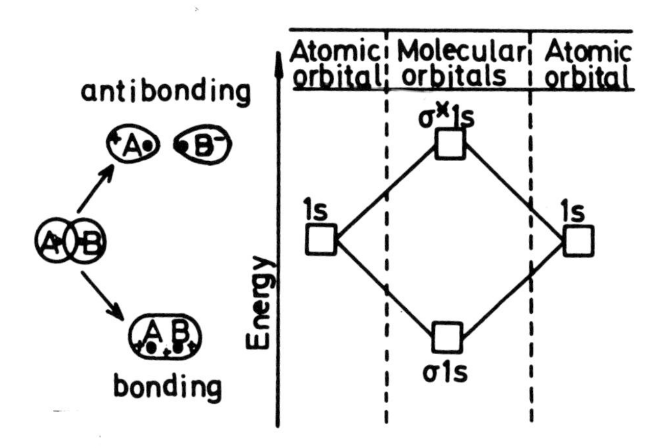

Fig. 3.3. The combination of two 1s atomic orbitals to give two molecular orbitals.

What is also shown in this diagram is the way that electrons are distributed in the molecular orbital. Generally the orbitals are filled in order of increasing energy, although there are many cases known in which the electrons prefer not to pair (low spin and high spin complexes (ref. 7)). One important factor will be the energy gap to the next available orbital.

For the more complex molecules such as transition metal complexes, molecular orbitals can best be set up by considering a number of steps. Taking an octahedral complex as an example, (a) consider how the unhybridised orbitals of the central atom can interact with the ligand orbitals. Just as in the valence bond approach, these are the s, p and $d_{x^2-y^2}$ and d_{z^2} orbitals, and they can be considered to interact with combinations of ligand molecules. (b) These orbitals can be combined in the same way as the simple molecules i.e. by taking linear combinations.

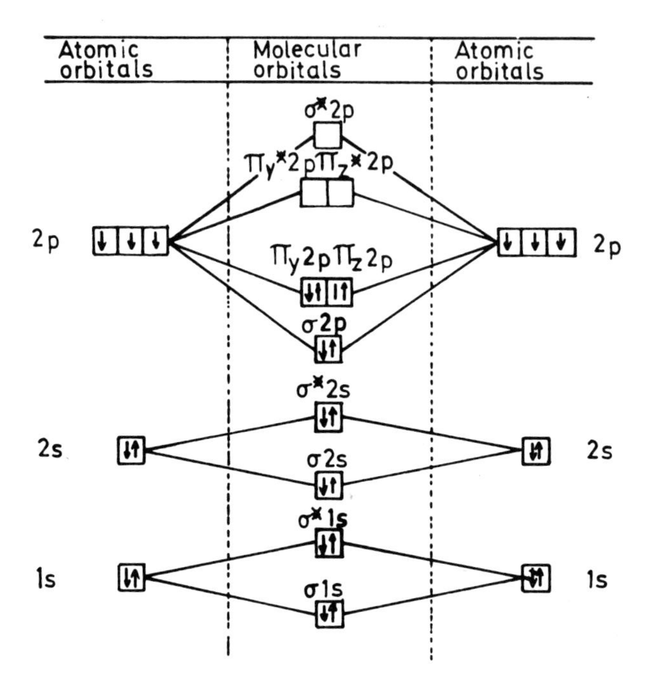

Fig. 3.4. Electronic configurations of two separate nitrogen atoms and of the nitrogen molecule.

(c) The next stage is to construct the molecular orbital of the complex. This is done by taking a linear combination of orbitals together with some measure of the degree of overlap of the bonds. The orbitals are combined from (i) the central metal atom and (ii) the ligand group orbital with the same symmetry. This results in a diagram as shown below. (Fig. 3.5).

Note that, for the time being, the d_{xy}, d_{yz} and d_{zx} (known, as a shorthand, as the t_{2g} orbitals) are not involved in bonding, and the fate of these will be considered later.

(d) The final stage is to apportion the available electrons in accordance with the Pauli exclusion principle and Hund's first rule: this may mean that electrons are placed in anti-bonding orbitals.

As stated above, the t_{2g} orbitals have not been considered, as a result of the fact that their spatial location (Fig. 3.1) is unsuitable for bonding, at least at first sight. In fact, this is often a gross simplification, since the t_{2g} orbitals may be located in a good position to interact with ligand orbitals, and particularly with p, d and anti-bonding π molecular orbitals. Thus, for example, carbon monoxide has the electronic configuration

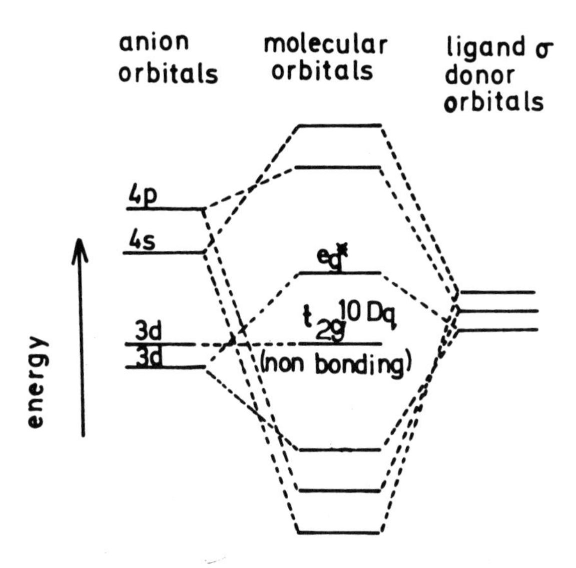

Fig. 3.5. Bonding scheme for an octahedral complex.

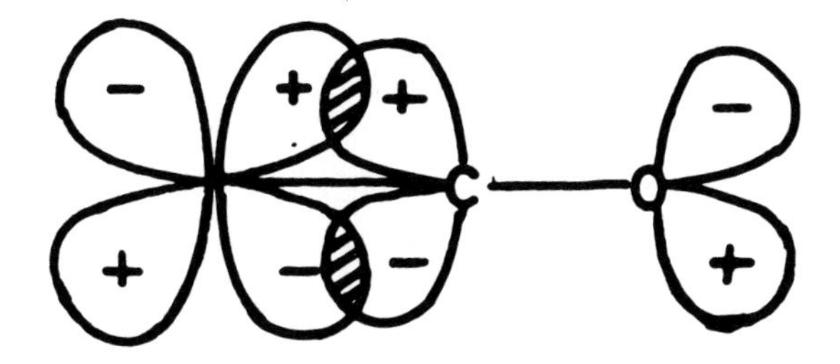

Fig. 3.6. Overlap of a d orbital of a metal with an empty anti-bonding π orbital of carbon monoxide.

$$(\sigma 1s)^2(\sigma^* 1s)^2(\sigma 2s)^2(\sigma^* 2s)^2(\sigma 2p)^2(\pi_y 2p)^2(\pi_z 2p)^2$$

leaving the $\pi^*_y 2p$ and the $\pi^*_z 2p$ orbitals empty. Spatially, as shown in Fig. 3.6, these are well located to interact with the t_{2g} orbitals of the central atom. (Fig. 3.6).

Similar types of interaction occur with olefin complexes, for example with silver.

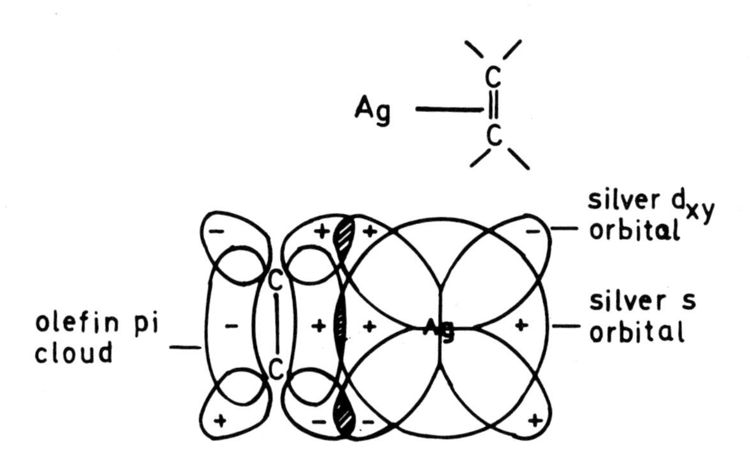

Fig. 3.7. Sigma and pi bonding in an olefin - silver ion complex.

There are, however, two types of π bond. If the π bonding orbitals are empty, the electrons from the central atom can be used in π bonding. In this case, electrons may be transferred from the ligand to the central atom via the σ bond, and back-transferred through the π bond. If, on the other hand, the π orbitals of the ligand are full, then electrons may be transferred to the central atom both via the σ bond and via the π bond. Thus, for example, for Pt-olefin complexes, back donation may occur, as shown in Fig. 3.8.

The molecular orbital theory does give a good qualitative description of the structure and properties of coordination complexes, but is very difficult to apply in molecules other than the simplest systems. Of much more use, in the sense that it can produce quantitative predictions of quite good accuracy, is the electrostatic theory and the related crystal field theory (refs. 2,8,9).

Application of the theory dates from the observation (ref. 2) that a simple electrostatic picture of complexes could be used to explain many of their properties. The parameters needed (the charge and size of the central atom and the dipole moments, polarisabilities and sizes of the ligands) were largely available, and insertion of values into ordinary potential energy equations of electrostatics gave surprisingly good predictions (ref. 10). Thus, for example, it can be easily demonstrated that linear, tetrahedral and octahedral complexes should be formed for coordination numbers of 2, 4 or 6, since these structures minimised the electrostatic repulsions of the ligands. However, using the simple theory, it is not possible to explain square planar complexes nor complexes containing virtually non-polar ligands such as CO.

Many of these problems can be resolved by including a correction due to the crystal field (refs. 2,8,9,11). The essence of the theory lies in the suggestion that five d orbitals, although equal in energy in a free gaseous metal atom, become differentiated in the presence of a 'ligand'. This 'ligand' may be other metal atoms (as in a crystalline solid) or a bonded anion (as in a complex). It is easy to see how this differentiation occurs by considering the formation of a complex. If, for example, six fluorine ligands approach a free metal atom along the x, y and z axes, then an octahedral complex will be formed. However the electrons arriving with the fluorine will interact with electrons in the atomic orbitals, the repulsion being strongest along the axis that the ligand approaches. As a result, the energy of electrons in the $d_{x^2-y^2}$ and d_{z^2} orbitals (pointing along the bond axes: see Fig. 3.1) will increase more than that of the t_{2g} orbital electrons, as shown in Fig. 3.9.

In contrast to the molecular orbital theory, where the difference in energy levels resulted from the formation of bonding and anti-bonding orbitals, the energy difference (10 Dq) arises from the octahedra of negative charges. The magnitude of 10 Dq (which may be measured spectroscopically (refs. 1,2)) depends on the central metal atom and its oxidation state and upon the ligand. For the same atom in the same oxidation state, it is possible to list the order of ligands corresponding to increasing values of 10 Dq, and this order usually holds regardless of the metal (although the value of 10 Dq will change, of course). In general, highest values of 10 Dq are observed for complexes in which the ligand can receive electrons back from the central metal atom by π bonding.

The differentiation in energy would be expected to be different for different arrangements of ligands, and this is found to be the case. Theoretical calculations of single electron energies of the d orbitals for a number of geometries are presented in Table 3.2 (refs. 2,9), and are shown schematically in Fig. 3.10.

The energy levels obey the rule that, if they are filled, the total energy is zero. For a given number of d electrons, the stabilisation energies may be calculated by filling up the levels in order, bearing in mind the possibility of spin pairing and non-pairing, as was the case for the molecular orbital theory.

Any comparison of the three theories must start from the knowledge that each of them is an approximation. Although the molecular orbital theory is, in principle, the most powerful of the three, it suffers from the disadvantage that quantitative calculations are very difficult. The valence bond theory has fallen into disrepute,

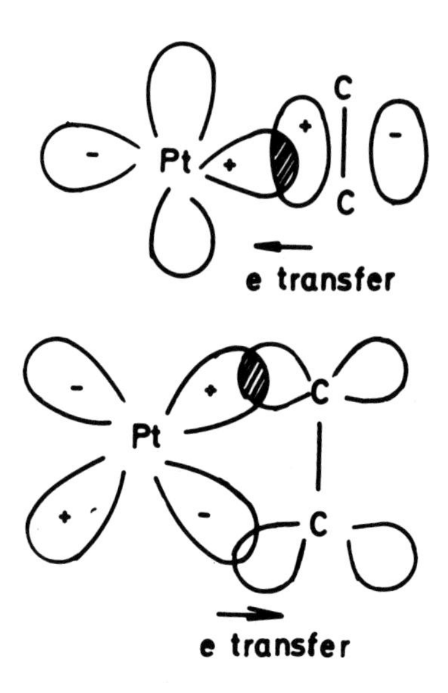

Fig. 3.8. Donation of electrons from the pi cloud of ethylene to an empty d s p^2 Pt orbital and to the anti-bonding pi^* orbital of the ethylene from the platinum.

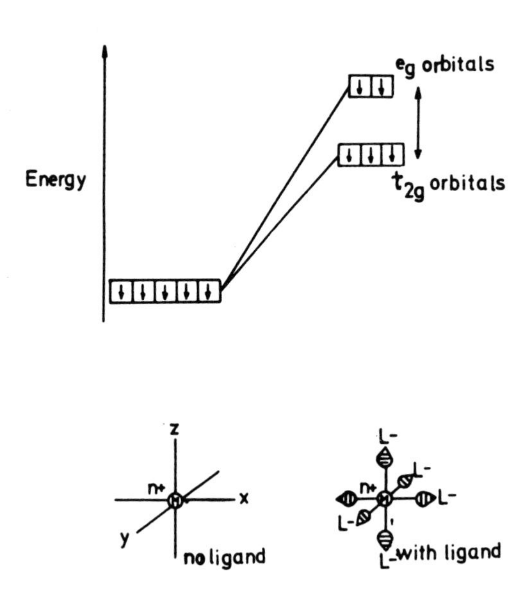

Fig. 3.9. Effect of octahedrally arranged ligand ions upon the energies of d orbitals of metal.

since it cannot be used to explain many of the properties of complexes: it does, however, offer a very useful visual image of bonding. For quantitative calculations there is no doubt that the crystal field theory is by far the most useful of the three. Using a relatively simple approach, it gives numbers which turn out to be almost correct. These can be used, as is discussed below, to provide detailed information on various aspects of catalysis.

TABLE 3.2

d-Orbital energy levels in crystal fields of different symmetries

Coordination No.	Structure	d_{zy}	d_{xz}	d_{yz}	$d_{z^2-y^2}$	d_{z^2}
1	...[1]	-3.14 Dq	0.57 Dq	0.57 Dq	-3.14 Dq	5.14 Dq
2	linear[1]	-6.28	1.14	1.14	-6.28	10.28
3	trigonal[2]	5.46	-3.86	-3.86	5.46	-3.21
4	tetrahedral	1.78	1.78	1.78	-2.67	-2.67
4	square planar[2]	2.28	-5.14	-5.14	12.28	-4.28
5	trigonal bipyramid[3]	-0.82	-2.72	-2.72	-0.82	7.07
5	square pyramid[3]	-0.86	-4.57	-4.57	9.14	0.86
6	octahedron	-4.00	-4.00	-4.00	6.00	6.00
7	pentagonal bipyramid[3]	2.82	-5.28	-5.28	2.82	4.93

Only electrostatic perturbations are considered.

1 Bonds lie along z axis.
2 Bonds in the xy plane.
3 Pyramid base in xy plane

One other approach to describing the properties of a solid deserves some mention, although it has been found to be of limited applicability in the context of catalysis. This is the electron band theory (refs. 12,13,14) which rests on the principle that, although each orbital corresponds to a definite energy level in an isolated atom, if atoms are assembled to form a crystal then energy levels lose their identity in an energy band. The theory has been very succesful in describing some bulk properties of solids, but is much less applicable to surfaces: since catalysis is a surface phenomenum, it is not surprising that applicability has been limited.

In a free atom, electrons may occupy s, p, d or f orbitals, and a unique description of the ground electron state can be written for any atom. On assembly into a crystal, atoms may be regarded as a lattice of positive nuclei (together with closed shell electrons) and valency electrons - which may roam throughout the crystal. Of course, they will be assigned to different energy levels which, for metals, may overlap as shown below. (Fig. 3.11).

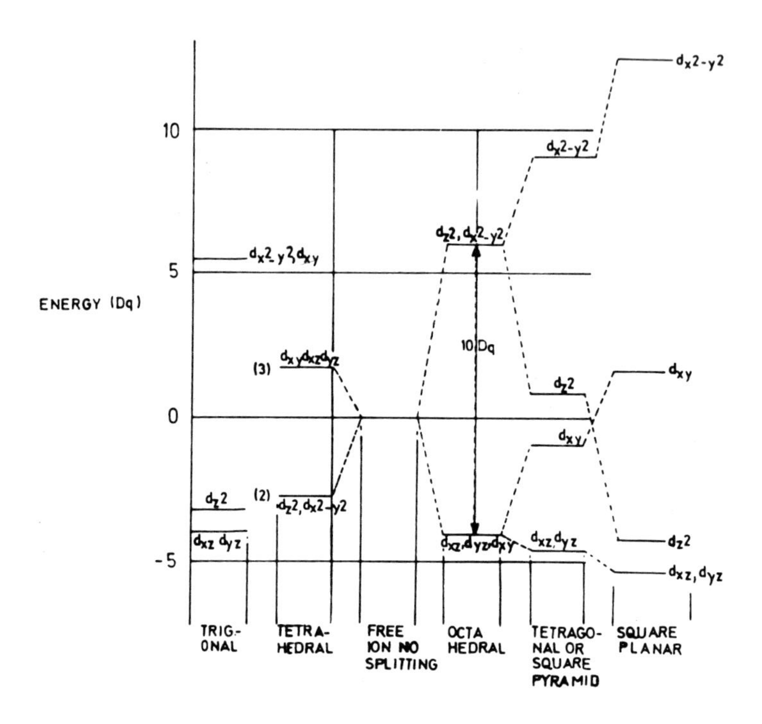

Fig. 3.10. Energy levels of d orbitals for different symmetries.

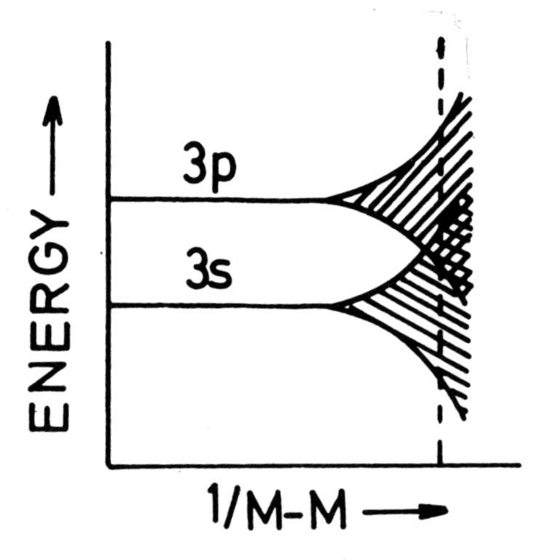

Fig. 3.11. Representation of the overlap of energy bands. M-M is the internuclear distance, and the dashed line corresponds to the internuclear distance obtaining in a metal.

As with other theories it is now necessary to ask whether the probability of finding an electron within a given band can be calculated. Indeed it can, although the calculation may be carried out in various degrees of sophistication. In the simplest form, electrostatic repulsion between electrons is ignored, and the central atoms (balancing positive charges) are assumed to be distributed uniformly. Treatment of this model by Fermi-Dirac statistics shows that the maximum electron energy at absolute zero of temperature is given by

$$E_{max} = \frac{P^2_{max}}{2m}$$

where P_{max} is the momentum in the highest occupied state and m is the mass of the electron. The number of electrons (n(E)) with energy between E and (E+dE) (density of states) can be shown to vary as $E^{\frac{1}{2}}$.

The next stage in refinement concerns the distribution of positive charge which will vary with the periodicity of the lattice. When this is taken into account, it is found that n(E) no longer increases regularly, but that there are zones of permitted energy which may, or may not, have energy gaps between them. This is shown in Fig. 3.12.

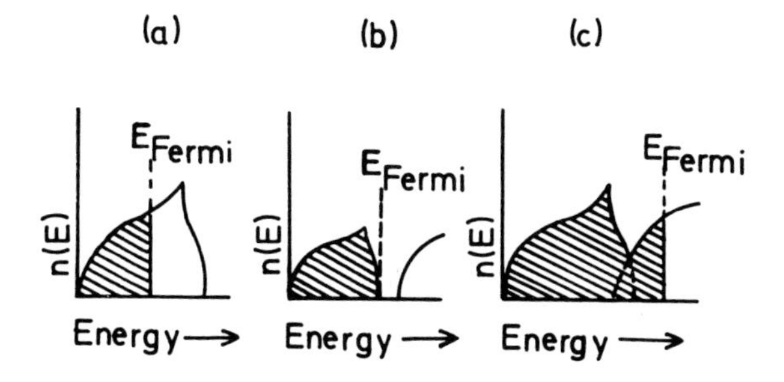

Fig. 3.12. Some n(E) versus E diagrams according to the electron band model.
(a) A half-filled s-band, for example, the s-band in alkali metals.
(b) An insulator.
(c) Overlapping bands in a metal.

Where the allowed energy bands overlap, the electron can roam throughout the lattice and the solid is a conductor. In catalytic terms, it is usually a metal. When the forbidden gap is very large, electrons pass between allowed zones only infrequently and the solid is an insulator. In catalytic terms, the most common examples of these are acidic catalysts such as silica-alumina. Where the gap is narrow, the solid is a semi-conductor, and n- or p-type semi-conductors are often used as oxidation catalysts.

Distributions of electron level densities can be measured experimentally by soft X-rays, and the results are often in quite good agreement with predictions based on the electron band theory. It is, however, of limited applicability in the context of catalysis by metals, and it is interesting to see why this should be so.

The most obvious reason lies in the fact that catalysis occurs at a surface which is, by definition, a discontinuity in the lattice. Not only that but the surface is also known (a) to be microscopically rough (b) to act as a concentration zone for impurities. We must then ask what size of ensemble must exist before we can apply to a collective theory and, perhaps of more importance, how do we know that impurity ions are not affecting the distribution of electron densities?

Less obvious, but of more importance, is the fact that the electron band theory does not take into account the nature of adsorption at the surface. Knowing that a gas can adsorb in more than one way, how can we expect a theory based on the bulk properties of a metal to give us accurate predictions?

The electron band theory has been considerably more successful in explaining catalysis on semi-conductors, and this is less surprising when we know something of the mechanism of reactions occurring. As a gross simplification, oxidation has been described as the transfer of an electron from the reactant to the catalyst, the transfer of an electron from the catalyst to oxygen, and the reaction of positively charged reactant with negatively charged oxygen to give products. Although this is much too simple, it does explain why a theory that describes the movement of electrons in the solid should have more success.

The electronic theory and its application to semi-conductors has been well described in several texts (refs. 15,16,17), and there is no need to go into details. It is useful, however, to consider the extension to catalysis and the usefulness of the theory to the catalyst designer.

There is no doubt that the electronic theory can be of use to the catalyst designer but it is of less use than would be expected. The most obvious case can be illustrated by the classic example of the decomposition of nitrous oxide (ref. 18). Stoichiometrically, the reaction may be written

$$2N_2O = 2N_2 + O_2 \qquad (1)$$

and, following the design procedure, when the reactions are written on the surface

$$N_2O_x + e^- \rightleftarrows N_2 + O^-_x \quad (2)$$

$$2O^-_x \rightarrow O_2 + 2e^- \quad (3)$$

$$O^-_x + N_2O \rightarrow N_2 + O_2 + e^- \quad (4)$$

where e^- is an electron coming from or going to the catalyst.

From a design viewpoint, the situation is clear: either reaction 2 is rate determining or reaction 3 or 4 is rate determining. It is necessary then to test a catalyst which can easily donate electrons (n-type semi-conductor) or one that can easily accept electrons (p-type semi-conductor). It is well known that the latter are, in fact, more active catalysts, but the point is that presentation of the problem in terms of electronic effects does lead to identification of active catalysts.

Heartened by this success, it would seem rewarding to extend the electronic argument to consider adsorption. Is it possible, for example, to obtain information as to the strength of bonding from application of the electronic theory? In the sense that it is possible to explain cumulative and depletive chemisorption (refs. 17,19,20) in terms of the theory, this application can be useful. The extension, however, - although developed widely by Volkenshtein-suggested that the strength of chemisorption can be related to the degree of charge transfer across the surface. In "weak" chemisorption, adsorbed species were suggested to remain neutral and electrons/holes of the semi-conductor did not contribute to the bond. "Strong" chemisorption, on the other hand, was suggested to involve electron transfer across the surface to or from the adsorbate. "Weak" and "strong" chemisorption may interconvert, and by the use of Fermi-Dirac statistics (ref. 21), the fraction of "weak" or "strong" chemisorbed species can be computed in terms of the distance from the Fermi level to the conduction band and to the valence band.

It is difficult to find fault with these concepts, since they are general and logical. As with metals, however, quantitative calculations are very difficult because of the nature of the surface. As a result, one is forced back to thinking of the strength of adsorption in terms of possible electron transfer, a situation not too dissimilar from the simple picture illustrated for the decomposition of nitrous oxide.

One strength of the electronic theory has been its ability to explain the action of dopers. The presence of small amounts of additive that can insert donor or acceptor levels in the forbidden gap is shown to change semi-conductivity and to influence catalysis. Indeed, it is possible to explain catalysis in these terms.

Thus, for example, Wentrcek and Wise (ref. 22) have studied the hydrodesulphurisation of butyl mercaptan on single crystals of MoS_2 doped with Co^{2+} (ref. 22). They found that the catalytic activity was proportional to the hole carrier density in the doped catalyst.

Similarly Sleight (refs. 23,24) has studied the oxidation of olefins over scheelite catalysts, in which it is possible to vary numbers of cations and the number of defects. Although describing the reaction in terms of adsorption on the scheelite structure, they relate the oxidation of propylene to the electronic structure of the catalyst. The electron bands for bismuth molybdate based solid were represented as

Bi 6 p
Mo 4 d
Bi 6 s — Fermi level
O 2 p

During the reaction the following steps were suggested

$$C_3H_6 + (MoO_4)^{2-} \rightarrow (C_3H_5.MoO_4)^{2-} + H^+ + e^-$$

$$(C_3H_5.MoO_4)^{2-} \rightarrow C_3H_4O + MoO_3 + H^+ + 3e^-$$

The electrons involved initially occupy the Mo 4 d states, but there is near degeneracy of these states with the Bi 6 p states. Re-oxidation of the catalyst can then occur at the Bi^{3+} surface ions.

$$O_2 + Bi^{3+} + 4e^- \rightarrow O^{2-} - Bi^{3+} - O^{2-}$$

The adsorbing cation is not oxidised itself, since the electrons were delocalised in the Bi 6 p band. The newly created oxygen anions diffuse away from the surface to continue the cycle.

Application of the concept of doping can be useful to the designer, but is more commonly used in connection with the identification of secondary components of the catalyst.

Overall, then, the electron band theory of catalysis is seen to be useful under specific circumstances, and particularly with respect to semi-conductors. Even here, however, it fails because it does not take into account possibilities of coordination at the catalyst surface. Where such possibilities are not important, it can succeed, and it must always be kept in mind during a design.

II. THEORIES OF BONDING AND ADSORPTION

The first aspect of catalysis that can be considered in the light of these theories is adsorption. Even here, application of the theories can be complex and is often best left to the specialist. It is valuable, however, to consider how the theories may be applied.

Perhaps the most simple case has been treated by Dowden (refs. 8,9) with the application of elementary electrostatic theory to ionic lattices. For a perfect (001) plane of magnesia, each surface ion lies at the centre of the base of a regular square pyramid of nearest neighbours, and the energies of the electron states of the surface atoms can be located by methods well known for bulk states. Thus, for example, the position of the valence level in the full O^{2-} band may be estimated as below:

a) Remove O^{2-} from the surface

$$O^{2-} \rightarrow O^{2-}_{gas} + V_o + \frac{4\alpha e^2}{r} - R$$

where V_o is a vacancy, α is the Madelung constant, e the electronic charge, r the interionic distance and R, R_o are the repulsive energies between O^{2-} and O^- (see below) respectively and their neighbours.

b) Convert O^{2-}_{gas} to O^{-}_{gas}

$$O^{2-}_{gas} \rightarrow O^{-}_{gas} + 2 + E$$

c) Replace O^{-}_{gas} in the lattice

$$V_o + O^{-}_{gas} \rightarrow O^{-} + \frac{2\alpha e^2}{r} + R_o - W$$

where W is the polarisation energy of the lattice due to the deficiency of charge on O^-.

Adding these together gives the result

$$O^{2-} \rightarrow O^{-} + e + \frac{2\alpha e^2}{r} + E + (R_o - R) - W$$

In other words the basic process of interest is being split into its component parts and then re-assembled to give an overall picture. Similar treatments can be applied to other reactions, as we shall see.

If the lattice is not perfect (as is usually the case) the presence of vacancies or defects cause both the ionisation energy and the electron affinity of the lattice to change. In magnesia, for example, a cation vacancy reduces both values by ca 10 eV. In these cases, the theoretical problems caused by deviations from ideality can be large.

Dowden treats the adsorption of oxygen, hydrogen and water in a similar manner, with the object of finding the species most likely to be adsorbed. Thus, for example, on stoichiometric magnesia, hydrogen could adsorb as H_2^+, H_2^-, H^+, H^-, H_2 or H, interaction occurring with Mg_m^{2+}, Mg_i^{2+}, O^{2-}, O^- or with lattice vacancies V_m or V_o. The subscripts m, i and o refer to the lattice metal, interstitial metal or oxygen respectively. Then, for example, if adsorption involves

$$O^{2-} + H + Mg_m^{2+} = O^{2-}\cdots\cdot\ H^{\delta+}\cdots\cdot Mg_m^{(2-\delta)+}$$

the reaction may be expressed as

$$H_{2g} \rightleftarrows 2H_g + 4.5\ eV$$

$$Mg_m^{2+} + O^{2-} \rightleftarrows Mg_m^+ + O^- + 5\ eV$$

$$Mg_m^+ + H_g \rightleftarrows MgH_m^+ - 2.1\ eV$$

$$O^- + H_g \rightleftarrows OH^- - 4.7\ eV$$

which, when summed, gives

$$\Delta H < 2.7\ eV$$

The numbers of the right-hand side of the above equations are obtained from the properties of the components applied, where necessary, in equations such as above.

A similar calculation may be carried out for all the alternatives to show that hydrogen, if adsorbed, will be only weakly chemisorbed: this is in agreement with experiment.

These calculations work quite well for ionic lattices, and their development proceeded alongside that of the band theory. Descriptions with considerable more detail are given elsewhere (refs. 8,9).

The next degree of sophistication in dealing with adsorption on solids involves use of the crystal field effect (refs. 2,9). If we consider the surfaces of solids, then we can regard adsorption as being the addition of a ligand. Thus, for example, in the sodium chloride structure, chemisorption would lead to the following coordination changes.

(100) face: square pyramidal $\longrightarrow$ octahedral
(111) face: trigonal $\longrightarrow$ tetrahedral $\longrightarrow$ square pyramidal $\longrightarrow$ octahedral
(110) face: tetrahedral $\longrightarrow$ square pyramidal $\longrightarrow$ octahedral

Now the physical properties of the oxides of the first long period are usually interpreted as involving a weak field and, using the results tabulated in Table 3.2, the crystal field stabilisation energies for these coordination changes can be calculated. The results are shown in Table 3.3 together with typical values of 10 Dq obtained from summarising the literature (ref. 8). Of course, these calculations contain some approximations, including the fact that some ions of higher valency should include some reference to the strong field approximation, and that it has been assumed no special effects are introduced to the crystal field at a surface. What is clear from the table is that the energy change exhibits a distinctive twin peak pattern between d^0 and d^{10}, regardless of the coordination change involved.

Dowden and Wells (ref. 8) went on to consider the crystal field contribution to the adsorption of gases on oxides, including the adsorption of hydrogen. Considering the gas to be adsorbed as a polarised molecule

$$H^{\delta-} \text{ ---- } H^{\delta+}$$

$$O^{2-} - M^{x+} - O^{2-}$$

they showed that the crystal field contribution had the form shown in Fig. 3.13. Multiplying the ordinate by the literature values of Dq gives the absolute contribution to the heats of adsorption, but these values should not be taken too seriously. The twin peaked pattern is, in fact, much more significant, since we shall see that it duplicates many catalytic activity patterns.

Similar patterns may be constructed for other modes of adsorption, and a similar pattern would be expected to hold on sulphides (refs. 8,9). Superficial application of the theory to an emergent edge dislocation on a sodium chloride crystal shows that the twin peak pattern is destroyed (ref. 9), but the theory is not really suited for such application.

Similar calculations may be carried out for other adsorbing gases with a fair degree of confidence (e.g. water (refs. 8,9)) or with little confidence (e.g. oxygen). Where it is possible to have some confidence in the calculations, a twin peaked pattern is always observed.

As a result, we can say that it is possible to apply the crystal field theory to adsorption. Such application can be expected to give us an indication of the form of adsorption at least of simple molecules, since we can calculate the relative energy gains for adsorption in different modes. In addition, the calculations lead us to suspect that the pattern of crystal field contribution as a function

TABLE 3.3

Crystal field stabilisation energies during coordination changes for weak field oxides (refs. 2,8,9)

d electrons	Ion	10 Dq Kcal	Trigonal → Tetrahedral	Tetrahedral → Square Pyramid	Square Pyramid → Octahedral	Tetrahedral → Octrahedral	Trigonal → Square Pyramid	Trigonal → Octrahedral	Square Planar → Square Pyramid
0	Ca^{2+}/Sc^{3+}	0	0	0	0	0	0	0	0
1	Ti^{3+}	50	+1.19 Dq	-1.90 Dq	+0.57 Dq	-1.33 Dq	-0.71 Dq	-0.14 Dq	+0.57 Dq
2	V^{3+}	48	+2.38	-3.80	+1.14	-2.66	-1.42	-0.28	+1.14
3	Cr^{3+}/V^{2+}	46/35	+7.36	-6.44	-2.0	-8.44	+0.92	-1.08	+4.56
4	Mn^{3+}/Cr^{2+}	53/39	+3.68	-7.36	+3.14	-4.22	-3.68	-0.54	+3.14
5	Fe^{3+}/Mn^{2+}	34/22	0	0	0	0	0	0	0
6	Co^{3+}/Fe^{2+}	53/30	+1.19	-1.90	+0.57	-1.33	-0.71	-0.14	-0.57
7	Co^{2+}	28	+2.38	-3.80	+1.14	-2.66	-1.42	-0.28	+1.14
8	Ni^{2+}	23	+7.36	-6.44	-2.0	-8.44	+0.92	-1.08	+4.56
9	Cu^{2+}	37	+3.68	-7.36	+3.14	-4.22	-3.68	-0.54	+3.14
10	$Cu^{+}Zn^{2+}$	0	0	0	0	0	0	0	0

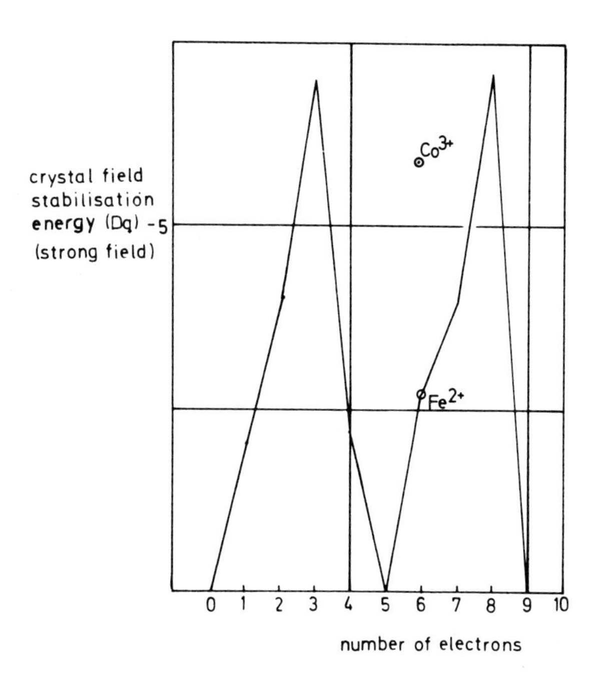

Fig. 3.13. Theoretically calculated crystal field stabilisation energies.

of the number of d electrons should be twin peaked, with minima at d^0, d^5 and d^{10}. The possible significance of this finding will emerge when we compare this result with catalytic activity patterns (see below).

Adsorbed molecules on solid surfaces must also include large covalent contributions in their bonding, and these must be considered in the light of the molecular orbital theory. The complexity of the calculations required will, except for the simplest molecules, increase by an order of magnitude as compared to those based on the crystal field theory.

Perhaps the most simple method of using the molecular orbital theory is in terms of the directional properties of the bonds, as was done in predicting the existence of σ and π bonds (see above). This was studied by Bond (ref. 25) in considering adsorption of gases on different faces, the probability of adsorption being related to the angle of e_g and t_{2g} bonds emerging from the surface. Although this approach is very useful, it suffers from distinct disadvantages. The first of them is theoretical: given that the emergence of orbitals can be predicted, how can the degree of occupancy of these orbitals be measured? Goodenough (ref. 26) has used magnetic measurements to study this problem for first row transition metals, but no quantitative information on the noble metals is available. The second problem is essentially practical. The models proposed by Bond are, by definition, on an atomic scale (refs. 2,25), and discuss the possibilities of adsorption on an atomic ensemble. In practice, surfaces are rough and may rearrange under reaction conditions: as a result, the correct type of ensemble may be a transient phenomenum or may result from surface roughness effects. This approach to adsorption must be regarded, then, as being a very useful visual aid, with little quantitative advantage.

It is possible to improve on this position, although the improvement requires a simplifying assumption (ref. 27). If catalysis is regarded as a chemical reaction which proceeds on a surface, then molecular orbital calculations could be applied if the catalyst is regarded as a perturbation of the electronic structure of the system. This could be done by regarding the catalyst as a single atom (ref. 28), as a line of atoms (ref. 29) as a two dimensional array of atoms (ref. 30) and as a three dimensional array (ref. 28). The latter, although most realistic, imposes severe demands on computer capacity. There is no doubt that the calculations are complex, and that the results have dubious value in representing the absolute energies associated with chemisorption. However, as with the crystal field approach, predictions of the relative probabilities of adsorption of different forms of adsorbed molecule can be very useful (ref. 28). In contrast to the crystal field approach, where a sufficiently interested designer could calculate the probability of adsorption, calculations based on the molecular orbital theory should not be undertaken without considerable forethought.

From the point of catalyst design, it is valuable to have some indication of the probability of adsorption in a particular form: this is particularly useful in writing possible reaction mechanisms. Applications of bond theory to adsorption can give some idea of probabilities of adsorption, but such calculations are not to be undertaken lightly. Given that the information is needed badly enough, application of the crystal field theory would appear to be possible. Application of the molecular orbital theory should be left to the expert, with no great hope of success. Where the information exists in the literature (refs. 28,31), no hesitation should be used in taking the easy way out.

III. BOND THEORIES AND CATALYSIS

It has been seen that the correlation between calculated values of the crystal field contribution to heats of chemisorption and catalytic activity - particularly for reactions involving hydrogen - tend to suggest that the crystal field theory could help to explain some aspects of catalysis. As would be expected from the nature of the systems, quantitative calculations are not easy. A crystal field approach does, however, provide a very useful qualitative guide as to what may be occurring. For clarity, it is useful to illustrate the idea in terms of a single complex, and then to consider the extension to a lattice. Although the mechanism is speculative, the oxidation of thiols on copper ions is a good example (ref. 32).

The overall reaction may be described as

$$2RS^- + Cu + \tfrac{1}{2}O_2 \underset{aqua}{=} RSSR + Cu + O^{2-}$$

The catalytic cycle has been suggested to involve copper complexes of varying valency and, since Cu^{I} favours two coordination and Cu^{II} favours four coordination, of varying coordination number:

$$(X - Cu^{I} - X)^{x-} + O_2 \rightleftarrows \left[\begin{matrix} X & & O \\ & Cu^{II} & \\ X & & O \end{matrix} \right]^{x-}$$

$$\left[\begin{matrix} X & & O \\ & Cu^{II} & \\ X & & O \end{matrix} \right]^{x-} + 2RS^- \rightleftarrows \left[\begin{matrix} X & & RS \\ & Cu^{I} & \\ X & & RS \end{matrix} \right]^{x-}$$

$$\left[\begin{matrix} X & & RS \\ & Cu^{I} & \\ X & & RS \end{matrix} \right]^{x-} \xrightarrow{fast} (X\text{-}Cu^{I}\text{-}X)^{x-} + RSSR$$

Describing this reaction sequence in terms originating from heterogeneous catalysis and coordination chemistry, oxygen adsorbs on the complex and the copper ion changes valency and changes preferred coordination number. Replacement of oxygen by two thiol anions (competitive adsorption) causes a change of valency back to Cu^{I} with the preferred coordination number of two. This leads to the elimination (desorption) of the product disulphide.

It should be possible to see this kind of behaviour on an energy diagram of the type shown in Fig. 3.10, and indeed it is. Consider a situation, divorced from the example above, where the original contains 4 d electrons and the ion is tetrahedrally coordinated. As shown below (Fig. 3.14) these electrons would be expected to be located in the d_{z^2}, $d_{z^2-y^2}$ orbitals. If, however, more electrons are fed into the system (for example, from a ligand) then these electrons either have to go into the d_{xy}, d_{xz} or d_{yz} orbitals of the tetrahedral complex, or the complex can change to octahedral coordination. In this case up to 6 electrons can be accommodated at lower overall energy (in the d_{xy}, d_{yz} and d_{xz} orbitals) as compared to the tetrahedral complex. At the same time, free coordination positions become available to which the ligand can be attached and "adsorption" takes place.

The same type of process could be envisaged as occurring in catalysis. Thus, for example, if two ligands attached to an octahedral complex containing six d electrons were able, in some way, to accept two electrons from the ion, then the coordination number would tend to decrease, releasing the ligands to react together and form products.

Of course the picture is oversimplified, but it illustrates the way in which the crystal field theory offers a very "visual" approach to catalysis which - at least on a qualitative scale - can relate adsorption and catalysis to coordination and valency changes.

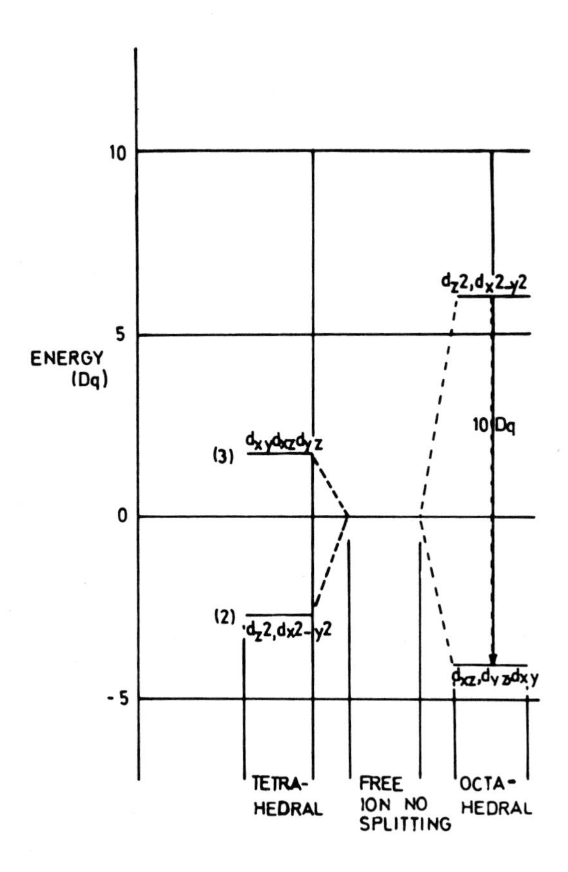

Fig. 3.14. Tetrahedral and octahedral complexes.

Extension of the concept to solid catalysts is made easier by the knowledge that the crystal field theory was first developed for such systems (refs. 1,2). Obviously a change in coordination number is much less likely when a metal atom is locked into a lattice, but the general picture still remains. Indeed, we can extend it to consider geometric effects in catalysis on two levels. Thus, for example, if we consider different faces of a crystal, then the number of coordination sites not involved in lattice binding will depend on the crystal face chosen. Thus, for a body-centred cubic metal crystal, we may represent the situation as:

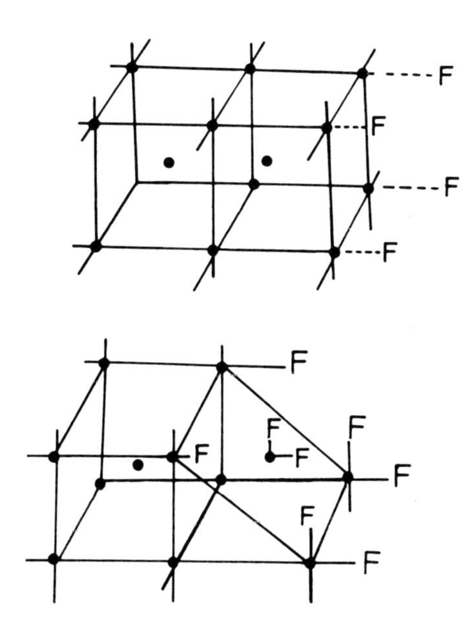

Fig. 3.15. Possible faces and coordination sites available for adsorption on a body-centred cubic crystal.

A slightly more detailed picture has been presented by Bond (ref. 25) using a molecular orbital approach. He considers the adsorption of unsaturated hydrocarbons on metal surfaces in terms of the emergence of e_g and t_{2g} orbitals from the surface. Although doubts arise as to the quantitative nature of the approach, the qualitative picture so obtained is invaluable in understanding adsorption and the mechanism of catalysis (see above).

IV. THEORIES OF BONDS AND CATALYST DESIGN

It is relevant to ask, at this stage, whether knowledge of theories of bonding can help the catalyst designer. The facile answer is negative since, for example, crystal field stabilisation energy effects can be represented merely as valuable activity patterns (Fig. 3.16). Indeed, it is a dedicated designer who could calculate the probability of adsorption on the basis of even the relatively simple crystal field theory. Much more likely is the situation where the designer searches the literature for experimental or theoretical evidence of possible forms of adsorption.

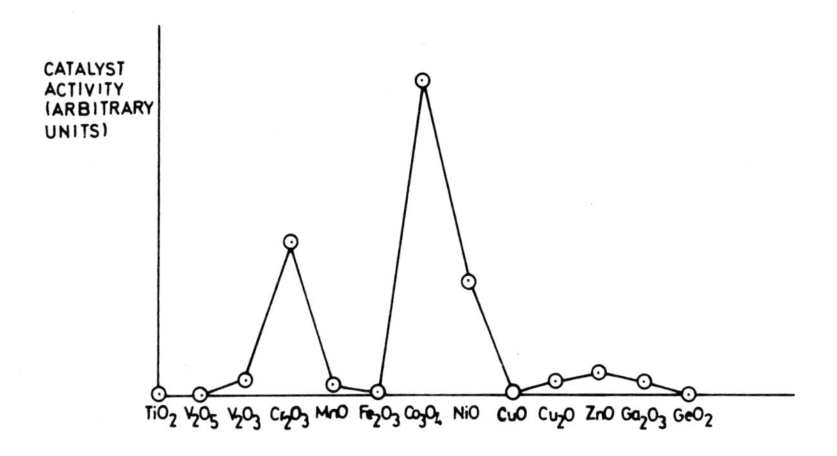

Fig. 3.16. Comparison of crystal field stabilisation energy and catalytic activity as a function of the number of d electrons.

The advantage of bond theories and mechanistic approaches based on them lies more in the background to the design. They do enable the construction of realistic mechanistic sequences (even if they are incorrect) and they do provide a theoretical background for many of the design steps. As such it is useful to have the information as a background, and to understand the basic reasoning behind some of the steps and observations made during the design.

If we consider the overall summary for the design of the primary constituents of the catalyst (Fig. 3.1), we see that an approach via the theories of bonds has helped to explain many of the steps (shaded area). The use of this information with respect to catalyst design is best illustrated by actual examples, and this will be done in part II of this book. There are, however, some aspects of the primary design which have not been covered (Fig. 3.1), and some areas in which aspects other than those obvious from application of bond theory can be expanded.

Perhaps the most important of these is the influence of geometric factors. From many studies we know that a given reaction may be structure sensitive or structure insensitive (ref. 33) and a rough guide to this was given in Chapter 2. We also know of the existence of the Multiplet theory (ref. 34), which considers the possibility of a catalytic reaction in terms of the match between the surface geometry and the molecules involved in catalysis. We have seen that different crystal faces can be expected to offer different numbers of coordination sites and, as a result, would be expected to have different catalytic activities. However, we also know that defects in the surface can be centres of catalytic activity, either as a result of geometric, electronic or adsorption effects.

In one sense, this knowledge highlights the empirical nature of the design. As was discussed in Chapter 2, one approach to design is to try to match lattice parameters with the reaction to be catalysed, and this is often successful. It must be emphasized, however, that this approach will not be successful unless the bulk of catalysis proceeds on a surface which is indeed described by the lattice parameters. If the reaction, in practice, proceeds only at defects, then the approach will fail. There is no way in which this can be predicted, and the only satisfactory method of assessing this is by experimental testing.

The approach to catalyst design on the basis of heats of adsorption does not emerge from application of theories of catalysis, although it has been discussed in Chapter 2. Little needs to be added. Where heats of adsorption are available, it can be a useful approach: regrettably this does not happen very often.

The final approach to design of the primary constituents is via consideration of activity patterns, and these are important enough to warrant a separate section.

V. ACTIVITY PATTERNS AND CATALYST DESIGN

It has long been known that it is possible to observe patterns of behaviour of catalysts if their activity is compared for similar reactions. There are usually sound theoretical reasons for this, and these are discussed elsewhere. From the point of view of catalyst design, however, these activity patterns can offer very useful pointers to potentially useful catalysts.

Various types of activity pattern have now been identified, ranging from simple comparisons of, for example, the minimum temperature at which catalysis is favoured (ref. 15) to more complex patterns in which the activity and selectivity of catalysts is compared (ref. 9) or in which the reactivity of related molecules on a given catalyst is studied (refs. 35,36). Some of the more widely applicable patterns are presented below, but it is useful to give one caveat at this stage.

Activity patterns exist for a reason. It may be that they reflect the heats of adsorption of reactants (ref. 37) or the ease of electron transfer between a reagent and the catalyst (ref. 15). In trying to relate an activity pattern for one reaction to a second (similar) reaction, there is the built-in assumption that the same processes are important. Although this is often correct, it need not necessarily be so.

Indeed, it need not be correct even over a range of different temperatures, as shown by comparison of the activity of metal oxides for cyclohexane and hydrogenation at 400°C and 450°C (Ref. 151). At the higher temperature, the activity pattern was in agreement with the predictions of the ligand field theory but, at only 50°C lower, this pattern became distorted.

Nonetheless, activity patterns are extremely useful guides for the catalyst designer. They must, however, be used with care.

It should be emphasized that any relationship, be it empirical or theoretical, can be of use to the catalyst designer. Activity patterns may be based on sound theoretical reasoning, such as the crystal field theory (refs. 9,38), on the position of an element in the periodic table (ref. 15), or even on the phase of the moon when the catalyst was made: in all cases, if they are shown to be of reasonably wide applicability, then they are of potential use. It does not matter that the same data can be used in different ways, provided that a recognisable pattern emerges.

Because of their importance, many activity patterns of general interest are presented in Chapter 7. In addition, however, the literature is full of more specific examples, of which a selection is presented in Table 3.4. As a first guide, the designer should turn first to a general activity pattern (Chapter 7) and then use the literature (e.g. Table 3.4) to seek activity patterns for reactions as similar to the reaction under study as is possible.

TABLE 3.4

Catalytic activity patterns

Catalyst	Reaction catalysed	Activity related to	Ref.
Reactions involving hydrogen			
Metals	methanation	metal	40
Pd, Rh, Ru, Pt, Ir	hydrogenolysis of heptane	% d character	41
Pt, Ir, Os, Pd, Rh, Ru	hydrogenolysis cyclopropane	comparison with ethane hydrogenolysis	42
Ni on SiO_2, Al_2O_3 and SiO_2/Al_2O_3	hydrogenolysis ethane	support	43
Ru, Rh, Pt, Ni, Co, Ir, Pd, Fe	hydrogenolysis and isomerisation of pentane	various mechanisms	44
Pt on SiO_2, C, Al_2O_3	isomerisation and hydrogenolysis of neopentane	dispersion	45
Group 8 metals	hydrogenolysis of ethane	% d character: atomic radius: dispersion	46,47
metals	hydrogenation	review	48
metal oxides	ethylene hydrogenation	crystal field theory	49
metal oxides	hydrogen adsorption	topochemistry	50
62 metals	homogeneous hydrogenation	redox potential and lipands	51
metals	various	review: crystallite size	52
metals	hydrogen adsorption	electronegativity: work function	53
Mo, Rh, Pd, Ag, Ta, W, Rh, Ir, Pt	hydrogen-deuterated benzene exchange	% d character	54
Pt, Rh, Pd, Ni	H_2-D_2O, C-C=C + D_2O	ΔH_{M-O}	55
rare earth metals + hydrides	o-p H_2 exchange	H_2 content in metals	56
d and f-shell metals and compounds	o-p H_2 exchange	atomic number: crystal field	57
rare earth metals	homomolecular H_2 exchange	atomic magnetic moment	58
Ni	hydrogenolysis	electronic and steric effects in hydrocarbons	59
Cu, Pt, Rh	hydrogenation of ketones	electronegativity of hydrocarbon	60
Cobalt molybdate	hydrogenation	structure of model compounds for coal	61

continued on p. 66

TABLE 3.4 (Cont'd)

Catalyst	Reaction catalysed	Activity related to	Ref.
Pd/C	hydrogenation	nature of poison present	62
Alloys			
Ru/Cu, Os/Cu on SiO_2	hydrogenolysis and dehydrogenation	composition	63
Cu/Ni	hydrogenolysis and dehydrogenation	composition	64
Cu/Ni	hydrogenolysis	composition	65
Pd/Ag	hydrogenation of nitrobenzene	electrical properties	66
Ni/Cu	hydrogenation	composition. Number of paired electrons	67
Pd/Rh films	CO oxidation	composition	68,69
Pd/Ag films	ethylene oxidation	composition	70
Pd/Rh films	ethylene oxidation	composition	71
Pt/Rh, Pd/Rh, Pt/Ir, Pt/Ru, Os/Pt	isomerisation	electron deficiency in d band	72
Pd/Ni/Cu	$N_2H_4 + H_2O = N_2 + 2H_2 + H_2O$	composition	73
Acid-Catalysed Reactions			
Various solids	various reactions	acidity	74
Molecular sieves	cracking	molecular size, porosity and acidity	75
Modified metal oxides	alkylation of aromatics	acidity, modifier, organic structures	76
Cracking catalysts	dealkylation and isomerisation	substitution in molecules: ΔH for hydride abstraction	77
Acid catalysts	dealkylation	acid strength	78
Cracking catalysts	cracking	thermodynamics of systems	36
Metal oxides	dimerisation	acidity	79
Mordenites	hydrogenisation and hydrocracking	sodium ion exchange	80
Metal sulphates	butene isomerisation	acidity	81
Rare earth oxides	butane dehydrogenation and cracking	f electrons, acidity	82

TABLE 3.4 (Cont'd)

Catalyst	Reaction catalysed	Activity related to	Ref.
Oxidation			
Various	gas and liquid	review	83
Transition metal oxides	adsorption	crystal field	84
Co_3O_4, MnO_2, ZnO, $TiO_2.Cr_2O_3$, V_2O_5, CuO	oxidation of hydrogen	redox of surface	85
Rare-earth oxides	oxidation of hydrogen	semiconductivity	86
Pt	total oxidation hydrocarbons	hydrocarbon structure	35
Metal oxides	total oxidation ethylene	energies of activation	87
Metal oxides	CH_4 total oxidation	oxygen exchange on surface	88
Ir, Pt, Pd, Ru, Rh	total oxidation hydrocarbons	hydrocarbon structure	89
Metal oxides	propene oxidation	no coincidence with O_2 exchange crystal field	90
Group IV metal oxides	HC=CH total oxidation	binding energy oxygen-surface	91
Rare earth oxides	propylene total oxidation	surface-oxygen bond	92
Ferrites	H_2, CH_4 total oxidation	surface-oxygen bond	93,94
Zeolites + metals	CH_4, propylene total oxidation	acidity and nature of metal	95,96
Metal oxides	total oxidation RH	structure of RH	97
NiO	lower olefin total oxidation	type and number of allylic hydrogens	98
Rare earth oxides	oxygen exchange	crystal field	99
Transition metal oxides	$CO_2^x + CO_2 = 2COO^x$	homomolecular exchange	100
IV period metal oxides	O_2 exchange	isotopic exchange	101
Oxides of Sn/Sb, Bi/Mo, Fe/Mo, Co/Mo and Fe	1-butene to butadiene		102
Metal oxides	H_2, CH_4 oxidation		103
Metal molybdates	NH_3, C_3H_6 oxidation	strength of oxygen bond with the surface	104
Various oxides	NH_3, C_3H_6, MeOH oxidation		105

continued on p. 68

TABLE 3.4 (Cont'd)

Catalyst	Reaction catalysed	Activity related to	Ref.
Metal oxides	CH_4, H_2 oxidation		106
Group IV metal oxides	benzene, allene oxidation		107
Metal oxides	i-butene, propylene, ethylene propane oxidation	strength of oxygen bond with the surface	40
Metal oxides	propylene oxidation		108
Co, Pb, Cd, Bi, Sb oxides	H_2 oxidation		109
Metal oxides	various		110
Metal oxides	n-butane oxidation		111
Metal oxides	various	and electrical properties: review	112
Rare earth oxides	CO oxidation	oxygen mobility on surface	113
Metal oxides	i-butene oxidation	oxygen mobility	114
Metal molybdates (group III)	acrolein oxidation	surface-oxygen bonding	115
Metal oxides	oxidation of acrolein	acidity of surface	116
Phosphates, sulphates, borates	ammoxidation of propylene	presence of oxygen centres	117
Sb/Tl oxides + dopers	i-butene to methacrolein	screening	118
bi-metal oxides	oxy dimerisation/aromatisation	screening	119
Metal oxides	propylene to either acrolein, benzene, acetone or CO_2	screening	120
Metal oxides	oxidative dealkylation alkylaromatics oxidation	screening	121
Group IV metal oxides	cumene oxidation: liquid phase	ROOH concentration	122
Semiconductors	hydrocarbon oxidation: liquid phase	semiconductivity	123
Metal oxides	tetralein oxidation: liquid phase	soluble vs insoluble catalysts	124
Various	various: mainly liquid	review: crystal field	125
Ions and oxides of transition metals	liquid cyclohexane oxidation	atomic number and oxidation state	126

TABLE 3.4 (Cont'd)

Catalyst	Reaction catalysed	Activity related to	Ref.
Metal oxides	gas-liquid hydrocarbon oxidation	surface bonding: structures of	127
Metal oxides	MeOH → HCHO	screening	128
Metal carbides	oxidation H_2, NH_3	oxygen-carbide bond strength	129
Metal oxides	oxidation of alkyl aromatics to aromatic aldehydes	carbon-catalyst bond strength	130
Sulphide catalysts	$H_2S + CO = H_2 + COS$	crystal field	151
Metal sulphides	hydrogenolysis of carbon disulphide	crystal field	152
ZnO, CuO, ZnS, Al_2O_3, MgO	recombination H·	defects	153
Metal sulphides	diethylsulphide cyclisation	sulphur mobility	154
3rd-6th period metal oxides + sulfides	organic sulphide cracking	charge transfer from sulphur to metal	155
Metal sulphides	diethylsulphide cyclisation	sulphur mobility and metal-sulphur bond strength	156
Various	various	good comparative review	157
Reactions involving nitrogen oxides			
Noble metals	NO + H_2	nature of metal	131
Semiconductor metal oxides	N_2O decomposition	ΔH_f of oxide	132
Metal oxides	N_2O decomposition	chemical nature of oxide	133
Metal oxides	NO decomposition	ΔH_{M-O}	
40 metal oxides	NO decomposition	effect of oxygen	155
40 metal oxides	NO decomposition	rate of desorption of oxygen	136
Rare earth oxides	N_2O decomposition	rate of desorption of oxygen	137
Ag	RH + NO	electron donation to the catalyst	138
Metal oxides	CO + NO	adsorption-desorption characteristics	139

continued on p. 70

TABLE 3.4 (Cont'd)

Catalyst	Reaction catalysed	Activity related to	Ref.
Reactions involving carbon			
Metals	carbon gasification	screening	140
Alkalis	carbon gasification	screening	141
Alkali/metal oxides	coal gasification by steam	screening	142
Metals	carbon gasification by hydrogen	spill over	143
Metals, oxides, carbides	$CO \rightarrow C$	screening	144
Refractories	$CH_4 \rightarrow C$	electron location	145
Ni + metal oxide promotors	$CH_4 \rightarrow C$	efficiency of promotor and position in periodic table	146
Ni	hydrocarbon-steam	molecular weight and structure of hydrocarbon	147
Hydration/dehydration			
Precious metals	demethylation of toluene by steam	bifunctional	148
100 catalysts	dehydration	multiplet theory	149
Metal sulphates	propylene hydration	valency of metal	150

REFERENCES

1 J.A. Duffy, General Inorganic Chemistry, Longman (1966).
2 F. Basolo and R.G. Pearson, Mechanisms of Inorganic Reactions, John Wiley, London, (1969)
3 F.A. Cotton and G. Wilkinson, Advanced Inorganic Chemistry, Interscience, London (1962).
4 C.A. Coulson, "Valence", Oxford University Press (1952).
5 L. Pauling, "The Nature of the Chemical Bond", Cornell University Press, Ithaca (1940).
6 J.H. Van Uleck, J. Chem. Phys., 3, 803-807 (1935).
7 F.A. Cotton and G. Wilkinson, Basic Inorganic Chemistry, John Wiley, London, p.364 (1976).
8 D.A. Dowden and D. Wells, Actcs 2nd Congres du Catalyse, Technip, Paris, p.1499, (1961).
9 D.A. Dowden, Catal. REv. Sci. Eng., 5, 1 (1972).
10 F.J. Garrick, Phil. Mag., 9, 131 (1930), 10, 71,76 (9130), 14, 914 (1932).
11 H. Bethe, Ann. Physik, 3, 133 (1929).
12 N.F. Mott and H. Jones, "Theory of the properties of metals and alloys", Oxford Press (1936).
13 J.A. Catterrall, J. Roy. Inst. Chem., 84, 311 (1960).
14 W.R. Trost, Canad. J. Chem., 37, 460 (1959).
15 O.V. Krylov, Catalysis by Non-metals", Academic Press, London (1970).
16 K. Hauffe, Angew Chem., 68, 776 (1956).
17 F.F. Volkenshtein, The Electron Theory of Catalysis on Semiconductors, Pergamon Oxford (1963).
18 S.J. Thomson and G. Webb, "Heterogeneous Catalysis", Oxford Chemical Texts (1968).
19 P. Aigran and C. Dugas, Z. Elektrochcm., 56, 363 (1952).
20 K. Hauffe and H.J. Engell, Z. Electrochem., 56, 366 (1952).
21 Th. Volkenshtein, J. Phys. Chem. USSR, 32, 2383 (1958).
22 P.R. Wentrcek and H. Wise, J. Catal., 51, 80 (1978).
23 A.W. Sleight and W.J. Linn, Ann. N.Y. Acad. Sci., 272, 22 (1976).
24 A.W. Sleight, W.J. Linn and K. Aykan, Chem. Tech., 235 (1978).
25 G.C. Bond, Disc. Faraday Soc., 41, 200 (1966).
26 T. Goodenough, Magnetism and the Chemical Bond, Interscience, London (1963).
27 G. Blyholder, Comput. Chem. Educ. Res. (Proc. Int. Conf.) 3rd, 189 (1976).
28 S. Beran and R. Zagradnik, Kin. and Cat., 18, 299 (1977).
29 P. Kadura and W. Gunther, Z. Chem., 12, 476 (1972).
30 I. Horiuti and T. Toya, Solid State Surf. Sci., 1, 1 (1969).
31 D.A. Dowden, Chem. Eng. Symp. Sci., No. 73, Vol. 63, 90 (1967).
32 C.J. Swan and D.L. Trimm, J. Appl. Chem., 18, 340 (1968).
33 J.H. Sinfelt, Catal. Revs. Sci. Eng., 9, 147 (1974).
34 A.A. Balandin, Adv. in Catalysis, 10, 96 (1958).
35 L. Hiam, H. Wise and S. Chaikin, J. Catal., 9, 272 (1968).
36 B.S. Greensfelder, H.H. Voge and G.M. Good, Ind. Eng. Chem, 41, 2573 (1949).
37 G.C. Bond, Catalysis by Metals, Academic Press, New York, (1962).
38 J.A. Busby and D.L. Trimm, J. Catal, In press (1979).
39 Y. Moro-Oka, Y. Morikawa and A. Ozaki, J. Catal., 7, 23 (1967).
40 M.A. Vannice, Catal. Rev. Sci. Eng., 14, 157 (1976).
41 J.L. Carter, J.A. Cusumano and J.H. Sinfelt, J. Catal., 20, 223 (1971).
42 R.A. Dalla Betta, J.A. Cusumano and J.H. Sinfelt, J. Catal., 19, 343 (1970).
43 W.F. Taylor, D.J.C. Yates and J.H. Sinfelt, J. Phys. Chem., 68, 2962 (1964).
44 E. Kikuchi, M. Tsurumi and Y. Morita, J. Catal., 22, 226 (1971).
45 M. Boudart, A.W. Aldag, L.I. Ptak and J.E. Benson, J. Catal., 11, 35 (1968).
46 J.H. Sinfelt and D.J.C. Yates, J. Catal., 8, 82, 348 (1967).
47 J.H. Sinfelt and D.J.C. Yates, J. Catal., 10, 362 (1968).
48 G.C. Bond, AIChE, Symp. Ser., 73, 3 (1967).
49 D.L. Harrison, D. Nicholls and H. Steiner, J. Catal., 7, 359 (1967).
50 G.E. Batley, A. Ekstrom and D.A. Johnson, J. Catal., 34, 368 (1974).
51 A.B. Fasman and Zh.A. Ikhsanov, Kinet. Catal., 8, 53 (1967).

52 J.H. Sinfelt, AIChE., Symp. Ser., 73, 16 (1967).
53 S. Trasatki, Trans. Faraday Soc., 1, 2, 229 (1972).
54 R.B. Moyes, K. Baron and R.C. Squire, J. Catal., 22, 333 (1971).
55 W.G. McNaught, C. Kemball and H.F. Leach, J. Catal., 22, 333 (1971).
56 I.R. Konenko, et. al., Kinet. Catal., 14, 163, 359 (1973).
57 I.R. Konenko, et. al., Kinet. Catal., 16, 363 (1975).
58 R.N. Zhavoronkova and L.M. Korabel'nikova, Kinet. Catal., 14, 844 (1976).
59 K. Koechloefl and V. Bazant, J. CAtal., 10, 140 (1968).
60 J. Simonikova, A. Ralkova and K. Kochloefl, J. Catal., 29, 412 (1973).
61 L.D. Rollmann, J. Catal., 46, 243 (1977).
62 Y. Fujii and J.C. Bailar, J. Catal., 52, 342 (1978).
63 J.H. Sinfelt, J. Catal., 29, 308 (1973).
64 J.H. Sinfelt, J.L. Carter and D.J.C. Yates, J. Catal., 24, 283 (1972).
65 T.J. Plunkett and J.K.A. Clarke, Trans. Faraday Soc., 1, 4, 600 (1972).
66 A. Metcalfe and M.W. Rowden, J. Catal., 22, 30 (1971).
67 G.D. Lyubarskii, E.I. Evzerikhin and A.S. Slinkin, Kinet. Catal., 5, 277 (1964).
68 R.L. Moss and H.R. Gibbens, J. Catal., 24, 48 (1972).
69 R.L. Moss, H.R. Gibbens and D.H. Thomas, J. Catal., 16, 181 (1970).
70 R.L. Moss and D.H. Thomas, J. Catal., 8, 151, 162 (1967).
71 R.L. Moss, H.R. Gibbens and D.H. Thomas, J. Catal., 16, 117 (1970).
72 T.J. Gray, N.G. Masse and H.G. Oswin, Proc. 2nd Intern. Congr. Catalysis, Paris, Technip, p.1697 (1961).
73 T.V. Lipets, Sh.L. Vert and I.P. Tverdovskii, Kinet. Catal., 10, 167 (1969).
74 J.E. Germain, Catalytic Conversion of Hydrocarbons, Academic Press, New York, 1969.
75 C.L. Thomas and D.S. Barmby, J. Catal., 12, 341 (1968).
76 Yu.I. Kozorezov, Kinet. Catal., 14, 1048, 1309 (1971).
77 I. Mochida and Y. Yoneda, J. Catal., 7, 386, 393 (1967).
78 Y. Yoneda, J. Catal., 9, 51 (1967).
79 V.A. Dzisko and M.S. Borisova, Kinet. Catal., 1, 130 (1960).
80 J.A. Gray and J.T. Cobb, J. Catal., 36, 125 (1975).
81 M. Misono, Y. Saito and Y. Yoneda, J. Catal., 9, 135 (1967).
82 Kh.M. Minachev and Yu.S. Khodakov, Kinet. Catal., 6, 74 (1965).
83 G.K. Boreskov, Kinet. Catal., 11, 1 (1970).
84 R. Killrack, J. Catal., 18, 314 (1970).
85 E.A. Mamedov, V.V. Popovskii and G.K. Boreskov, Kinet. Catal., 11, 799, 807, (1970).
86 T.T. Bakumento, Kinet. Catal., 6, 61 (1965).
87 B. Dmuchovsky, M.C. Frearks and F.B. Zienty, J. Catal., 4, 577 (1965).
88 T.V. Andrushkevich, V.V. Popovskii and G.K. Boreskov, Kinet. Catal., 6, 777, (1965).
89 N.W. Cant and W.K. Hall, J. Catal., 27, 70 (1972), 16, 220 (1970).
90 M.Ya. Rubanik, K.M. Kholyavenko, A.V. Gershingorina and V.I. Lazukin, Kinet. Catal., 5, 588 (1964).
91 V.V. Popovskii and Yu.D. Zverev, Kinet. Catal., 12, 529 (1971).
92 Kh.M. Minachev, D.A. Kondtrat'ev and G.V. Autoshin, Kinet. Catal., 8, 108 (1967).
93 G.K. Boreskov, V.V. Popovskii, N.I. Lebedova, V.A. Sazonov, T.V. Andrush-Kevich, Kinet. Catal. 11, 1039 (1970).
94 V.V. Popovskii, G.K. Boreskov, Z. Dzeventski, V.S. Muzykantrov and T.T. Shul'master, Kinet. Catal., 12, 871 (1971).
95 R. Rudham and M.K. Sanders, J. Catal., 27, 287 (1972).
96 S.J. Gentry, R. Rudham and M.K. Sanders, J. Catal., 35, 376 (1974).
97 V.V. Popovskii, Kinet. Catal., 10, 872 (1969).
98 T. Uchijima, Y. Ishida, N. Uemitsu and Y. Yoneda, J. Catal., 29, 60 (1973).
99 L.A. Sazonov, G.A. Mitrofanova, L.V. Preobrazhenskaya and S.V. Moskvina, Kinet. Catal., 13, 705 (1972).
100 G.K. Boreskov, L.A. Kasatkina and V.G. Amerikov, Kinet. Catal., 10, 79 (1969).
101 A.P. Dzizyak, G.K. Boreskov and L.A. Kasatkina, Kinet. Catal., 3, 65 (1962).
102 G.K. Boreskov, S.A. Ven'yaminov, N.N. Sazonova, Yu.D. Pankrat'ev and A.N. Pitaeva, Kinet. Catal., 16, 1253 (1975).

103 G.K. Boreskov, Kinet. Catal., 14, 2 (1973).
104 F. Trifiro, P. Centola and I. Pasquon, J. Catal., 10, 86 (1968).
105 F. Trifiro and I. Pasquon, J. Catal., 12, 412 (1968).
106 V.A. Sazonov, V.V. Popovskii and G.K. Boreskov, Kinet. Catal., 9, 255 (1968).
107 O.N. Kimkhai, V.V. Popovskii, G.K. Boreskov, et. al., Kinet. Catal., 12, 322 (1971), 13, 815 (1972).
108 Y. Moro-Oka and A. Ozaki, J. Catal., 5, 116 (1966).
109 V.V. Popovskii, G.K. Boreskov, V.S. Muzykantov, V.A. Sazonov and S.G. Shubnikov, Kinet. Catal., 10,683 (1961).
110 K. Tanaka and K. Tamaru, Kinet. Catal., 7, 219 (1966).
111 V.A. Levin, T.V. Vernova and A.L. Tsailingold, Kinet. Catal, 13, 454 (1972).
112 G.K. Boreskov, Kinet. and Catal., 8, 878 (1967).
113 L.A. Sazonov, E.V. Arkamonov and G.N. Mitrofanova, Kinet. Catal., 12, 329 (1971).
114 V.M. Zhiznevskii and E.V. Fedevich, Kinet. Catal., 12, 1073 (1971).
115 M.N. Yakubovich, Ya.B. Gorkhoratskii, T.G. Alkhazov and K.Yu. Adzhamov, Kinet. Catal., 17, 1310 (1977).
116 Yu.V. Belokopytov, K.M. Kholyavenko and M.Ya. Rubanik, Kinet. Catal., 14, 1124, (1974).
117 L.V. Skalkina, I.K. Kolchin and L.Ya. Margolis, Kinet. Catal., 9, 84 (1968).
118 V.M. Zhiznevskii, E.V. Fedevich, D.K. Tolopko and I.M. Sulima, Kinet. Catal., 12, 374 (1971).
119 K. Ohdan, et al., Kogyo Kagaku Zasshi, 73, 292 (1970).
120 T. Seiyama, M. Egashira, M. Iwamoto, Rep. Sov. Jap. Seminar on Catalysis, I, 35 (1971).
121 C.J. Norton and T.E. Moss, Ind. Eng. Chem. Process. Des. Dev., 3, 23 (1964).
122 N.P. Evmenenko, Yu.B. Gorokhovatskii and M.V. Kost', Kinet. Catal., 12, 1275 (1971).
123 L.L. Ioffe and N.V. Klimova, Kinet. Catal., 4, 682 (1963).
124 A. Mukherjee and W.F. Graydon, J. Phys. Chem., 71, 4232 (1967).
125 Emanuel: Eighth World Pet. Congr. IV, 408.
126 E.S. Gould and M. Rado, J. Catal., 13, 238 (1969).
127 Ya.B. Gorokhovatskii, Kinet. Catal., 14, 62 (1973).
128 D. Klisurski, Kinet. Catal., 11, 221 (1970).
129 N.I. Ilchenko, Kinet. and Catal., 18, 26 (1977).
130 Yu.I. Pyatnitskii, G.I. Golodets and T.G. Skorbilina, Kinet. Catal, 17, 125 (1976).
131 T.P. Kobylinski and B.W. Taylor, J. Catal., 33, 376 (1974).
132 A.K. Vijh, J. Catal., 31, 51 (1973)
133 Y. Saito, Y. Yoneda and S. Makishima, Actcs 2nd Intern. Congr. Catalysis, Paris, 1937.
134 T.M. Yur'eva, V.V. Popovskii and G.K. Boreskov, Kinet. Catal., 6, 941 (1965).
135 E.R.S. Winter, J. Catal., 34, 431, 440 (1974).
136 E.R.S. Winter, J. Catal., 22, 158 (1971).
137 E.R.S. Winter, J. Catal., 15, 144 (1969).
138 V.M. Belousov, I.Ya. Mulik and M.Ya. Rubanik, Kinet. Catal., 10, 687 (1969).
139 M. Shelef, K. Otto and H. Gandhi, J. Catal., 12, 361 (1968).
140 R.T. Rewick, P.R. Wentrcek and H. Wise, Fuel, 53, 274 (1974).
141 K. Otto and M. Shelef, Proc. VI Internat. Congr. Catalysis, London, p.1082 1976.
142 W.P. Haynes, S.J. Gasior and A.J. Forney, Adv. Chem. Ser., 131, 179 (1974).
143 A. Tomita and Y. Tamai, J. Catal., 27, 293 (1972).
144 R.P. Todorov, D.K. Lambiev and St. M. Mechkova, Kinet. Catal., 15, 611 (1974).
145 V.M. Grosheva and G.V. Samsonov, Kinet. Catal., 7, 782 (1966).
146 V.V. Veselov and P.S. Pilipenko, Kinet. Catal., 12, 939 (1971).
147 M.R. Arnold, K. Atwood, H.M. Baugh and H.D. Smyser, Ind. Eng. Chem., 44, 999 (1952).
148 G.L. Rabinovich, G.V. Dydykina, G.N. Maslyanskii, and M.I. Dement'eva, Kinet. Catal., 15, 847 (1974).
149 A.A. Baladin, Zh. Fiz. Khim., 31, 745 (1957).
150 Y. Ogino, J. Catal., 8, 64 (1967).
151 J.M. Criado, J. Catal., 55, 109 (1978).

152 K. Fukuda, M. Dokiya, T. Kameyama and Y. Kotera, J. Catal., 49, 379 (1977).
153 B.P. Fish, D.A. Dowden and D. Wells, Unpublished results reported in reference 2.
154 V.V. Styrov, Kinet. Catal., 9, 101 (1968).
155 T.S. Sukhareva and A.V. Mashkina, Kinet. Catal., 12, 1183 (1971).
156 A.V. Mashkina, Kinet. Catal., 12, 95 (1971).
157 A.V. Mashkina, T.S. Sukhareva and G.L. Veitsman, Kinet. Catal., 13, 223 (1972).
158 O. Weisser and S. Landa, Sulphide Catalysts: their properties and applications, Pergamon, Oxford, and Vieweg, Braunschweig, 1973.

CHAPTER 4

DESIGN OF THE SECONDARY COMPONENTS OF A CATALYST

I. INTRODUCTION

The term secondary components of a catalyst is used in this text as indicating any minor component which is added to a catalyst to improve selectivity or activity. Basically, the need for secondary components emerges only from experimental testing of primary components. This shows that, for whatever reason, the catalyst is not behaving as is desired and, as a result, secondary components are added to improve the situation. One of these, a spacer used to reduce catalyst sintering, is discussed in Chapter 5. Others, such as dopers, additives, poisons etc., are dealt with in this Chapter.

As was stated earlier, there are two ways of approaching the design of secondary components. The first can be described as the application of scientific common sense to the problem: the second is more scientific but is more time consuming. As such it is usually used only when it has been established that the catalyst is of considerable interest.

The first approach is best illustrated by example. Thus, for example, in considering the design of a catalyst for hydrocarbon isomerisation, it could be found that products originating from cracking reactions are significant. These could be minimised either by reducing the acidity of the catalyst or by reducing the temperature. In the former case, acidity could be reduced by poisoning with alkali (a secondary component), or by choosing a catalyst of lower acidity.

Similarly, for the design of a methanation catalyst, it was suspected that removal of water from the active site could accelerate the reaction. The chemistry of oxygen deficient oxides suggested that steam could be chemisorbed via the reaction

$$MO_{x-1} + H_2O \rightarrow MO_x + H_2$$

with an order of activity (ref. 1):

$$UO_2/U_3O_8 > MoO_2/MoO_3 > WO_2/WO_3 > Pr_2O_3/Pr_4O_{10} > Ce_2O_3/CeO_2 > CrO_2/Cr_2O_3$$

As a result, a small amount of urania was added to the catalyst and methanation was, indeed found to accelerate.

Again, in the oxidative dehydroaromatisation of olefins, carbon dioxide was found to be an unwanted side product (ref. 2,3). Whatever else is involved, the production of carbon dioxide requires more oxygen than any other product. As a result an oxygen repelling additive was added to the catalyst, with the net effect that the production of carbon dioxide decreased and the selectivity increased (ref. 3).

Obviously there are as many applications of this kind of reasoning as there are catalyst designs to do. As a result, it is probably best to use the examples in the second part of this book to illustrate different applications.

The second approach to the design of secondary components is based on a study of the reaction mechanism, in the expectation that this will lead to the recognition of minor components that will improve performance. This is sometimes, but not always, justified. Studies of the mechanism may be carried out with respect to the reaction and/or with respect to the catalyst.

II. SECONDARY COMPONENT DESIGN VIA MECHANISTIC STUDIES

The second route to the design of secondary components of a catalyst involves knowledge of the mechanism of the catalytic reaction. In many senses it is a classical argument that the understanding of the mechanism of catalysis must lead to improvements in the catalyst, an argument that has been disproved as often as it has been proved! Since, however, such understanding can lead to improvements then, if the catalyst is of sufficient interest, the effort may be worthwhile. The problem is that such mechanistic studies can take considerable time (of the order of years), and no guaranteed advantage can result. As a consequence, this type of approach is treated with caution by industrial research management.

In terms of the arguments already presented, the picture is fairly clear. The design of the primary components rests on the construction of a reaction sequence on the surface: in most circumstances this is not based on fact, but on probabilities. Thus, for example, the literature may show that it is probable that a reactant will adsorb in a given form on a given surface, or that it is probable that a particular solid will catalyse a particular reaction sequence. Any studies of any stage of the suggested mechanism must result in a much firmer basis for design and usually in a firmer basis for the selection of secondary components.

Various approaches to the study of the reaction mechanism have been well described (refs. 4,5,6) and a detailed survey would probably complicate the concept of design more than it would clarify. Perhaps the most important point to remember is that it is necessary, where possible, to get as close to the conditions of the catalytic reaction as possible. Many modern techniques give extremely valuable information, but this may be suspect because, for example, the pressure range applicable is some orders of magnitude less than that relevant to the catalytic reaction under study (refs. 7,8). As a result, some older techniques may (or may not) still give more useful information. In this connection, the application of isotopic labelling techniques (ref. 4) is certainly less fashionable but can be extremely useful.

One other problem arises with the use of mechanistic studies in the context of design. This is that the situation may be complicated by interrelated changes in a complex catalyst. What one would really like to do would be to vary one parameter - and only one parameter - of a catalyst at a time, in order to study its effect on activity and selectivity. In recent years, several attempts have been made to reach this situation, and these are of interest in the present context.

The general approach to the problem is to try to set up a "framework" catalyst, in which the catalytic species of interest can be locked into a given environment: this may be geometric or electronic. Changes can then be made in the "framework" to induce desired alterations in the catalyst.

The success of this approach depends on the system chosen. The use of one of the most obvious frameworks (alloy catalysts) has probably led to more problems than it has solved (refs. 9,10,11,12), but solid solutions of oxides have proved profitable to study (refs. 13-17). In addition, it has proved possible to substitute catalytic species into a crystalline unit cell, an environment which has proved to be an excellent framework (refs. 13-18): care must be taken, however, to ensure that the framework itself does not affect catalytic activity.

It is rewarding to consider different types of system individually, since their advantages and disadvantages vary considerably.

(a) Alloy catalysts

Both as a result of theoretical (refs. 9,10,11,12) and industrial (refs. 19,20, 21,22) interest, considerable attention has been focused on alloy catalysts for a long time. In the past, theoretical interest has focused on their properties in connection with electron band theories of catalysis (Chapter 3): this is now changing, as will be discussed below. Industrially, bimetallic or multimetallic reforming, hydrogenation and oxidation catalysts are of considerable interest (refs. 11,19).

From the point of view of catalyst design, alloy catalysts are of interest in their own right and because they can show two distinct effects:

(a) The influence of geometry, called the 'ensemble' effect.

(b) The influence of electronic interaction, called the 'ligand' effect.

The former case can be regarded as the effect of dilution of one metal by another. Both Kobozev (ref. 22) and Dowden (refs. 19,23) have speculated on the role of ensembles of surface atoms (see above). Thus, for example, adsorption of a molecule may require one, two or more adjacent surface atoms. Carbon monoxide may adsorb linearly or bridge-bonded (ref. 24) while the adsorption of oxygen on silver as O^{2-} has been suggested to require four adjacent silver ions (ref. 25). Similarly, a catalytic reaction may require an ensemble of several atoms. Methanation (refs. 26,27) and hydrogenolysis (refs, 11,28) have both been suggested to require four or more adjacent atoms. If a catalytically active metal is diluted by alloying with an inactive metal (such as Ni-Cu alloys (refs. 11,29)), then the change of occurrence of such ensembles must be reduced. We will see later that a direct correlation between activity and dilution would not be expected, since one component of an alloy tends to enrich the surface (refs. 11,30,31). As a result, the surface composition may be very different from the bulk.

Interest in electronic factors in alloys was initiated on the basis of predictions of the electron band theory (Chapter 3). The advent of modern techniques (refs. 32,33) have shown that complete mixing of electrons in an alloy is unlikely (refs. 34,35): instead the components of the alloy retain much of their individuality, although this may be modified by the presence of the other component (refs. 34,35). It is these modifications which are of more interest in modern thinking about electronic effects. In many ways, the concept is similar to the crystal field approach (Chapter 3) in that an atom in the surface is viewed as being attached to ligands, which can be gas, metal A or metal B. The electron density on an atom will depend upon interactions with near neighbours, and changing the composition of an alloy will obviously change electron energy distributions (refs. 36,37). Again, however, surface enrichment by one component of an alloy will mean that surface effects will be different from bulk effects.

Alloy catalysts are both a blessing and a curse for the catalyst designer. They are, of course, of great interest in their own right and, at least on paper, they offer many possibilities of theoretical interest. Some of these are of definite practical interest: others could be of interest, but it is difficult to know what to do about them.

The ensemble effect is certainly of interest. Thus, for example, studies have shown that some reactions require an ensemble of sites (refs. 26,27,28) and, if such reactions are undesired, they may be minimised by alloying. Ponec (ref. 11), working with Ni/Cu alloys, managed to distinguish between reactions in which the effect of alloying was approximately proportional to the alloy surface composition ($RH-D_2$ exchange, hydrogenation and dehydrogenation of RC=CR' or RC=O) and those in which the effect on selectivity was much greater than that to be expected on the grounds of dilution alone. This latter category, in which an ensemble is required, included C-C bond breakage, C-C-C bond rearrangement, methanation, $R-NH_2$ disproportionation etc.

More careful work of this type is needed, but the problems are very real - not only in measuring catalytic activity/selectivity but also in determining surface structures. Some general rules emerge from the mass of information available on the latter (11):

a) When an alloy contains two equilibrated phases, then the alloy with the lower surface energy (higher rate of diffusion, lower energy of sublimation) will concentrate at the surface (refs. 38,39).

b) In one-phase alloys, the surface tends to be enriched by the component of lowest surface energy (refs. 38,39,40).

c) The surface composition is affected by the ambient gas. The component with the higher heat of adsorption of the ambient gas tends to migrate to the surface (refs. 39,40,41).

d) For evaporated film alloys, the Tamman temperature of equilibration has to be ca. 0.3 for equilibration to be attained in a reasonable time (ref. 42).

These rules help to predict surface enrichment, but give no quantitative descriptions which would be necessary to study ensemble effects. As a result, design of catalysts could involve

(i) A literature search for quantitative information of alloy surfaces in question.

(ii) Measurement of the surface composition of alloys.

(iii) Trial and error testing based on the possibility of an ensemble effect.

It is obvious that course (ii) is undertaken with great reluctance.

The ligand effect ought to be of interest to the catalyst designer, but the main problem lies in prediction. The calculations can be very complex (Chapter 3), always provided that one knows the nature of the surface in the neighbourhood of a given atom. As a result, it is rare that the approach is actually used in design.

Part of the problem is seen to lie in the nature of the bulk and the surface of an alloy. It might be thought that some difficulties could be avoided by use of metal solid solutions (refs. 9,19). The problems that arise, however, are almost exactly the same as those inherent to alloys. Electronic interaction between near neighbours is possible, and there is no certainty, even in dilute solution that clusters of a minority component will not be formed on the surface.

It would seem that alloy catalysts produce almost as many problems as they solve. Nonetheless, the concept behind their use as a design tool is valid and they are, indeed, very important catalysts in their own right (ref. 19). What is needed is some less complicated framework that can be used to study the reaction mechanisms.

(b) Metal cluster catalysts

Since many catalysts involve transition metals supported on oxidic supports, perhaps the most obvious way to look at the catalyst is to remove the influence of the support. In recent years it would appear that this is possible, since methods of preparing the so-called metal cluster catalysts (ref. 43) have been developed. These consist of clusters of metal atoms (containing at least two atoms) which may include one or more metallic elements (refs. 43,44). They would appear to have promise as framework catalysts a) because small clusters of mono or bi-metallic solids may be prepared and tested in the absence of support effects and b) because, if the cluster is small enough, ligand (or electronic) effects originating from anything else but very near neighbours will be absent.

Much of this promise remains unfulfilled for the catalyst designer, for various reasons. This can easily be seen from the nature of the problem. In designing secondary components, we wish to study the mechanism of the reaction and how it is affected by a given change in the catalyst. It is certainly true that the effect of the absence of support can be studied using these systems, but we should then ask whether the metal clusters are truly representative of the metal particles on the support. By definition, small metal clusters must have high surface energy, and this will depend upon methods of preparation (refs. 43,44,45). Supported metals, on the other hand, although localised in small particles, have probably equilibrated during preparation. As a result, some difference in surface energy (and possibly configuration) can be expected between supported metals and metal clusters.

The second problem lies in the fact that the designer wishes to change a parameter and to study the effect of this on catalytic activity/selectivity. Although we begin to understand how metal clusters may be prepared (refs. 43,44,45), we still do not know enough about their preparation and characterisation to induce any desired change, although we may move between known metal clusters (e.g. examine the activity/selectivity of $M_R M'_B$ and $M_C M'_D$, where $M_A M'_B$ and $M_C M'_D$ are known clusters). However, from the viewpoint of catalyst design, we wish to know the effect of a given change on the <u>overall</u> process i.e. the support, although left unchanged, may influence the results in different ways, if a different intermediate is produced from different metals.

As a result, metal cluster catalysts must be considered to have considerable potential for understanding the mechanism of catalysis, and, as a consequence, for use in the design of secondary components. At the moment, insufficient knowledge is available to realise this potential, except for particular combinations of metals.

The next refinement of this concept would seem to be to study the effect of single metal atoms on a support. Indeed, these can be prepared (as metals in zeolites (ref. 46) or metals on ion exchange resins (ref. 47)) and the catalysts are found to be active. Under these circumstances, however, the exact opposite effects are found: the support has an influence on the metal atom far beyond that experienced with normal catalysts. This may be due to the size limitations set up by the molecular sieve (ref. 46), to enormous gradients of potential that may exist within the cage of a molecular sieve (ref. 48), to the nature of the ion exchange resin itself (ref. 47) or to interaction between the metal and the resin/sieve (ref.48). In any case it is rare that the behaviour of a metal in such an environment is typical of the isolated or supported metal and, as a design tool (although not as catalysts in their own right) these systems have very limited applicability.

(c) Metal oxide solid solutions

One of the most widely applicable framework catalyst systems involves a series of solid solutions of metal oxides (ref. 13), in which the properties of a guest metal ion can be strongly influenced by a host metal oxide lattice. The concentration of the guest ion is never large, but - by variation of the concentration and the nature of the guest ion as well as by varying the host lattice - it is possible to study the effect of different factors on catalytic activity. These include:

(i) By varying the nature of the host lattice, it is possible to change the local coordination environment of the guest ion, allowing the study of crystal field effects on the catalytic behaviour of the guest ion.

(ii) By keeping the host lattice constant and varying the nature of the guest ion (but not its concentration), it is possible to study the intrinsic activity of different guest ions under otherwise constant conditions.

(iii) By varying the concentration of guest ions, it should be possible to study the effect of their electronic and geometric interaction on catalytic activity.

Generally the solid solutions involve host oxides that are relatively inactive as catalysts and form a homogeneous solution with the guest oxide. This can be very important since, for example, if one seeks the effect of a given parameter on catalytic activity in the context of design, then it is not desired to complicate the picture with what can be loosely described as "support" effects. Distinction between effects due to the "catalyst" and the "support" can be complicated in such solid solutions (refs. 13,14,15).

The exploitation of the systems in the context of the design of secondary components of a catalyst can now be considered.

(i) Changes in the local coordination environment.

In terms of design, the basic question revolves around the recognition of the best coordination environment which favours catalytic acitivity and selectivity. If it is possible to recognise such a geometry, the possibility of obtaining it and maintaining it in a conventional catalyst is a second - and perhaps more difficult - question.

Some idea of the possibilities open to study can be obtained by inspection of Table 4.1 (ref. 13), which lists host lattices and their coordination. Some care must be taken with this table. Thus, for example, if the guest ion has a valcency different to that of the host lattice, then electrostatic balance must be maintained (for example, by adding Li^{+} to balance charge on adding Mn^{4+} to MgO). Indeed, the amount of Mn^{2+}, Mn^{3+} and Mn^{4+} will be strongly influenced by the amount of Li^{+} present (ref. 49).

It is seen from Table 4.1 that octahedral coordination in the host lattice is easily obtained, but tetrahedral coordination is usually achieved by use of ZnO (which may have catalytic activity of its own). Spinels may involve either octa-

TABLE 4.1

Local co-ordination environments in host metal oxides

Local coordination	Host oxide	Interactions	Examples
Octahedral	MgO	M-O-M	Cr^{3+} (50), Mn^{2+}-Mn^{4+} (49), Fe^{2+}-Fe^{3+} (51)
			Co^{2+} (14,15) Ni^{2+} (52), Cu^{2+} (53)
Octahedral	α-Al_2O_3	M-O-M	Cr^{3+} (16)
		M-O_2-M	V^{3+} (54)
Octahedral	perovskites	M-O-M	see next section
Tetrahedral	ZnO		Cu^{2+} (54), Co^{2+} (14,15)
Octahedral (B site)/	$MgAl_2O_4$	(B)M-O-O-M(B)	Ni^{2+} (55), Co^{2+} (56), Cr^{3+} (57)
Tetrahedral (A site)		(B)M-O-M(A)	V^{3+}-V^{4+} (13)

hedral or tetrahedral coordination. In theory, then, it is easy to study the effect of local coordination geometry on catalysis, and, indeed, several valuable studies have been completed (ref. 13). Complications do arise, however, in that the concentration of guest ion can have a significant effect on catalysis. A general activity pattern has been recognised in which specific activity is very high at high dilution but drops as the concentration of guest ion approaches 10-15%: at higher concentration the activity rises again (refs. 13,54). An explanation of this behaviour has been advanced in terms of electronic interactions in the catalyst (ref. 13), but it is clear that the effect of the coordination environment will be most significant at low dilution, and could be affected by other factors at higher concentrations of guest ion.

(ii) The intrinsic activity of guest ions.

Evaluation of the intrinsic activity of metal ions is of considerable theoretical importance (ref. 59), but is of less significance in the context of catalyst design. Thus, for example, application of the crystal field theory suggests a twin peaked activity pattern (Chapter 3), and this is confirmed by experimental testing. It must be asked, however, exactly how dependant are the experimental results on factors other than the crystal field (e.g. the presence of defects etc.). If metal ions can be locked in exactly the same environment, as in a solid solution, then no such doubts arise. The preparation of solid solutions of transition metal ions shows that the twin peaked pattern is, indeed, experimentally verifiable (ref. 13).

For the purposes of design, this approach is perhaps most useful for establishing the role of individual metal ions present in a complex catalyst. Thus, for example, in a catalyst such as copper promoted zinc oxide-chromia (ref. 60), the approach could be used to distinguish the role of each metal ion.

(iii) Electronic and geometric interactions

In theory, it should be possible to study the dependence of catalytic activity on electronic and geometric interactions by varying the concentration of guest metal ion. In practice, this is not yet possible, for a variety of reasons. As stated above, one major problem is that we do not fully understand the dependence of activity on concentration (ref. 13) and it seems, at least in some cases, that a different reaction mechanism may be important at low and high concentrations of guest ion (ref. 13).

The second problem is that of characterisation. In considering the conversion of propylene to benzene (Chapter 2) it was suggested that the reaction could involve two adsorbed intermediates on one metal ion centre or one adsorbed intermediate on one ion reacting with another adsorbed intermediate on another ion. It would obviously be of interest to distinguish these possibilities, but then we come up against the difficulty of how to recognise the presence of one or two neighbouring metal ions in the surface. The preparation technique can be adjusted to favour one or the other in a solid solution (refs. 13,61), but experimental techniques are generally not advanced enough to characterise clearly the resulting catalyst in the detail required.

As a result, the possibility of using solid solutions to study possible electronic and geometric interactions in catalysis must be regarded as having considerable potential for the future, rather than as being a useful tool for the present.

Although the use of solid solutions would seem to be hedged with uncertainties, there is no doubt that the systems can be extremely useful. Vickerman has presented an excellent review of their use in the context of simple reactions (ref. 13), but it is interesting to consider a more complex reaction where the application of framework catalysts has proved invaluable. This is the oxidation of prop1lene to acrolein over bismuth molybdate based catalysts (ref. 62).

(d) Specific examples of oxide solid solution catalysts applied to design

Studies of one particular unit cell framework have proved very successful. As stated, the catalyst of interest involved bismuth and molybdenum oxides (refs. 62, 63), and the scheelite structure was used as the framework (refs. 18,64). The unit cell of a scheelite is shown in Fig. 4.1.

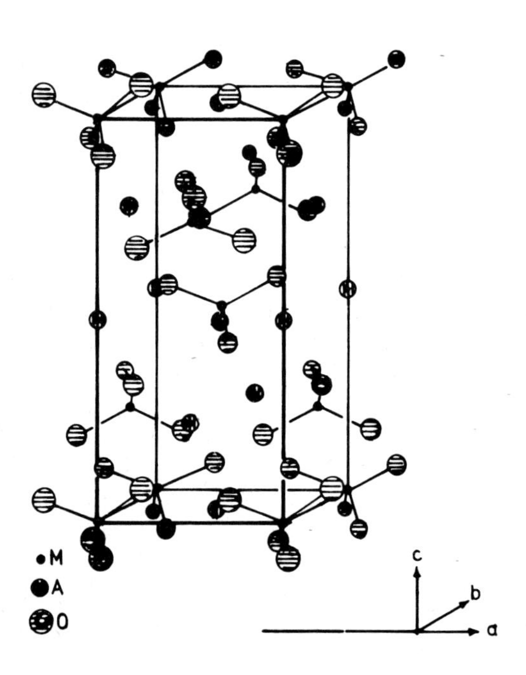

Fig. 4.1. The unit cell of a scheelite.

The general formula for scheelite oxides is AMO_4, the M cation being tetrahedrally coordinated. About one hundred compounds are known with this structure, and typical examples are given in Table 4.2 (ref. 18).

TABLE 4.2
Scheelite structure oxides
General structure AMO_4

	1+	2+	3+	4+	5+	6+	7+	8+
A	Li,Na,K, Rb,Cs,Ag, NH_4	Ca,Sr,Ba Cd,Pb,Gu	Bi,rare earths	Zr,HF,Th, Ce,U				
M		Zn	Ga,Fe	Ge,Ti	V,As,Nb,Ta Mo	Mo,W,Cr,S	Re,Tc,Ru	Os

Large concentrations of A cation vacancies can be introduced (up to one-third of the A ion content (ref. 65)) and, although low, some oxygen vacancies or interstitial ions must be present since the oxygen ion mobility can be high (ref. 18). Sleight (ref. 18) has reported the phase diagram for at least one scheelite $(Pb_{1-3x}Bi_{2x}\phi_x)MoO_4$.

The allylic oxidation of olefins over metal oxides has been studied in some detail (refs. 18,64). The overall reaction is known to go via an allylic oxidation route, in which a pi adsorbed olefin is involved (ref. 66), e.g.

$$H_3C\text{-}CH{=}CH_2 \;+\; O^{2-}\text{-}M\text{-}O^{2-} \;\rightleftarrows\; \begin{array}{c} CH_3\text{-}CH{=}CH_2 \\ \downarrow \\ O^{2-}\text{-}M\text{-}O^{2-} \\ \downarrow \\ CH_2\text{-}CH\text{-}CH_2^{\ominus} \\ \cdots\cdots\cdots\cdots \\ \downarrow \\ HO^{-}\text{-}M\text{-}O^{2-} \\ \downarrow \\ CH_2\text{-}CH\text{-}CH_2 \\ \cdots\cdots\cdots\cdots \\ \downarrow \\ HO^{-}\text{-}M^{-1}\text{-}O^{2-} \end{array}$$

$$CH_2{=}CH.CHO \;\leftarrow\; HO^{-}\text{-}M^{-1}\text{-}O^{2-}$$

Over a catalyst such as $BiMoO_4$, there is considerable uncertainty as to the role of Bi, of Mo, of defects and of oxygen. Much of this uncertainty can be resolved by the application of isotopic tracer techniques. Thus, for example, the existence of the pi-allyl adsorbed species was confirmed by the use of carbon tracers (ref. 66) and, more recently, it has been established that oxygen in acrolein always comes from the $(Bi_2O_2)_n^{2+}$ layer (ref. 67) while replacement of oxygen in the lattice probably involves the $(MoO_2)_n^{2+}$ layer. There are, however, many questions that cannot be resolved by isotopic tracers.

Thus, for example, it is impossible to determine the role of defects by isotopic labelling, and this is one area where framework catalysts such as scheelites come into their own. Use of the scheelites showed that the role of the defect appears to be to promote the formation of the allyl radical (C). Since this is rate-determining (ref. 66) defects are very important to the mechanism. Bismuth, on the other hand, was suggested to have an important role in rapidly replacing the oxygen at an active site. Sleight argued that trivalent bismuth would be stable on the surface, since it can orientate its lone pair of electrons in the direction of the gas. Molecular oxygen, seeking electron density on the surface, will be attracted to these electrons and the surface Bi^{3+} temporarily takes on the bulk coordination. The adsorbed oxygen anions can then be passed onto the active site on demand (ref. 68). Sleight and Linn have showed that the electronic structure of the catalyst is consistent with this picture, in view of the small energy gap between the Bi6p and Mo4d bands.

This work, carried out before the isotopic studies used to show that the acrolein oxygen originates from the bismuth (ref. 67), is important in the present context largely as an excellent example of the use of a framework catalyst. The accuracy of the mechanism proposed is less important than the concept of a framework in which it is possible to change one parameter at a time. Given the limitations of components summarised in Table 4.2, it would seem that the scheelites offer a valuable framework system for the study of other reactions.

Another type of framework catalyst that has received a great deal of attention (and particularly from the Bell laboratories) are the perovskites (refs. 70-74). These have been used as frameworks for transition metal ions, in particular. The crystal structure of the unit cell is shown in Fig. 4.2, and the transition metal ions are in octahedral coordination at the corners of the cube. The central ion may be Ca, Sr, Ba, Na, K, Rb, Bi, Pb, La or a rare earth metal.

Perovskites are of interest as catalysts in their own right, since Ru, substituted into the lattice, is used as an automobile exhaust clean-up catalyst (refs. 72,73). The solid behaves as if Ru was isolated in a matrix which, although it retains the Ru in situ, does not interfere with the catalytic properties of the metal.

As opposed to the scheelites, perovskites do not really offer the same degree of control of one parameter in the catalyst, since the strength of the metal-oxygen bonds dictate the involvement of lattice oxygen in the catalytic reaction

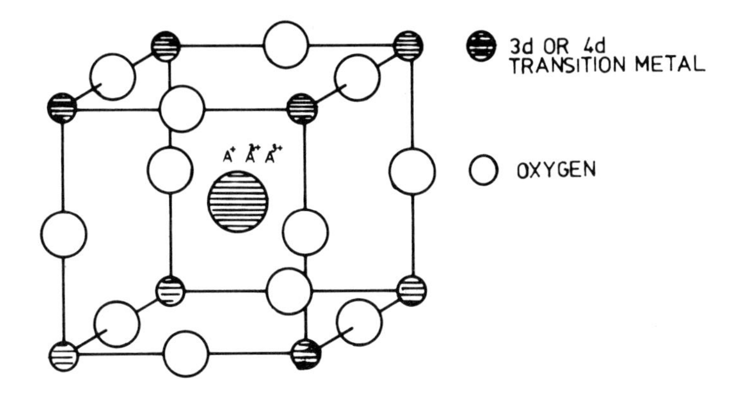

Fig. 4.2. The unit cell of a perovskite.

(refs. 72,73). This was shown elegantly by Voorhoeve, Remeika and Trimble (refs. 72,73), who studied the oxidation of carbon monoxide and the reduction of nitric oxide on transition metal perovskites. In the latter case, the evidence points to a reduction-oxidation mechanism (ref. 73).

$$\text{M-O-M} + \text{CO} \rightarrow \text{M-V}_0\text{-M} + \text{CO}_2$$
$$\text{M-V}_0\text{-M} + \text{NO} \rightarrow \tfrac{1}{2}\text{N}_2 + \text{M-O-M}$$

Where V_0 is an oxygen vacancy. In this case the metal-oxygen bonding will be important in dictating the activity of the catalyst.

In the oxidation of carbon monoxide, on the other hand, the efficiency of the catalysts seems to be dictated by the symmetry and energy of electron orbitals near to the Fermi level. Given the perovskite formula (AMO_3), changes in the A cation valency can be expected to result in corresponding changes in the B cation. Correlations of the activity of the catalysts with such substitutions (ref. 73) suggests that it is the electron orbitals energies which are important.

This finding emphasizes both the strength and the weakness of framework catalysts. There is no doubt that correlations of the reducibility of NO with M-O bonding are best made with a series of catalysts in which the framework remains constant. On the other hand, the mechanism of the reaction may be changed in a different situation. The conclusion is that framework catalysts can be very useful indeed but, as with other systems, one must be sure that some unforeseen event is not important.

Palmierites have also been used as framework catalysts (ref. 75): these are mixed metal oxides of the general composition $A_3B_2O_8$. The B site is tetrahedrally coordinated, and may be occupied by V, Mn, Cr, P, S or As. The three A atoms can have two different coordinations and may be occupied by Ba, Sr, Pb, Rb, K, Na, NH_4 and Tl. A diagram of the crystal structure is given by Longo and Clavenna (ref. 75).

In catalytic terms, the palmierites are of interest in that the B cation can be stabilised with respect to valency. Thus, for example, the V^{5+} state is stabilised to 1200°C in Ba_3VO_8, instead of the easy reduction of V^{5+} observed, for example, in V_2O_5. As with the scheelite system, mobility of oxygen ions was high, at least for A = Ba. This framework catalyst is, as a result, more suitable for studies of the catalytic activity of stabilised cations, care being taken that the A cation does not interfere with the reaction under study.

III. SUMMARY

It is difficult to generalise about the design of secondary components of catalysts because the problem is particular, and depends on results of testing of primary component catalysts. Two approaches can be distinguished, however. The first involves the application of scientific common sense. The second involves a study of the mechanism of reaction, in the hope that this will lead to improvements in the catalyst. Considerable care must be taken to ensure the system is worth the degree of attention that will be necessary if the second approach is adopted.

In addition to traditional means of studying reaction mechanisms, considerable advantage can result from the use of "framework" catalysts, in which the dependence of catalytic activity on only one parameter can be ascertained. The various examples discussed above show that distinction between the desired effect and spurious effects (due, for example, to the activity of the framework itself) can be difficult, but that studies of these analogous framework catalysts can be very revealing in the context of the design of secondary components of a catalyst.

REFERENCES

1 D.A. Dowden, C.R. Schnell and G.T. Walker, Proc. IV Internat. Congr. on Catalysis, Moscow, 201 (1968).
2 L.A. Doerr and D.L. Trimm, J. Catal., 26, 1 (1972).
3 M. Goldwasser and D.L. Trimm, J. Appl. Chem. Biotechnol., in press (1979).
4 M. Ozaki, Isotopic Studies of Heterogeneous Catalysis, Academic Press, New York, 1977.
5 J.G. Germain, Catalytic Conversion of Hydrocarbons, Academic Press, New York, 1969.
6. J.M. Thomas and W.J. Thomas, Introduction to the Principles of Heterogeneous Catalysis, Academic Press, N.Y., 1967.

7 L. Lee, Characterisation of Metal and Polymer Surfaces, Vols. 1 and 2, Academic Press, New York, 1977.
8 J.R. Anderson, Structure of Metallic Catalysts, Academic Press, New York, p. 395, 1975.
9 W.M.H. Sachtler and R.A. Van Santen, Adv. Catal., 26, 69 (1979).
10 W.M.H. Sachtler, Catal. Revs. Sci. Eng., 14, 193 (1976).
11 V. Ponec, Catal. Revs. Sci. Eng., 11, 41 (1975).
12 R.L. Moss and L. Whalley, Adv. Catal., 22, 115 (1972).
13 J.C. Vickerman, Catalysis Vol. 2: Specialist Periodical Report on the Chemical Society, London, p.107 (1978).
14 A. Cimino, M. Lo Jacono, P. Porta and M. Valigi, Z. Phys. Chem., Frankfurt, 70, 166 (1970).
15 M. Lo Jacono, A. Sgamellotti and A. Cimino, Z. Phys. Chem., Frankfurt, 70, 179 (1970).
16 F.S. Stone and J.C. Vickerman, Trans. Farad. Soc., 67, 316 (1971).
17 P. Porta, F.S. Stone and R. Turner, J. Solid State Chem., 11, 135 (1974).
18 A.W. Sleight and W.J. Linn, Annals N.Y. Acad. Sci., 272, 22 (1976).
19 D.A. Dowden, Catalysis Vol. 2: Specialist Periodical Report of the Chemical Society, London, 1 (1978).
20 J.H. Sinfelt, Adv. Catal., 23, 91 (1973).
21 T.J. Gray, N.G. Masse and H.G. Oswin, Act. 2nd Congr. Internat. Catal., Paris, 1697 (1961): Technip.
22 N.I. Kobozev, Acta Physicochem., USSR., 9, 895 (1938).
23 D.A. Dowden, Proc. 5th Internat. Congr. Catal., Amsterdam, 621 (1972).
24 G.A. Martin, M. Primet and J.A. Dalmon, J. Catal. (in press) (1979).
25 P.A. Kilty, N.C. Rol and W.M.H. Sachtler, Proc. 5th Internat. Congr. Catal., Amsterdam, 929 (1972).
26 G.C. Bond and B.D. Turnham, J. Catal., 45, 128 (1976).
27 M. Araki and V. Ponec, J. Catal., 44, 439 (1976).
28 G.A. Martin, Compt. Rend. Acad. Sc. Paris, t284, ser. C, 479 (1977).
29 D.H. Seib and W.E. Spicer, Phys. Rev., B2, 1676 (1970).
30 H. Verbeek and W.M.H. Sachtler, J. Catal., 42, 257 (1976).
31 R. Bouwman, G.J.M. Lippits and W.M.H. Sachtler, J. Catal., 25, 350 (1972).
32 C.D. Gelatt and H. Ehrenreich, Phys. Rev., B10, 398 (1974).
33 C. Norris and H.P. Myers, J. Phys. Metal. Phys., 1, 62 (1971).
34 D.H. Seib and W.E. Spicer, Phys. Rev. Lett., 20, 1441 (1968).
35 S. Hufner, G.K. Wertheim, R.L. Cohen and J.H. Wermick, Phys. Rev. Lett., 28, 488 (1972).
36 G.M. Stocks, R.W. Williams and J.S. Faulkner, Phys. Rev., B4, 4390 (1971).
37 P.O. Nilsson, Phys. Kondens. Mat., 11, 1 (1970).
38 F.L. Williams and M. Boudart, J. Catal., 30, 438 (1973).
39 R. Bouwman and W.M.H. Sachtler, J. Catal., 14, 127 (1970).
40 R. Bouwman and P. Biloen, Surface Sci., 41, 348 (1974).
41 R.L. Moss, H.R. Gibbens and D.H. Thomas, J. Catal., 16, 117 (1970).
42 R. Bouwman and W.M.H. Sachtler, Surface Sci., 24, 350 (1971).
43 G.A. Ozin, Catal. Rev. Sci. Eng., 16, 191 (1977).
44 J. Hulse and M. Moskovich, Surface Sci., 57, 125 (1976).
45 H. Huber and G.A. Ozin, Inorg. Chem., 20 (1978).
46 J.G. Firth and H.B. Holland, Nature, 212, 1036 (1966).
47 F.J. Wolf, Separation Methods in Organic Chemistry and Biochemistry, Academic Press, New York, 1969.
48 R. Kellerman and K. Klier, Surface and Defect Properties of Solids, Vol. 4, Specialist Periodical Reports, Chemical Society, London, 1 (1975).
49 A. Cimino, M. Lo Jacono, P. Porta and M. Valigi, Z. Phys. Chem., Frankfurt, 59, 134 (1968).
50 M. Valigi, Z. Phys. Chem., Frankfurt, 97, 241 (1975).
51 M. Valigi, F. Pepe and M. Schiavello, J.C.S. Faraday, 1, 71, 1631 (1975).
52 A. Cimino, M. Lo Jacono, P. Porta and M. Valigi, Z. Phys. Chem., Frankfurt, 55, 14 (1967).
53 M. Schiavello, F. Pepe and S. de Rossi, Z. Phys. Chem., Frankfurt, 92, 109 (1974).

54 A. Callaghan, M.J. Rossiter and F.S. Stone, Trans. Farad. Soc., 62, 3463 (1966).
55 P. Porta, F.S. Stone and R. Turner, J. Solid State Chem., 11, 135 (1974).
56 S. Greenwald, S.J. Pickart and F.H. Grannis, J. Chem. Phys., 22, 2597 (1954).
57 J.C. Vickerman, Trans. Farad. Soc., 67, 665 (1971).
58 T.A. Egerton, F.S. Stone and J.C. Vickerman, J. Catal., 33, 299, 307 (1974).
59 D.A. Dowden, Catal. Revs. Sci. Eng., 5, 1 (1972).
60 V.M. Vlasenko, V.L. Chernobrivets, N.K. Lunev and A.I. Melchevskii, React. Kinet. Catal. Lett, 6, 195 (1977).
61 Proc. Internat. Symp. on Preparation of Catalysts, Ed. B. Delmon, Elsevier, Amsterdam, 1976.
62 J.M. Peacock, M.J. Sharp, A.J. Parker, P.G. Ashmore and J.A. Hockey, J. Catal., 15, 373, 379, 387, 398 (1969).
63 P.A. Batist, B.C. Lippens and G.C.A. Schuit, J. Catal., 5, 55, 64 (1966).
64 A.W. Sleight, W.J. Linn and K. Aykan, Chem. Tech., 235 (1978).
65 L.H. Brixner, A.W. Sleight and M.S. Licis, J. Solid State Chem., 5, 247 (1972).
66 H.H. Voge and C.R. Adams, Adv. in Catal., 17, 151 (1967).
67 H. Miura, T. Otsubo, T. Shirasaki and Y. Morikawa, J. Catal., 56, 84 (1979).
68 T. Takahashi and H. Iwahari, J. Appl. Electrochem., 3, 65 (1973).
69 R.J.H. Voorhoeve, J.P. Remeika, P.E. Freeland and B.T. Matthias, Science, 177, 353 (1972).
70 R.J.H. Voohoeve, J.P. Remeika and D.W. Johnson, Science, 180, 62 (1975).
71 R.J.H. Voorhoeve, J.P. Remeika, L.E. Trimble, A.S. Cooper, F.J. Disalvo and P.K. Gallagher, J. Solid State Chem., 14, 395 (1975).
72 R.J.H. Voorhoeve, L.E. Trimble and J.P. Remeika, in The Catalytic Chemistry of Nitrogen Oxides, General Motors Research Symp., Oct. 6-8, 1974, Ed. R.L. Klimisch and J.G. Larson, Plenum Press, New York, 1974.
73 R.J.H. Voorhoeve, L.E. Trimble and J.P. Remeika, Ann. N.Y. Acad. Sci., 272, 3 (1976).
74 F.J. Morin and T. Wolfram, Phys. Rev. Lett., 30, 1214 (1973).
75 J.M. Longo and L.R. Clavenna, Annal. N.Y. Acad. Sci., 272, 45 (1976).

CHAPTER 5

CHOICE OF SUPPORT MATERIALS

The previous chapters have shown that catalytically active materials are often expensive, and that their activity depends on factors such as surface area, porosity, geometry of the surface, resistance to deactivation etc. In an effort to optimise these factors, it is common practice to disperse the active ingredient of many catalysts on the surface of "inactive" solids, called supports or carriers, or to intersperse the active ingredient with small particles of an "inert" solid, called a dispersant, spacer or a stabiliser. Although these materials are essentially diluents, they play an important multifunctional role in dictating catalytic activity. This may include chemical reaction with the catalyst, and they are designated as "inactive" only to distinguish them from bifunctional catalysts, in which the "support" plays a major role in catalysis. Most of this chapter is not concerned with bifunctional catalysis.

Although the use of supports is extremely widespread, it is an area which, in some respects, has not received the attention which it should command. This results from the many factors which are affected by the support, which may be listed

1. Economic
(a) To reduce cost by extending an expensive catalyst.

2. Mechanical
(a) To give mechanical strenth
(b) to optimise bulk density
(c) to provide a heat sink or a heat source
(d) to dilute an overactive phase.

3. Geometric
(a) to increase the surface area of a catalyst
(b) to optimise the porosity of a catalyst
(c) to optimise crystal and particle size
(d) to allow the catalyst particles to adopt the most favourable configuration.

4. Chemical
(a) to react with the catalyst either to improve specific activity or to minimise sintering

(b) to accept or donate chemical entities, possibly via a spillover mechanism.

5. Deactivation

(a) to stabilise the catalyst against sintering

(b) to minimise poisoning.

In that the support is required, in a perfect world, to fulfil all of these functions, overall optimisation can be complex and can lead to incompatible requirements. Thus, for example, optimal porosity usually leads to a catalyst which is too weak mechanically. This is particularly important in that a major potential cause of catalyst failure is a result of mishandling during loading of the reactor.

Before considering these factors individually, it is useful to consider briefly the type of materials that are used as supports. Since a solid is totally inactive, a support is usually chosen for its stability towards temperature over the range of operating conditions and for its surface area and porosity. Generally, the support is not catalytically active over the range of operating conditions, but if there is some small unwanted activity, this can be eliminated by, for example, selective poisoning.

The most stable range of supports includes such solids as alumina, magnesia, silicon carbide and alumino-silicates such as synthetic mullite ($3Al_2O_3.2SiO_2$). The synthetic materials are usually calcined to high temperatures, and are available as coarse powders. Their porosity is low and their mechanical strength is high, but their surface area is generally low (ca. 0.5 m^2g^{-1}). They are particularly useful for high temperature operation or for use in a fluidised bed, with the proviso that - since the catalyst is on the outside of the support - attrition should not be too much of a problem.

Even these refractory solids are susceptible to attack, for example by alternate high and low temperature cycles or by chemical reaction - for example, with steam. We shall see that one of the major factors in the choice of a car exhaust catalyst support is the low coefficient of thermal expansion, and of a steam reforming catalyst is the avoidance of material, such as silica, which volatilizes in steam at high temperature.

By far the most widely used range of supports do possess some chemical or catalytic activity, and any choice of support must be made with this in mind. Thus, for example, such a support would be perfectly acceptable provided there was no reactant with which they could interact, or that the support-catalysed reaction was favoured under reaction conditions far removed from these favouring the desired reaction (e.g. at much higher temperature), or that any support activity could be removed by poisoning. In that these relatively inactive supports are available with a wide range of surface area and porosity, the advantages of their use far outweigh these disadvantages.

Nearly all of the insulating oxides are useful as supports, although - on economic grounds - alumina and silica are most widely used. However, the final choice depends on the chemical properties of the oxide, which can be correlated with their ionic potential (ref. 1). Oxides such as alumina, silica, zirconia and thoria have a high ionic potential and tend to be acidic, while calcia and magnesia, with low ionic potential, are basic. These properties are either of no importance, or they can be removed by poisoning, or they can favour the reaction. Thus, for example, the removal of carbon by steam is favoured by basic salts, and magnesia or calcia are obvious candidates for use in the steam reforming reaction.

Many naturally occurring materials belong in this group, such as pumice, asbestos, calcined clays (e.g. bentonite, sepiolite) and diatomaceous earths such as kieselguhr. As a result of the wide variation in structure, the solids offer a wide range of surface area and porosity, although synthetic versions of the same materials may be preferred in that they offer a more closely defined range of properties (see Table 5.1).

TABLE 5.1

Typical support materials

Surface area	Porosity	Materials
Low surface area $< 1\ m^2g^{-1}$	essentially non-porous	ground glass Alundum (α-Al_2O_3) silicon carbide
	porous	kieselguhr pumice
High surface area $> 1\ m^2g^{-1}$	essentially non-porous	natural silica-alumina carbon black titania zinc oxide
	porous	natural clays synthetic silica-aluminas alumina magnesia activated carbon silica asbestos

Semiconducting oxides are also widely used as catalyst supports, despite the fact that their potential activity is high. Thus, for example, chromia, titania and zinc oxide are widely used, despite the fact that they have intrinsic activity of their own. Although their resistance to sintering is high, they are generally used for low-temperature reactions, where their reactions with reactants or with the catalyst (alloying) is minimal.

One good example of a catalyst support in which the advantages overcome the disadvantages is the use of carbon, as active or amorphous carbon or as graphite. The support is destroyed in oxidising atmospheres (except in a liquid medium or at low temperatures), but the range of properties available make it a very useful support for reactions such as hydrogenation. Thus, for example, active carbons can be produced in a granular or a powder form with "surface areas" of between 100 and 1000 m^2g^{-1} (ref. 2), and have little catalytic activity of their own, except for some chlorination or oxidation reactions (ref. 3). On the other hand, carbons can also be produced with molecular sieve properties (ref. 4), in which high or low surface areas can be combined with a closely defined pore system (8-10 Å diameter).

One other feature of these supports, which is both welcome and unwelcome, is their high adsorptive capacity. This is unwelcome in that exposure to contaminated atmospheres can lead to deactivation of the catalyst. However, this self-same property may result in a much wider use of the support for pollution-prevention catalysts. The basic problem with pollutants is that they are present in low concentrations. Passage of large quantities of air through an activated carbon operating as an adsorption bed could lead to concentration of the pollutant: subsequent increase in temperature would then lead to removal of the concentrated pollutant over a catalyst either suspended on the activated carbon or immediately downstream of the adsorbent bed.

In the case of carbon, we also see a good example of incompatibility. The supports can be made cheaply from coal or from wood but, not surprisingly, such carbons have earned a bad reputation for irreproducibility. Alternatively, the carbons may be made from single precursors under carefully controlled conditions. The carbons so obtained have perfectly reproducible properties, but the cost rises. The choice obviously depends on the eventual use of the catalyst.

One other advantage of carbon supports, which is often ignored, is that their activity depends on preparation. High-temperature carbons tend to have basic, hydrophobic surfaces, while low-temperature carbons have acidic hydrophilic surfaces. Where these properties may be incompatible with desired surface areas or porosities, doping during the course of preparation is very easy.

Of all the support materials, metals have been most infrequently used, as a result of the fact that most of them can be active catalysts. However, they do come into their own under conditions where it is necessary to remove large quantities of heat from the catalyst. Perhaps the most widely known example of this is in the U.S. Bureau of Mines Synthane process, where the catalyst, Raney nickel, is deposited directly onto the surface of a metallic heat exchanger (ref. 5).

In considering the individual factors which dictate the choice of support, it must be remembered that the final choice depends on a weighing of these factors in the context of the use to which the catalyst is to be put. Many requirements are in opposition, as we shall see by considering two of these together, mechanical strength and geometric considerations.

I. THE TEXTURE AND STRENGTH OF THE SUPPORT

If the rate of a catalytic reaction is dictated by the rate of the chemical reaction on the surface, then the observed activity will be a function of the surface area of the catalyst (specific activity). In practice, however, the overall rate of reaction is usually affected by mass or heat transfer, when the porosity and geometry of the catalyst pellet becomes of increasing importance. As a result, the choice of support depends on the surface area of catalyst which can be made available to the reactants and on the porosity of the catalyst.

It is possible to quantify these factors to some extent, as is done in Table 5.2. The basis of this quantification depends on factors that affect mass and heat transfer in the catalyst (refs. 6,7).

For a catalytic reaction

$$A \rightarrow B$$

the rate may be expressed as

$$\frac{dB}{dt} = k\, A^{\alpha} \eta$$

where α is the order of reaction and η is the effectiveness factor. If the reaction is controlled by the rate of the chemical reaction, then $\eta = 1$. If the reaction is mass transfer controlled, then η is less than one (for an isothermal pellet) and is proportional to the inverse of the Thiele modulus (refs. 6,7). The effect of this on the kinetic parameters is that the activation energy reduces from E to E/2 and the reaction order changes from α to $(\frac{\alpha+1}{2})$(ref. 6). As a consequence, for the simple reaction above, the observed rate is slower (mass transfer control) than would be expected from the concentrations and activation energies.

For a reaction network

$$A \rightarrow B$$
$$C \rightarrow D$$

this fact can be turned to advantage. Consider the situation where the rate constant for the first reaction is large and that for the second reaction is small.

The selectivity of the network can be written

$$\frac{d\,B}{d\,D} = \frac{k\,A^{\alpha}}{k^1\,C^{\gamma}}$$

where the reaction is kinetically controlled.

If, on the other hand, the network is mass transfer controlled, then the selectivity must be written

$$\frac{d\,B}{d\,D} = \frac{k\,A^{\alpha}\eta}{k^1\,C^{\gamma}\eta^1} = \frac{k\,A^{\frac{\alpha+1}{2}}D_{eff}}{k^1\,A^{\frac{\gamma+1}{2}}D^1_{eff}}$$

where D_{eff} is the effective diffusion coefficient.

Under these circumstances, it is easy to see (since $k \gg k^1$) that, if B is required, it is better to work under chemically controlled conditions while, if D is required, mass transfer control is preferable.

Having identified conditions under which mass transfer control could be preferable, it is necessary to ask how this could be achieved. Bulk diffusion is dependent on the thickness of the boundary layer of fluid around the outside of the pellet, but Knudsen diffusion is predominant when the mean free path of the molecules is less than the pore diameter (ref. 6). At atmospheric pressure the mean free path of a molecule of diameter 2×10^{-8} cm is about 10^{-5} cm, and catalysts with a pore diameter of ca. 100 Å will induce Knudsen diffusion. If, on the other hand, the pores are too narrow, entry of fluid into the pore could be restricted, and surface diffusion could predominate (ref. 8). Under these circumstances, the arguments above will not apply, since the mechanism of surface diffusion (mobile adsorption) is totally different. As a result, it is probably better to ensure that the pore diameter is at least 50 Å. As a result of these arguments, quantification of desired pore size is possible (Table 5.2) and, since the porosity of the catalyst is very dependent upon that of the support, some guidelines to the selection of the latter may be set up.

The basis of selection of the thermal conductivity of the support is considerably simpler, since the catalyst pellet will be effectively isothermal (although it may be at different temperatures to the fluid (ref. 7)). Calculation of the actual temperature of a pellet can be difficult, particularly as it may be dependent on the position in the bed. However, it is relatively easy (ref. 6) to show that the maximum temperature difference between the pellet and the fluid is given by

TABLE 5.2

The reaction produces	Temperature control* is		Diffusional effects are		Surface area **	Porosity**	Thermal conductivity
	desirable	unnecessary	desirable	undesirable			
Terminal products such as CO_2, CH_4 etc.	✓		✓		medium	medium : maxima in PSD is 50-100 A	high
		✓		✓	high	High, provided temperature is not too high low if temperature rise is very large	any value
Two products concurrently, one of which is desired	✓		✓		medium	medium : maxima in PSD at 50-100 A	high
	✓			✓	medium	low porosity or very wide pores	high
Two products consecutively, the first of which is desired	✓		✓		medium	medium : maxima in PSD at 50-100 A	high
	✓			✓	medium	low porosity or wide pores	high
A product, but there is a potential poison in the feed or produced	✓		✓		medium	medium. Pores must be such either not to allow the poison to enter the pores or to avoid pore blocking by poison accumulation	
				✓	Medium		high
	✓						
A very high temperature rise	✓				low	non porous	

*For any reason: it is assumed that the catalyst is stable at the highest temperature liable to be observed

**Consistent with desired strength

$$\Delta T = \frac{-D\Delta H}{K} C_S$$

where ΔH is the heat of reaction, D is the diffusion coefficient, K the thermal conductivity and C_S the concentration of reactant at the exterior surface of the pellet. This estimation is useful not only to estimate the effect of temperature on rate, but also to give some indication as to whether sintering (see below) will be important or not.

Thus, from the point of view of mass and heat transfer, some assessment of the desired properties of the support may be made. It is possible, however, to control the distribution of active phase on the support to some extent (ref. 9) and this, too, will influence activity and mass and heat transfer. The concept may be discussed generally, and illustrated by a specific sample.

Any non-porous support may have rough or smooth surfaces on which the active catalyst can be deposited from a melt or from solution. Obviously, there are very real advantages when the surface of the support is rough, in that the catalyst can be keyed to the support. This can be particularly important with active catalysts which are mobile under reaction conditions, as, for example, is the case with vanadia-based catalysts for the oxidation of sulphur dioxide or of naphthalene. If the catalyst-support bond is not strong enough, adhesion can be enhanced by a glue. The most common versions of these (starch, albumin, polyvinyl alcohol, etc.) can be removed by combustion or by evaporation, although this may lead to changes in porosity (see later). It is possible to use a cement, but care must be taken to avoid encapsulation of the catalyst or chemical reaction to form, for example, a less active spinel.

Where the support is porous, the degree of availability of the catalyst can be controlled during preparation. The impregnation of the support with dissolved salts, for example, offers two extreme cases. Dissolved salts which are rapidly and strongly adsorbed by the support tend to concentrate near the pellet exterior, while weakly adsorbed salts will be distributed homogeneously through the pellet. The choice of method will obviously depend on diffusion limitations in the reaction to be catalysed: if diffusion limitations are high, then the reactants will not penetrate to the interior of the pellet, and any catalyst in this zone will be wasted. There are, of course, many preparation methods for these different catalysts. Thus, for example, preparation of a platinum colloid in solution leads to deposition on the exterior of a pellet, while impregnation with a solution of a platinum salt - which can subsequently be converted to platinum - will lead to homogeneous distribution of the metal.

Where an intimate mixture of support and catalyst is required, aqueous pastes or dry powders can be blended before granulation, to be followed - if necessary - by leaching of one of the components (e.g. Raney nickel). Alternatively, the precursors of the two constituents can be co-precipitated from solution (ref. 10). However, the chemistry of co-precipitation can be very complex (ref. 11), with the two salts precipitating at different pH to give encapsulation of one component by another. The development of a large scale process based on a catalyst prepared by co-precipitation should, therefore, be preceded by a careful investigation of the preparation method.

The problem with any detailed study of methods of obtaining optimal surface area is that little information is of general applicability and that individual systems require individual attention. However, some of the methods can be illustrated by using the example of a widely used catalyst (Pt/Al_2O_3) which has been studied in some detail by Zaidman et al. (refs. 12-16).

The investigation was initiated from the observed fact that a saturation value of catalyst can be reached on any support, such that the catalytic activity no longer increases as the weight of active component increases. In some cases this may be due to monomolecular coverage being exceeded or, for example, that pores in the support become blocked by the active component. Alternatively, as for example with alumomolybdenum catalysts (refs. 17,18) the maximum in activity corresponds to the formation of compounds by the reaction of support and catalyst (see later).

Accepting the first two postulates, it should be expected that the saturation value should depend on the degree of dispersion of the catalyst. This, in turn, has been shown to have a linear dependence on the reciprocal of surface concentration, at least for Pt/Al_2O_3 and Ag/corundum (ref. 13)

$$\frac{1}{\gamma} = 1 + K\, C_{sur}$$

where γ = dispersion, the fraction of active component on which gas can be absorbed,

k_t = crystallisation constant, which characterises the increase in particle size with surface concentration,

C_{sur} = surface concentration, the weight of the catalyst divided by the surface area of the support.

Crystallisation constants have been reported for a number of systems (ref. 19) and it is possible to calculate the saturation values for various catalysts knowing the degree of dispersion. As shown in Table 5.3, good correlation is observed with experimental observation, for a variety of catalysts.

Thus, at least for some catalysts, it is possible to relate dispersion to surface concentration and to calculate saturation. However, the relationship does

TABLE 5.3

Optimal catalyst loading

Conc^n.of active component (wt.%)	Support surface area (m^2g^{-1})	Preparation T (^oC)	Reaction	Optimal conc^n. of active component (wt.%)	Reference
Cr_2O_3/alumino silicate: $K_t = 12.8 \times 10^{-3}\ m^2g^{-1}$ at 500^oC					
1.4 - 25	260	550	ethylene polymeris^n.	3.8 - 46	20
1.4 - 25	560	550	" "	9.1	20
1.5 - 10	300	300-600	" "	4.4 - 5.8	21
0.5 - 20	300	400	" "	3.8 - 5.8	22
1 - 7	300	-	" "	3.8	23
0.1 - 10	300	400	" "	3.8	24
Cr_2O_3/Al_2O_3: $K_t = 4 \times 10^3\ m^2g^{-1}$ at 500^oC					
10 - 50	200	550	dehydrog^n. butane	18-23	25
	200	550	" isopentane	30	26
Pd/Al_2O_3: $K_t = 14 \times 10^3\ m^2g^{-1}$ at 300^oC					
.16 - 5.9	200	50	hydrog^n.cyclohexene	0.2 - 2	27
Ni/Ce_2O_3: $K_t = 2.6 \times 10^3\ m^2g^{-1}$ at 350^oC					
35 - 50	160	350	hydrog^n. benzene	15.25	20
Pt/Al_2O_3: $K_t = 7.9 \times 10^3\ m^2g^{-1}$ at 500^oC					
0 - 2	100-240	500	dehydrog^n.cyclohexane	> 2	28

depend on crystallisation constants, which may be controllable by experiment (see later) but cannot be predicted. Once again, the need for experimentation on individual systems becomes obvious.

Although the optimisation of surface area is an important factor, it has been shown to be related to other factors which affect the texture and strength of a catalyst. Thus surface area and porosity are closely related, and we can extrapolate further in that porosity and mechanical strength are also interrelated. To ensure long life, a catalyst needs a stable structure that is strongly bound together, and this is certainly not the case if the porosity is too high.

The natural porosity of supports is difficult to control in a systematic fashion. Zeolites or carbon molecular sieves possess most of their internal surface areas within pores, which have a critical diameter for the entry of gas of ca. 5-10 Å. Some gamma aluminas have a pore diameter distribution in the 100-200 Å region, while some foamed aluminas have only a few macropores in their structure. Given this natural pore size distribution it is possible, within limits, to alter the distribution. The most common method of doing this is by heat. Thus Walker has reported that the surface area and pore size of carbon molecular sieves is dependent upon the pyrolysis temperature (ref. 29), the available surface area increasing to a given preparation temperature and then decreasing as pores are shut off by heating to higher temperatures. Similarly, Elo and Clements (ref. 30) have reported that heat treating silica-aluminas results in a decrease in pore volume and surface area, but without marked changes in pore radius. Heating in steam, on the other hand, increased the surface area and increases the pore size, probably due, at least in part, to the increased mobility of silica in steam.

Pore diameters can also be decreased by careful precipitation of material in the pore mouth. As with the deposition of carbon on a zeolitic molecular sieve, this can be unwanted but, in some cases, can lead to desirable changes. Thus, for example, greater distinction can be made between the adsorption of oxygen and nitrogen on carbon molecular sieves if carbon is carefully deposited in the pore mouth by the controlled pyrolysis of organic fuels (ref. 31).

In use, however, most supports have a bimodal pore size distribution, arising from the fact that particles of support material are compressed together in a pellet to give inter and intra particle pores. Many catalysts have a continuous network of transitional pores (20-200 Å) and of macropores (> 200 Å), their relative contribution to the total pore volume being adjusted to suit the end use. The macropores have a useful part to play, in that they provide a pathway for gaseous reagents to the centre of the pellet and hence help to remove mass transfer limitations.

The development of these pore structures can be controlled to some extent. In some cases sufficient is known about the preparation to control pores by chemical methods: larger pore sizes can be obtained in silica gel, for example, by forming the gels at high pH. Alternatively, powdered support materials may be mixed with an organic material such as graphite before pelleting: subsequent oxidation removes the graphite and leaves macropore channels. Perhaps of most importance is the fact that formation of pellets can be carried out by different means, which can result in large differences in porosity. It is here, however, that the effects of strength and porosity are most obviously in opposition.

One of the most common means of forming a catalyst granule is by pelleting in a press. The catalyst-support mixture is prepared and ground to optimal size and density, mixed with a die wall lubricant (such as graphite) and fed to a press. Although the resulting properties of the pellet depend on such factors as amount of diluent, pelleting pressure, etc., there are limiting factors. Thus, there is a broad general relationship between the strength and hardness of inorganic solids and their melting points (ref. 32). High melting point solids have high hardness and strength, while low melting point solids are ductile and soft. Since pelleting machines are generally fabricated from steel, only materials up to a Moh's hardness of ca. 4 can be satisfactorily pelleted, showing that a mixture of a refractory oxide and graphite could not be pelleted unless a third component is added that conforms to the criteria for pelletability.

The pelleting operation is particularly important in determining the resistance of the pellet to stresses occurring before and during reactor loading. During reactor loading, for example, the minimum drop for failure has been shown to be related to the ultimate tensile strength(s) of the pellet and to the length:diameter ratio (ref. 32)

$$\text{drop for failure (m)} \sim 0.075 \, \frac{s \times d}{l}$$

where s is expressed in kg cm^{-2}. A long thin pellet is obviously much weaker than a short fat pellet. If the desired geometry is intrinsically weak, then addition of a cement (or further calcining after pelletising) can strengthen the pellet, although this is almost certain to have some effect on the catalytic activity of the solid.

Some variation in porosity is also possible using other methods of pellet preparation, extrusion, in particular, being widely used. Again, however, the effect of different parameters on the strength and porosity of the final pellet is rather empirical, theoretical treatment of the process being complicated (ref. 33).

Although it is difficult to write in general terms concerning the preparation of pellets which have optimal dispersion of catalyst, optimal porosity and optimal strength, the preceding discussion should indicate the factors that are important in considering any particular example. The remaining factors, the optimisation of crystal and particle size and the generation of the most favoured configuration of catalyst are perhaps best discussed in terms of chemical interaction between the catalyst and the support.

II. CHEMICAL INTERACTION

One of the more important areas in which the support has a positive role to play is in chemical interaction with the catalyst. As distinct from bifunctional catalysis, where the support acts as a co-catalyst, we should consider, here, interactions between the support and the catalyst that result in a modification to the to the catalyst. Although important, the amount of information is limited as a result of the fact that suitable means of detection of such interaction have been developed only recently.

The kind of effect under discussion can perhaps most clearly be seen by reference to Table 5.4. The selectivity of supported platinum catalysts to the isomerisation of hydrogenolysis of neopentane was found to depend on the support (ref. 34). Careful investigation showed, in fact, that it was not the support itself, but the effect of the support on the "concentration" of (111) faces in the platinum that was responsible for the changes observed. This is an example of the first kind of interaction, where the geometry of the catalyst is dependent upon the support, at least to some extent. Although insufficient information is available to present reasonable correlations, it is possible to discuss the factors that could be important.

TABLE 5.4
Selectivity as a function of support

Support	Dispersion[1]	Selectivity[2]	Pre-treatment
γ Al_2O_3	73	0.55	reduced at 425^oC in H_2 for 10 h
η Al_2O_3	64	0.41	" " 425^oC " " " 10 h
γ Al_2O_3	73	1.5	" " 500^oC " " " 10 h
η Al_2O_3	64	1.5	" " 500^oC " " " 10 h
SiO_2	17	0.29	" " 500^oC " " " 10 h
Spheron	12	27	reduced at 500^oC: evacuated at 900^oC: reduced at 500^oC in H_2 for 10 h
η Al_2O_3	7.6	2.4	sintered in H_2 at 650^oC
-	0.03	9.1	heat in H_2 and O_2: reduce at 500^oC in H_2 for 10 h

1 Dispersion calculated as Pt on surface/Pt total

2 Selectivity = rate isomerisation neopentane/rate hydrogenolysis neopentane: 300^oC:RH:H_2 = 0.1

Of these the most obvious is that a least one of the catalysed reactions should be surface-sensitive (see Chapter 3). Having established this, preparation methods must also be important in determining catalyst crystallite size and the extent of catalyst-support interaction. Unfortunately, this is, again, a particular problem: typical compounds investigated include Ni-Al_2O_3, Ni-Cr-Al_2O_3 (ref. 35), Pd-Al_2O_3 (ref. 36) and Pt-Al_2O_3 (ref. 37).

The next question is the extent to which support effects influence the catalyst structure. An indicative paper on the orientation of copper, silver and gold has been published (ref. 38). This shows that, as the thickness of a deposit increases, the preferred orientation ot the metal changes from (111) and (111) twinning to (110) and/or (211) orientation as a result of secondary twinning. Of course, these orientations will be stable only as long as local reorganisation is unimportant (see later).

Similar geometric effects are important in oxides, but here it is necessary to descend a scale in magnitude. Our understanding of the effects rests mainly on the E.S.R. work of Kazansky and others, and it is useful to discuss such interactions in some detail.

The understanding of catalytic oxidation is complicated by the fact that different adsorbed species may be present on the catalyst. Considerable discussion has been focused on the importance of O_2^-, O^-, etc. as oxidising agents for particular reactions (ref. 39). In 1972 Shvets and Kazansky (ref. 40) set out to study the formation of O_2^- and O^- on a range of supported catalyst containing Ti, B and Mo ions.

Considering VO_x/SiO_2, the sequence of the formation of various adsorbed oxygen species during oxidation of a partially reduced catalyst could be presented as follows

$$O_{2\ gas} \rightarrow O_{2\ ads} \rightarrow O_2^- \rightarrow 2\ O^- \rightarrow 2\ O^{2-}$$

where O^{2-} is the oxygen of the oxide lattice. If this is so, then O_2^- anion radicals would be formed before O^-, and O^- would be expected to be more stable at higher temperature. This is indeed the case, and their presence has been observed at temperatures up to 300^oC (ref. 41); they are, however very reactive.

Oxygen is capable of adsorbing as the O_2^- anion radical on V/SiO_2 as a result of electron transfer from tetrahedrally coordinated V^{4+} ions to oxygen molecules (refs. 42,43). This electron transfer does seem to occur preferentially with the tetrahedral coordination, as can be seen from the examination of other V^{4+}-containing catalysts. Thus, for example, the adsorption of oxygen on partially reduced V/γAl_2O_3 (which contains V^{4+}) does not produce O_2^-, and the V^{4+} coordination is either a square pyramid or an octahedron shortened by the formation of a vanadyl bond (refs. 53,44). The vanadyl complexes $(V=O)^{2+}$ are unable to donate electrons to oxygen. Similarly, no O_2^- anion radicals are produced on the surface of unsupported V_2O_5 (ref. 46). Again, if the tetrahedral coordination of V/SiO_2 is brought to octahedral by adsorption of H_2O or NH_3 (refs. 43, 47), oxygen adsorption no longer results in the formation of O_2^- (ref. 40).

Thus, it would seem that at least one of the factors affecting the formation of adsorbed oxygen radicals is the coordination of the transition metal surface ions for which, in the first case, the structure of the support is responsible. Of course, examples have been selected that illustrate the point, and the situation is more complex than this in practice. Thus, for example, V/MgO or V/ZrO_2 catalysts do not have tetrahedral configuration (ref. 45), and yet anion-radicals are formed - as a result of the fact that the support can participate in electron transfer (ref. 40). However, these papers do show that geometric effects, on a molecular scale, caused by interaction of the catalyst and the support, can have a major effect on catalyst activity. Similar effects have been observed for supported CoO and MoO_3 catalysts (refs. 48,49) and - despite the fact that insufficient information is available for a range of catalysts - the effect can be expected to be quite general.

When experimental information is not available, it is possible to calculate the probabilities of formation of different adsorbed species. This is not easy and can only be expected to give a comparison between, for example, the calculated energies of bonds formed between the adsorbate and different surface complexes, but it can give a useful guideline. Calculation procedures have been discussed by Dowden (ref. 50) and by Van Krevelen and Chermin (ref. 51).

Generally speaking, there is always a possibility of geometric effects on a molecular scale or of electron transfer effects when the catalyst and support can interact. In some cases, this is obvious: thus, for example, the formation of spinels is well known (ref. 52). In other cases, this is much less obvious, and sophisticated techniques such as E.S.R. or photoelectron spectroscopy are necessary to detect such interaction. Several general guides as to the possibility of compound formation are given in the literature (e.g. refs. 2,6,32), although these must be treated only as approximations, in that small traces of metal ions have been found to promote the formation of compounds in systems where it would not normally occur (ref. 37). For the catalyst designer, support-catalyst interactions may be important, and reference to these general guides can give a first approximate answer to the question of what importance is attached to the effect.

The second kind of support-catalyst interaction occurs in alloys or in solid solutions of two metals, where a catalytically inactive metal (i.e. a support) could affect the structure of the active catalyst. This has been discussed in connection with alloy catalysts, and further consideration is unnecessary.

The third kind of interaction is also discussed elsewhere, in that it usually involves deactivation of a catalyst as a result of chemical reaction between the catalyst and the support. Perhaps the most widely known example of this is the formation of spinels (ref. 52), with subsequent loss of catalytic activity.

The fourth kind of catalyst-support interaction rests on the ability of some adsorbed gas to move across the surface from the catalyst to the support. This phenomenon, known as spillover, can be very important in the context of hydrogenation reactions.

Most of the quantitative studies of spillover have been carried out with systems specifically designed to show the effect. Schwabe and Bechtold (ref. 53) have studied the reduction of silver sulphide by hydrogen spillover from platinum, to show that the adsorbed molecule can migrate by up to 1000 Å from the adsorption site. Similarly, Levy and Boudart (ref. 54) have studied the kinetics and mechanism of spillover, using the reduction of WO_3 to H_xWO_3 by hydrogen spillover from supported or unsupported platinum. They show that co-catalysts can have a major effect, as a result of the fact that at least some of the hydrogen spills over as a proton.

In systems where spillover can be expected to have a significant effect on catalytic activity, the results are much less quantitative. Thus, for example, spillover of hydrogen on supported rhodium has been reported (refs. 55,56) and on supported platinum (ref. 57), but no details of mechanism or availability of spilled-over gas are given. One exception to this comes from work by Neikam and Vannice (ref. 58), who studied spillover from platinum onto a number of inorganic oxides and zeolites. They found that, in order for spillover to occur, a bridging compound was necessary in the system. This bridging compound facilitated spillover, and studies of various materials showed that large organic aromatic compounds were particularly effective, with hundreds or thousands of hydrogen atoms migrating through a single bridge. This work is particularly interesting, in that the type of reactions occurring on these catalysts do tend to deposit coke on the solids, and this coke should be of a structure such that efficient bridging should result.

Spillover effects can be expected to be particularly important in reactions such as hydrogenation or dehydrogenation, where an efficient source or sink for hydrogen facilitates reaction. The literature is still ambiguous as to which oxides favour spillover (thus, for example, Altham and Webb (ref. 57) report spillover on Pt/Al_2O_3, while Neikam and Vannice (ref. 58) report no significant spillover on the same catalyst), but the catalyst designer should attempt to consider this kind of catalyst support interaction in his deliberations.

III. DEACTIVATION

It is an unfortunate fact that all catalysts deactivate under reaction conditions, albeit at very different rates. The catalyst designer must always seek to maximise the time that the catalyst can operate economically. Deactivation is reversible or irreversible, and both are undesired - the former because regeneration means taking the catalyst off line.

Deactivation can occur by catalyst poisoning, which may be reversible in some cases, or by catalyst sintering, which is generally irreversible. The choice of a support and a spacer can have a major effect on sintering, and this is a major factor in selecting a support.

In that catalysts and supports are high-surface-area solids with a distinct porosity, there will always be a thermodynamic driving force to decrease free energy, i.e. to minimise surface area. The fact that all catalysts do not immediately sinter is due to kinetic limitations, which depend on the nature and physical arrangements of the solids and on the mechanism by which the solids rearrange. Although the chemical nature of the solids and the ambient atmosphere affect the rate of sintering, temperature is usually the dominant factor. Thus, at sufficiently low temperature, the solid retains its structure for long periods of time. As the temperature increases, possibly as a result of the fact that heat is being supplied by the catalysed reaction, surface diffusion becomes important, leading initially to the smoothing of very unstable surfaces, and eventually resulting in the production of faceted or spherical particles (see later). Growth of particles occurs where as-deposited catalyst particles are in contact with each other, or where transport across the surface or through the gas phase is possible.

At still higher temperatures, volume diffusion becomes important, and gross changes in the structure of the solid start to become more apparent. Eventually, at very high temperatures, evaporation of the solids becomes important, but this is rare in practice (see, however, the volatilization of noble metals from platinum/rhodium gauzes used to oxidise ammonia (ref. 59)).

The most commonly used indicator that sintering could be important was suggested, on empirical grounds, by Tamman (ref. 60), who noted that the onset of sintering occurred at temperatures, α_V, near to one half the melting point of the solid, $T_m(K)$:

$$\alpha_V \sim 0.5\ T_m$$

This was modified to some extent by Huttig (ref. 61) and other workers, who noted that sintering resulted initially from surface diffusion and that sintering started at a surface temperature, α_S, given by

$$\alpha_S \sim 0.3\ T_m$$

These relationships offer a very useful, but approximate, guide to the possible importance of sintering. Thus, for example, they reveal that metallic silver catalysts should sinter above ca. 350°C (α_V = 480°C, α_S = 320°C) while alumina (α_V = 600°C, α_S = 400°C) or silica (α_V = 855°C, α_S = 570°C) should be stable up to ca. 500°C. In fact, it is a common feature of all supports that sintering does not occur readily until temperatures in excess of those used during catalysis. Where the melting point of complex solids is not known, a reasonable guess may be made from lattice energies or ionic potentials, which tend to vary in the same way as melting points.

Although these relationships are very useful indicators, it is often worthwhile for the catalyst designer to pay more attention to the prevention of sintering, and it is useful to review briefly how this may be done.

One major problem results from sintering due to phase transformations, particularly in the support. Thus, for example, in the case of alumina, it is well known that various crystal forms may be produced, and Gitzen (ref. 62) has produced a very useful guide to the temperature stability of these compounds.

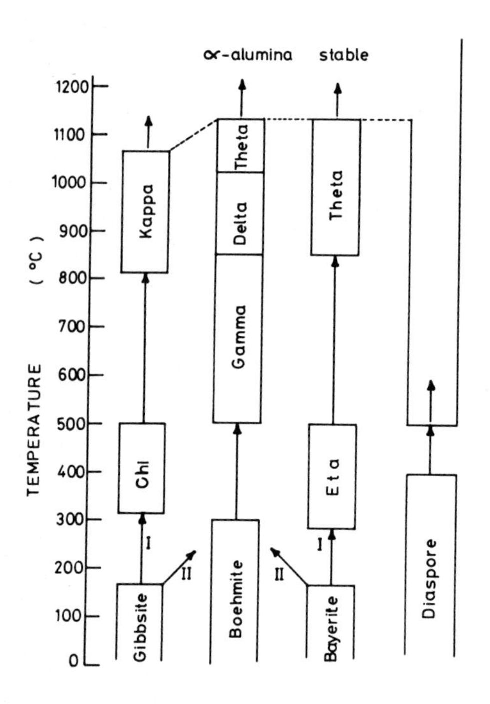

Fig. 5.1. Approximate temperature stability limits of alumina and aluminium hydroxides
I - favoured by dry small crystals
II - favoured by wet coarse crystals.

Considering the unit cell dimensions of the most common forms of alumina (Table 5.5) it is not surprising that large changes in surface area and porosity can occur when the crystal forms interconvert (refs. 62,63,64,65):

TABLE 5.5

Unit cell dimensions of aluminas (ref. 62)

Alumina	Crystallinity	a(Å)	b(Å)	c(Å)
η	cubic	7.90	-	-
θ	monoclinic	5.63	2.95	11.86
γ	tetragonal	7.95	7.95	7.79
δ	tetragonal	7.97	7.97	23.5

Although Fig. 5.1 shows the basic complexity in the alumina system, the situation becomes even more complicated in that the temperatures of transition are markedly affected by impurities - which may or may not include the catalyst. Thus, for example, low concentrations (ca. 2%) of group I and group II oxides have been shown to have an appreciable effect on the surface area of alumina calcined at 1100°C, by catalysing the conversion of γ- to α-alumina (ref. 66), and the presence of platinum at less than 3 wt.% reduces the temperature of phase transformations of alumina significantly (ref. 65).

It has also been well established that the nature of the support and the catalyst, as well as the ambient atmosphere, can have a marked effect on sintering. What is not generally realised, however, is that the relative importance of each factor depends on the mechanism of sintering and that preventive methods must be adjusted to suit the most important mode of sintering. Thus, we must first distinguish the mechanism of sintering and then apply appropriate remedies.

One of the most useful methods of distinguishing mechanisms of sintering has been developed by metallurgists, and involves model catalysts operating under reaction conditions. The model catalysts are prepared by evaporating known thickness of metal onto the appropriate support: under reaction conditions, these reorganise initially at a defect such as a grain boundary (ref. 67). Two factors are important: the shape of the grain boundary groove that develops and the time needed for the groove depth to extend to the metal-support interface (t_i) (ref. 68). Both are easily recognisable by electron microscopy (ref. 68). These measurements are important in that theoretical relationships have been developed which show that different mechanisms of metal reorganisation predict measurable differences in these parameters and in the activation energy of the process - which can be measured from the dependence of t_i on temperature (ref. 69).

It is important to measure metal reorganisation under reaction conditions, in that the rate of the process depends on the temperature and the ambient gas.

Thus, for example, silver sinters easily in oxygen but is stable in hydrogen; with nickel the reverse is true (ref. 70). This effect of ambient gas is widespread and, for different reasons, affects most of the mechanisms of reorganisation. Thus, for example, the surface self-diffusion coefficient is very responsive to impurities, of which adsorbed gas is one example (ref. 71). Oxygen accelerates the local reorganisation of silver and copper, since it affects the surface self-diffusion coefficient (ref. 68). However, chemisorbed oxygen tends to stabilise distributed copper (ref. 72), presumably as a result of the fact that individual particles tend to come to equilibrium within themselves, and the driving force to minimise surface energy by going to still larger particles is lessened. Again, metal sintering by evaporation-condensation is markedly affected by gas. Carbon monoxide reacting with nickel to form volatile nickel carbonyl is the classic example, but the action of water in greatly enhancing vapour-phase transport of oxides such as silica or beryllia (ref. 73) is perhaps of more importance.

Growth of particles is hindered by distributing them widely upon a second high-surface-area surface, and this is the major role of a support. Although catalytic activity may be maximised by the presence of very small particles (ref. 74), rates of sintering are substantially decreased by an increase in particle size (ref. 75). The as-deposited (or the equilibrium) size of the particles on a support is therefore controlled by these two opposing factors. The process can be understood in a quantitative manner from the application of mathematical models. Thus Flynn and Wanke (ref. 76) have suggested a model in which individual metal atoms (or molecules formed from the catalyst plus a reactant) move from a metal particle to the surface of the support, migrate over the support and are captured by a second particle or frozen by a drop in temperature. It is shown that smaller metal crystallites equilibrate with higher concentrations of metal atoms on the support than do larger particles, resulting in the growth of the larger particles at the expense of the smaller. The rate of sintering is found to be very dependent on the particle size distribution, with an increase in the rate of sintering as the width of the dispersion increases. The rate of sintering is also shown to increase as the metal loading increases.

These models are particularly useful in that they can provide a theoretical explanation for many reported empirical facts. Thus, for example, many workers have reported that sintering may obey a power-rate law of the form

$$\frac{dD}{db} = -KD^n$$

where D is the dispersion and K and n are constants.

Flynn and Wanke show that their model is capable of predicting variations of n between 2 and 13, depending on conditions (ref. 76). In addition, the factors which control redispersion of a catalyst are discussed; redispersion, which is obviously of enormous interest if it leads to regeneration of the catalyst, has been observed in a few systems.

Pulvermacher and Ruckenstein (ref. 77) have extended these arguments to differentiate between the diffusion of the particle across a metal surface and the subsequent sintering of the colliding particles. As would be expected, the homogeneity and the magnitude of the surface energy of different supports is predicted to have a major effect on sintering.

That this is so has been experimentally determined by Raissian and Trimm (ref. 78), although, as with most catalytic phenomena, the situation is complicated. Studies of the sintering of silver catalyst supported on a variety of non-porous supports, using a wide range of experimental techniques, show that the surface energy of the support does have a significant effect on the activation energy of sintering by surface self-diffusion. However, photoelectron spectroscopy also reveals that very small traces of impurities migrate from the support to the surface of the metal, and these too can be expected to influence the surface self-diffusion coefficient. Similar observations have been made by Ruckenstein (ref. 79) who records that the miscibility of supported ruthenium-copper or osmium-copper depends both on configurational changes occurring in small assemblies of atoms because of their smallness and also on strain energy caused by differential expansion or surface energy changes between the support and the catalyst. The differential expansion effect has been shown to be particularly important in the initial sintering of catalysts brought to operating conditions for the first time (ref. 68).

The other major cause of catalyst deactivation is poisoning, and here the role of the catalyst support is usually less important. If the catalyst is coked, and regeneration necessitates removal of carbon by steam or by oxygen at high temperature, then the support must obviously resist sintering under both deactivation and regeneration conditions. However, the support can play a more positive role.

One good example of this is with steam reforming, which will be discussed in more detail at a later stage. In this reaction, carbon formation is undesirable, and one obvious precaution is to avoid activity in the metal and the support (refs. 32,80). Indeed, since alkalis are known to catalyse the reaction between carbon and steam (ref. 80), there may be very real advantages to adding alkali to the support. Such additives are used by I.C.I. in their catalyst formulations (ref. 32), although there are several problems to overcome. The first of these is that potassium hydroxide volatilizes in steam and is carried downstream - to the detriment of subsequent pipework, which is rapidly corroded. To overcome this, the alkali is combined with alumino silicates in the original catalyst, to form complex compounds such as $KAlSiO_4$. Under steam reforming conditions, these complexes slowly decompose to release low concentrations of potash, which catalyse carbon removal. I.C.I. catalyst 46-1 contains ca. 7% K_2O, which is combined with alumino silicates, and the catalyst is active over a period of three years or longer (ref. 74).

I.C.I. also add magnesia to their catalyst, with the object of combining with the alumino silicates released at the same time as the potash. There is, however, good evidence that the magnesia plays a more important role. Thus, Rostrup-Nielsen (ref. 81) reports that the specific activity of steam reforming catalysts is very low when alkali is present. He shows that coke formation is depressed by enhanced adsorption of steam on the catalyst, and this may be achieved by the inclusion of either alkali or magnesia in the catalyst formulation (ref. 81). A similar role has been assigned to urania, which is also added to supports (ref. 80). It is probable that the arguments over the relative usefulness of alkali additives arises primarily from the different uses of the catalyst. I.C.I. and Haldor Topsoe are primarily interested in a high-temperature catalyst, useful for steam reforming methane to produce hydrogen for the production of ammonia. British Gas, on the other hand, require a low-temperature catalyst to be used for the steam reforming of naphtha to produce methane.

Doping of the support to avoid deactivation is not common, with the exception that unwanted acidic or basic properties may be removed. There is, however, one other class of catalyst in which the support plays a major role in avoding poisoning - the molecular sieves. These are a very important class of catalysts in which the crystal structure of the support is such that pores are formed with an extremely well-defined entrance diameter. Thus, for example, various aluminosilicates can be produced which contain a cage with an entrance of 5-10 Å. Metal ions can be exchanged into these cavities, to give extremely good distribution of metal with minimal chances of sintering (ref.(82). Since the molecules are also extremely acidic, the catalysts have found wide application as industrial cracking catalysts (ref. 83).

In the present context, these "supports" are particularly interesting, in that the size limitation at the pore entrance may stop large poison molecules from reaching the catalyst. Thus, for example, Rudham (ref. 82) has described a range of zeolite X compounds exchanged with a range of transition metal ions, and Firth (ref. 84) has shown these catalysts to be effective in sulphur-containing atmospheres which would normally poison the metal. The size limitation is particularly useful for many of the poison molecules present in petroleum feedstocks, which tend to be high-molecular-weight polycyclic aromatics containing heterocyclic sulphur or nitrogen.

The high acidity of the zeolites does, on occasion pose a problem, in that unwanted carbon builds up and tends to block the pores. One useful molecular sieve catalyst under these conditions can be produced from carbon (ref. 85). Although they have the inherent reactivity of any carbon, they possess minimal acidity and do not tend to carbonise.

Various additives have been added to supports in an effort to minimise deactivation. These are probably best discussed in the light of particular examples, as given in Part 2 of this book.

IV. SUMMARY

The practical significance of this body of information to the catalyst designer is that if offers some useful information and some caveats for the future - if and when the detailed information becomes available. It is an unfortunate fact that detailed quantitative attention has been paid to the action of supports only comparatively recently, due, at least in part, to the recent development of the necessary tools. There is, in the literature, a wealth of data applicable to individual catalysts or supports, but little attention has been paid to the underlying reasons for the effect noted. It is only when this information can be assigned to the occurrence of particular basic effects that a generalised predictive picture will emerge: it is unfortunate that insufficient detail is given, in most cases, to make such an attribution.

However, we can develop a questionnaire for the design of a support, and assign an approximate order of importance to the various roles of a support - even though this may depend more on the availability of information than in the importance of the effect. It should be emphasized that this questionnaire does <u>not</u> cover the possibility that the support might be required to be catalytically active.

1. Important questions where information is liable to be available:

(a) What are the operating conditions of the catalyst?

(b) What is the desired surface area of the catalyst and the support?

(c) What is the desired porosity of the support?

(d) Which supports meet these specifications?

(e) What is the mechanical strength of the support?

(f) Would the operating conditions (temperature, oxidising or reducing gas) affect the support?

(g) What is the cost of different supports?

(h) At what temperature, judged from the Tammann or Huttig temperature, can sintering be expected?

(i) Given that a favoured support does not meet these requirements totally,

i. Is it possible to modify the pore structure or surface area?

ii. Is it possible to improve mechanical properties by addition, for example, of a cement?

iii. Is it possible to poison selectively any unwanted chemical properties?

iv. Is it possible to combine the use of a support and a spacer to minimise sintering?

(j) Are the thermal properties of the support satisfactory?

(k) Can catalyst preparation be adapted to suit a particular support and a particular reaction?

(l) Are there any obvious chemical reactions between the support and the catalyst or the support and the gases?

(m) What would be the influence of the support on potential poisoning reactions?

(n) What is the cost of the better supports?

2. Questions for which some information may be available:

(a) Is the reaction to be catalysed a demanding reaction?

(b) If so, can the geometry of the catalyst be influenced by the support?

(c) What is the mechanism of sintering of the catalyst or of the support? Can preventive measures be taken?

(d) What dispersion can be expected on a given support?

(e) Is the reaction to be catalysed liable to be affected by spillover?

3. Questions which should be asked, but for which information is liable to be minimal:

(a) Is there any catalyst-support interaction on a molecular scale?

(b) What is the effect of support on catalyst reorganisation, with respect to both surface energy and impurities?

It should be stressed that a perfectly adequate design can be formulated on the basis of the first group of questions alone: indeed, this is how a decision is usually made. However, if available, answers to the second and third group of questions will allow the design of a more active and longer-lived catalyst.

REFERENCES

1 O.V. Krylov, Catalysis by Non Metals, Academic Press, New York, 1970.
2 S.J. Gregg and K.S.W. Sing, Adsorption, Surface Area and Porosity, Academic Press, London, 1967.
3 E.S. Dokukina, O.A. Golovina, M.M. Sakharov and R.M. Aseeva, Kinet. Catal., 7, 580 (1966).
4 D.L. Trimm and B.J. Cooper, J. Catal., 31, 287 (1973).
5 R.R. Schechl, H.W. Pennline, J.P. Strakey and W.P. Haynes, Amer. Chem. Soc. Div. Fuel Chem. Prepr., 21, 2 (1976).
6 J.M. Thomas and W.J. Thomas, Introduction to the Principles of Heterogeneous Catalys, Academic Press, New York, 1967.
7 J.J. Carberry, Chemical and Catalytic Reaction Engineering, McGraw Hill, 1976).
8 J.R. Dacey, Ind. Eng. Chem., 56, No. 6, 27 (1965).
9 E.R. Becker and T.A. Nuttall, Preprint, Second Internat. Sympos. Scientific Bases for the Preparation of Heterogenous Catalysts, Louvoin-la-Neuve, Sept. 1978.
10 M.S. Borisova, V.A. Dzisko and L.G. Simonova, Kinet. Catal., 15, 425 (1974).
11 A.G. Walton, "The Formation and Properties of Precipitates, Interscience, New York, 1967.
12 N.M. Zaidman, Kinet. Catal., 13, 906 (1972).

13 N.M. Zaidman, U.A. Dzisko, A.P. Karnaukhov, N.P. Krasilenko, N.G. Koroleva and G.P. Vishnyakova, Kinet. Catal., 9, 709 (1968).
14 N.M. Zaidman, U.A. Dzisko, A.P. Karnaukhov, L.M. Kefdi, N.P. Krasilenko, N.G. Koroleva and I.D. Ratner, Kinet. Catal., 10, 313 (1969).
15 N.M. Zaidman, V.A. Dzisko, A.P. Karnaukhov, N.P. Krasilenko and N.G. Koroleva, Kinet. Catal., 10, 534 (1969).
16 N.M. Zaidman, N.P. Krasilenko, L.M. Kefdi and I.D. Ratner, Kinet. Catal., 11, 604 (1970).
17 T. Sukeno, M. Mimura, H. Nomura and K. Zasshi, J. Chem. Soc. (Japan), 68, 1838 (1965).
18 T.N. Asimolov and O.V. Krylov, Kinet. Catal., 11, 1028 (1970).
19 N.M. Zaidman, Kinet. Catal., 13, 1012 (1972).
20 B.A. Lipkind, T.G. Plachenov, E. Ya. Paramonkov, A.S. Semenova, B.G. Feddrov and S.V. Vainshtein in Scientific Fundamentals of Selecting and Manufacturing Catalysts, Nanka, (Novosibirsk) p.281, (Russian) 1964.
21 A.V. Topchiev, B.A. Krentsel', A.I. Perel'man and T.V. Rode, Izv. Akad. Nauk. SSSR, Otd. Khim Nauk, 1079 (1959).
22 F.M. Bukanaeva, Yu. I. Pecherskaya, V.B. Kazanskii and V.A. Dzis'ko, Kinet. Catal., 3, 258 (1962).
23 V.R. Gurevich, M.A. Dalin and L. Ya. Vedeneeva, Azerb. Khim. Zh., 6, 37 (1963).
24 Yu I. Grurakov, L.P. Ivanov, L.Ya. Al't, A.L. Gel'bshtein and V.F. Aunfrienko, Kinet. Catal., 9, 352 (1968).
25 E.S. Houdry, U.S. Pat. 2945823 (1960), 3180903 (1965).
26 B.A. Dadashev, A.V. Sarydzhanov and G.G. Aluramedov, Azerb. Neftyanoe Khozyaistvo, 12, 35 (1965).
27 G.M. Maxted and S. Ali, J. Chem. Soc., 9, 4137 (1961).
28 N.M. Zaidman, V.A. Dzis'ko, A.P. Karnaukov, N.P. Krasilenko, N.G. Koroleva and G.P. Vishuyakova, Kinet. Catal., 9, 862 (1968).
29 P.L. Walker, T.J. Lamond and J.E. Metcalf, Proc. Conf. Ind. Carbon Graphite, 3rd, S.C.I. London, p.7 1966.
30 A. Elo and P. Clements, J. Phys. Chem., 71, 1078 (1967).
31 S.V. Moore and D.L. Trimm, Carbon, 15, 177 (1977).
32 Catalyst Handbook, Wolfe Scientific Texts, London, 1970.
33 J.J. Benbow, Chem. Eng. Sci., 26, 1467 (1971).
34 M. Boudart, A.W. Aldag, L.D. Ptak and J.E. Benson, J. Catal., 11, 35 (1968).
35 M.S. Borisova, V.A. Dzisko and Yu.O. Bulgakova, Kinet. Catal., 12, 344 (1971).
36 N.T. Kulishkin, A.V. Mashkina, N.E. Buyanova, A.P. Karnaukhov, I.D. Ratner and L.M. Plyasova, Kinet. Catal., 12, 1361 (1971).
37 Ya. Neinska, V. Penchev and V. Kanazierev, Kinet. Catal., 14, 667 (1973).
38 K.K. Kakati and H. Wilmen, J. Phys. D., Appl. Phys. 6, 1917 (1973).
39 J.E. Germain, Intra-Science Report, 6, 101 (1972).
40 V.A. Shvets and V.B. Kazansky, J. Catal., 25, 123 (1972).
41 V.A. Shvets, V.M. Vorotinzev and V.B. Kazansky, Kinet. Katal., 12, 678 (1971).
42 V.A. Shvets and V.B. Kazansky, Kinet. Katal., 12, 935 (1971).
43 L.L. Van Reijen and P. Cossee, Disc. Farad. Soc., 41, 277 (1966).
44 M.Ya, Kon, V.A. Shvets and V.B. Kazansky, Kinet. Catal., 14, 339 (1973).
45 V.M. Fenin, V.A. Shvets and V.B. Kazansky, Kinet. Katal., 12, 1255 (1971).
46 A.I. Machenko, V.B. Kazansky, G.B. Parijsky and V.M. Sharapov, Kinet. Katal., 8, 353 (1967).
47 V.M. Vorotinzev, V.A. Shvets and V.B. Kazansky, Kinet. Katal., 12, 678 (1971).
48 V.M. Vorotinzev, V.A. Shvets and V.B. Kazansky, Kinet. Catal., 12, 1108 (1971).
49 O.V. Krylov and L.Ya. Margolis, Kinet. Catal., 11, 358 (1970).
50 D.A. Dowden, Cat. Revs. Sci. Eng., 5, 1 (1972).
51 D.W. Van Krevelen and H.A.G. Chermin, Chem. Eng. Sci., 1, 66 (1951).
52 D.A. Dowden, Ind. and Eng. Chem., 44, 977 (1952).
53 U. Schwabe and E. Bechtold, J. Catal., 26, 427 (1972).
54 R.B. Levy and M. Boudart, J. Catal., 32, 304 (1974).
55 S.E. Wanke and N.A. Dougharty, J. Catal., 24, 367 (1972).
56 J.U. Reid, S.J. Thomson and G. Webb, J. Catal., 29, 421 (1973).
57 J.A. Altham and G. Webb, J. Catal., 18, 133 (1970).
58 W.C. Neikam and M.A. Vannice, J. Catal., 27, 207 (1972)

59 A.S. Darling, Internat. Metall. Rev., 91, (1973).
60 G. Tammann and Q.A. Mansuri, Z. anorg. allg. Chem., 126, 119 (1923).
61 G.F. Huttig, Disc. Farad. Soc., 8, 215 (1950).
62 W.M. Gitzen, Alumina as a Ceramic Material, Amer. Cer. Soc., Columbia, (1970).
63 L.M. Kefeli and L.M. Plyasov, Kinet. Catal., 6, 1080 (1965).
64 G.S. Vinnikova, V.A. Dzis'ko, L.M. Kefcli and L.M. Plyasov, Kinet. Catal., 9, 1331 (1968).
65 G.C. Bailey, V.C.F. Holm and D.M. Blackburn, Amer. Chem. Soc. Div. Petrol. Chem. Preprints, 2, 329 (1957).
66 V.N. Kuklira, G.A., L.M. Plyasova and V.I. Zherkov, Kinet. Catal., 12, 1133 (1972).
67 W.W. Mullins, Trans. Met. Soc. AIME, 218, 354 (1960).
68 A.E.B. Presland, G.L. Price and D.L. Trimm, Progr. in Surf. Sci., 3, 63 (1972), Pergamon, London.
69 W.W. Mullins, Phil. Mag., 6, 1313 (1961).
70 M. Moayeri, Ph.D. thesis, Univ. of London (1974).
71 N.A. Gjostein, in Surfaces and Interfaces, ed. T.T. Burke, N.L. Reed, V. Weiss, Syracuse Univ. Press (1967).
72 J.A. Allen and J.W. Mitchell, Disc. Farad. Soc., 8, 309 (1950).
73 O. Glemser and H.G. Wendlandt, in Advances in Inorganic Chemistry and Radiochemistry, 5, 215 (1963), Academic Press, New York.
74 D.A. Dowden, I. Chem. E. Symp. Sci., 27, 18 (1968).
75 R.L. Coble and J.E. Burke, in Progress in Ceramic Science, 3, 197 (1963), Pergamon, London.
76 P.C. Flynn and S.E. Wanke, J. Catal., 34, 390, 400 (1974).
77 B. Pulvermacher and G. Ruckenstein, J. Catal., 35, 115 (1974).
78 M. Raissian, D.L. Trimm and P.E. Williams, Farad. Trans., 1, 72, 925 (1976).
79 G. Ruckenstein, J. Catal., 35, 441 (1974).
80 J. Rostrup-Nielsen, Steam Reforming Catalysts, Teknisk Forlag A/S, Copenhagen, (1975).
81 J. Rostrup-Nielsen, J. Catal., 33,184 (1974).
82 R. Rudham and M.K. Sanders, J. Catal., 27, 287 (1972).
83 J.E. Germain, Catalytic Conversion of Hydrocarbons,Academic Press, London, (1969).
84 J.G. Firth and H.B. Holland, Nature, 212, 1036 (1966).
85 B.J. Cooper and D.L. Trimm, Chem. Comm., 1969 (1970).

CHAPTER 6

EXPERIMENTAL TESTING

The one factor that is seen to be of paramount importance in the design of catalysts is the experimental testing which must be carried out. Reference to the general schemes discussed in Chapter 2 shows that catalyst testing is necessary at various stages in the design, ranging from simple comparisons of activity and selectivity of several possible catalysts to the more detailed investigations of the kinetics of reaction over one or two catalysts. Since the objectives of the testing are closely linked to the complexity of the test, it seems desirable to devote some attention to this area.

On the assumption that a new catalyst is required for a novel reaction, experimental testing can be applied at several points:

(i) When the paper design has suggested several possible primary constituents, a comparison of activity and selectivity can lead to selection of catalysts worthy of further consideration.

(ii) The roles of secondary components of a catalyst must be established by experimental testing.

(iii) Given that one or two catalysts are sufficiently interesting, the detailed kinetics of reactions must be established.

(iv) The life and the resistance of these catalysts to potential poisons must be tested.

(v) Large scale testing of possible catalysts must be carried out.

(vi) New, used and deactivated catalysts must be tested on a continuing basis.

(vii) New, used and deactivated catalysts must be characterised, with respect to factors such as composition, surface area and porosity, mechanical strength etc.

Obviously these tests are applied at different stages in the development of a catalyst, and a general flow diagram, as shown in Fig. 6.1, can be used to relate the experimental programme.

Before considering possible methods of catalyst testing it is rewarding to consider briefly what should be tested. The characterisation of catalysts has been well described (refs. 1,2) and will not be considered here. There is a case, however, to regard other tests as being a particular form of measurements of kinetics. Thus, for example, if a catalyst is active it is necessary to obtain some idea of the activity and selectivity as a function of contact time, of temperature, of reactant concentration/pressure and of surface area/porosity.

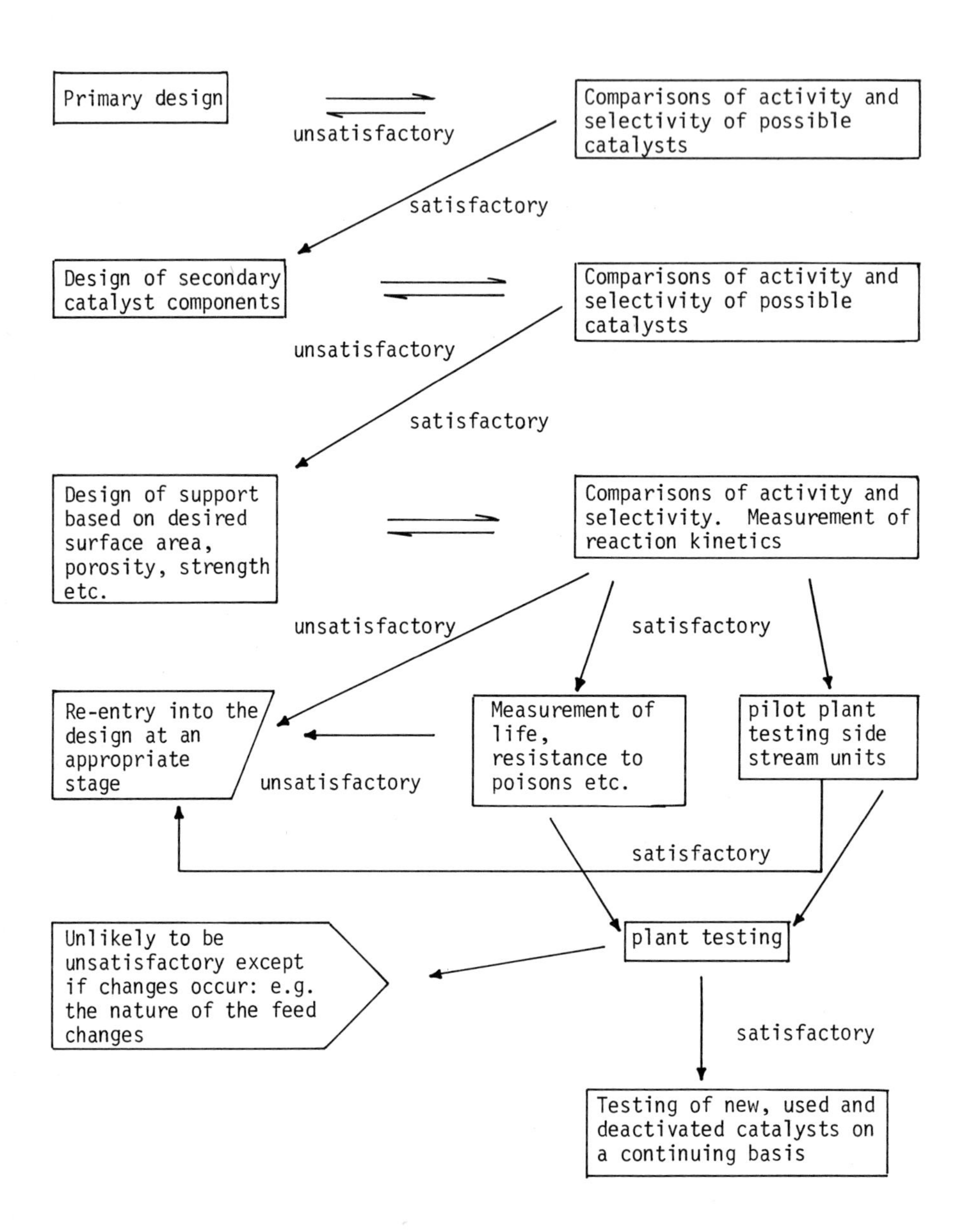

Fig. 6.1. Stages in the testing of catalysts.

This is as true for simple comparisons of activity and selectivity as it is for the much more detailed and accurate measurements required to establish the kinetics of reactions over the catalyst. Some of the problems associated with the measurements can, as a result, be easily identified from consideration of the rate equation. Experimental tests, and the forms of reactors in which these can be carried out, must be devised with these potential problems in mind.

A reaction of the general type

$$A + B \rightarrow C \qquad (1)$$

can be expected to proceed with the gain or loss of heat at a rate which can be expressed in the form of an equation

$$\text{rate} = k \times c \times (A)^{\alpha}(B)^{\beta}(C)^{\gamma} \times \eta \qquad (2)$$

where k is the rate constant, η is the effectiveness factor, c is the concentration of active sites and α, β and γ are the apparent orders of reaction. In this equation, the rate is considered as the change of concentration with time, and it is in the measurement of this that the first problems are apparent. Thus, for example, if we consider a tubular reactor filled with catalyst (Fig. 6.2)

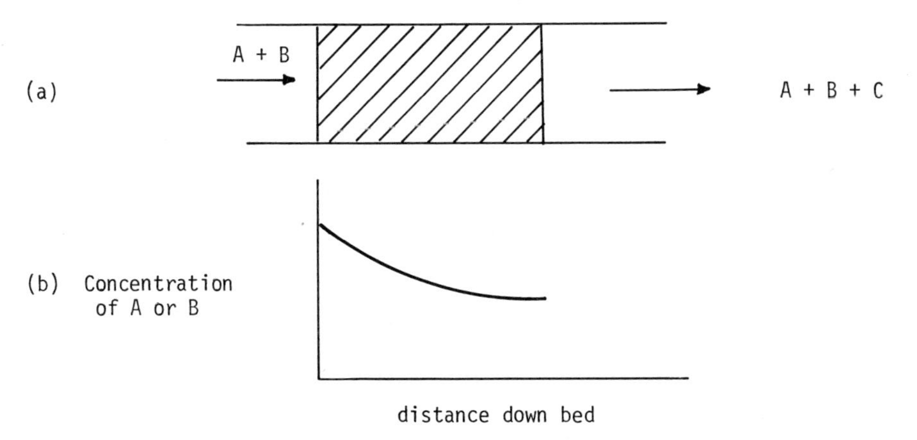

Fig. 6.2. Representation of a tubular reactor filled with catalyst (a) and of the reactants down the tube (b).

then the concentration of a reactant can be expected to be of the general distribution shown, while the temperature in the bed - unless the reactor is maintained at a given value - will reflect the heat of the reaction involved.

The difficulties introduced by the change of concentration down the bed, by possible inhibition of reaction by the products and by temperature variations are well known, and are reflected in the use of differential and integral reactors (ref. 3). In the former case, the initial rate approach (ref. 3) is used to measure kinetics under conditions where the concentration change (and the temperature) is small between inlet and outlet. There are, however, some residual problems in the measurement of time which, although they may not be important, can complicate the interpretation of results.

In absolute terms the time connected with the conversion of reactants is the time that the reactants remain in contact with the catalyst. In practice, however, this is not easy to measure and may, in fact, be less useful than other measurements of time. This can easily be seen from brief consideration of methods of measuring true residence times and of the processes that can occur in the catalyst bed.

The residence time (and residence time distribution) of reactants in a bed can be established by introducing a pulse (or step function) of a tracer into the inlet and by measuring the distortion in the pulse as it emerges from the outlet (refs. 3,4). The distortion will reflect the flow regime in the reactor, the importance of diffusion into and out of catalyst pores and the adsorption of the gas on the surface. Of these, the only factor truly pertinent to the reaction is the time of adsorption - which will vary as the nature of the gas and the nature of the catalyst. This is difficult to distinguish from the overall residence time and, even if this is possible, should - theoretically - be measured for every system.

Of course, this is impractical, but the situation is saved by the knowledge of what is really required. In fact, it is usually desired to know the conversion that can be effected by a given mass (or volume or surface area) of a catalyst when exposed to a reactant passing at a given flow rate. As a result, time is usually expressed as a contact time, defined as the mass (or volume) of catalyst divided by the flow rate adjusted to standard conditions. The rate of a reaction, then, is defined as the change of concentration occurring in the catalyst bed, divided by the contact time and by the surface area of the catalyst.

The difficulties in measuring the concentrations of reactants and the temperature in the reactor are closely related to the problems involved in measuring the change in concentration. These are solved in different ways, and it is useful to consider some of these in detail.

I. PRELIMINARY TESTING

In the context of catalyst design, the objectives of preliminary testing can be stated easily. Given that the paper design has suggested a number of possible catalysts, it is desired to ascertain which of these catalysts (if any) is most active/selective for the reaction under study. In discussing different test procedures that have been suggested, it should be remembered that some degree of subjectivity is involved in the choice of system.

Before considering different reactors, it is worth considering the reactions themselves. On occasion, it can seem worthwhile to test analogous reactions. Thus, for example, if a catalyst is sought for, say, the hydrogenation of an expensive olefin, then it could be suggested that the activity of catalysts for the hydrogenation of a similar (but cheaper) olefin would be indicative. Similarly, if thermodynamics indicate that experimental testing should be carried out at high pressure for the reaction

$$A + B = C$$

then it could be argued that the decomposition of C to A and B at low pressure would give an analogous comparison. Very often such comparisons are valid but, on occasion, they are not. Use of an analogous reaction can be rewarding but, in all cases, the greatest care should be taken in their use.

The most widely used test system is the microreactor, in which about 10 g of catalyst are maintained at temperatures as near isothermal as possible. Reactants are fed to the system and the exit gases are analysed: a combination of gas chromatography and mass spectrometry would seem to offer the most reliable means of product identification. The operation and design parameters for such a microreactor system have been reported in detail (refs. 2,3,4). For screening purposes, activity and selectivity are established at a near optimum contact time over a range of reactant concentrations and reactor temperatures. Typically, three reactant concentrations and three temperatures are sufficient for these approximate comparisons.

Several other types of reactor have been suggested for screening, and each have different advantages. At first the single pellet reactor or the string of pellets reactor (refs. 5,6,7) would seem to have advantages because it is necessary to prepare less catalyst. This may, however, be an illusory advantage, in that it is often as difficult to prepare one pellet of catalyst as it is to prepare 10 g of catalyst. In addition, the conversion observed over a single pellet (considered together with the fact that considerable by-passing may be observed) is usually such that product analysis may become a major problem. As a result, the single pellet (or string of pellets) reactor would seem to have advantages only in very particular circumstances (refs. 5,6,7).

It has also been argued that preliminary testing may be best carried out in a small reactor of the type to be used at a later stage: thus, for example, a small fluidised bed catalytic reactor could be more suitable for a reaction which is known to be very exothermic. Here, again, the advantage may be illusory, in that the advantage to be gained in these preliminary approximate tests has to be set against the disadvantage of setting up and operating such a reactor. With the exception of particular cases, the tubular microreactor would seem to be just as suitable a way of comparing catalysts for a given reaction.

II. TESTING OF THE EFFECT OF SECONDARY COMPONENTS

The choice of testing system vis-a-vis the study of secondary components of the catalyst depends on the complexity with which the reaction is being studied. If it is desired only to ascertain the effect of secondary components on the overall activity and selectivity, then a tubular microreactor is usually suitable. If it is desired to study the detailed mechanism of the reaction in order to optimise the catalyst composition, then other types of reactor can have considerable advantages.

These occur from the desire to recognise possible intermediates in a reaction. With a tubular reactor, short residence times can be achieved by reducing the amount of catalyst (which may cause major by-passing) or by increasing flow rate (which may cause by-passing or may result in a change of flow regime). As a result, special types of reactor may be more suitable.

The first of these is a pulse reactor, in which a pulse of reactants is introduced to the catalyst bed (ref. 8): the pulse may contain the mixed reactants transported in a stream of inert fluid, or may contain one reactant transported in a stream of the other reactant. In most cases, results obtained from the former geometry are more meaningful and are easier to interpret. The effect of the pulse is easily seen, since a close definition of time is obtained. Contact time is determined by the flowrate of the carrier, but the time of contact of the reactants with the catalyst is dictated by the size of the pulse and the flowrate. As such, the reactor system is particularly useful for the investigation of a catalyst where the activity may be rapidly changing (ref. 9).

There are, however, difficulties with the operation of this system. The first of these results from the fact that the catalyst is not at the steady state. Under normal conditions, the surface of a catalyst is conditioned to some extent. This may involve the adsorption of equilibrium amounts of reactants of the partial poisoning of the surface, for example by the deposition of polymeric species on the more active sites. If the gas passing over the surface is different to the reacting species normally used, then the equilibrium will be different and the observed behaviour of the catalyst could differ.

The second difficulty is related to another type of reactor that has been used to study reaction mechanisms. As a pulse of mixed reactants passes through a bed of porous material, chromatographic separation of the different species may occur. This may be desirable or not. If subsequent reactions between an intermediate and the reactants can occur, then separation of the intermediate can allow identification. In the pulse reactor, however, separation of reactants may stop reaction altogether. As a result, a clear distribution must be made between a pulse reactor operating non-chromatographically (ref. 9) and chromatographically (ref. 10). Both reactors may be useful, under different circumstances, for the study of reaction mechanisms. Indeed, the reactors have also been used to study reaction kinetics (see below) but the disadvantages listed above make the interpretation of results difficult.

As a general rule, then, the tubular microreactor is useful for the study of reaction mechanisms, but other types of reactor may be more useful under specific circumstances.

III. MEASUREMENT OF REACTION KINETICS

Inspection of the literature would seem to indicate that there are almost as many reactors to measure the kinetics of a catalytic reaction as there are reactions! However most of these fall into general classes, some of which are more useful than others: this can easily be seen by consideration of what is required. As discussed above, the main necessity is to measure rates under conditions of known concentration and temperature. For the purposes of reactor design, data is required which can be scaled up and which is not affected by spurious factors. This means that rate data should not be affected by mass and heat transfer, nor by effects such as reactions in the gas phase. Different approaches to this necessity require different reactors.

(a) Tubular reactors

By far the most widely used reactors are tubular, being either differential or integral in their mode of action. Given that a reaction sequence may be written

$$A + B \rightarrow C \rightarrow D \qquad (3)$$

then the variation of concentration, time and temperature down the bed has been discussed above. With a differential tubular reactor, it is usual to adopt the initial rate approach. A constant ratio of reactants are fed to the reactor and a plot is constructed of conversion versus contact time. If the contact time is short, the conversion in the bed will be small, no inhibition by products can be expected and the temperature of the catalyst bed (which would be affected by heat liberated or consumed during reaction) remains constant. In this region, to all and purposes the plot of conversion versus contact time will be linear, and the tangent to the slope at zero contact time will give the rate corresponding to the initial concentrations of reactants. As a guide line, plots of conversion versus contact time are usually linear up to ca. 10% conversion. Of course, the reactor must be kept as isothermal as possible, and this is usually done by using a narrow tube immersed in a liquid heating medium.

The basic assumption behind measurement of kinetics in tubular reactors is that the gas passes through the reactor as a plug (ref. 2). This plug flow assumption may fail because of longitudinal diffusion effects, of transverse diffusion effects or of temperature gradient in the reactor (refs. 2,3,4). If the ratio of the reactor length to particle size is of the order of 100 or more, then longitudinal diffusion will be insignificant (ref. 4). Transverse diffusion tends to work in favour of plug flow, since it tends to even out the concentration gradients across the reactor. However the packing of the reactor is looser near the reactor wall (refs. 4,11) and velocity gradients can be observed there. As a rough guide, the deviation from the flat profile assumed in plug flow is not more than 20%, provided that the tube diameter is at least 30 x the particle diameter (ref. 11). In practice, this is rarely possible in a laboratory scale microreactor used to measure kinetics. The reasons for this lie in the necessity of isothermality: the wider the reactor, the more chance of temperature gradients across the bed. The dependence of rate on the exponential term ($-E/RT$) means that temperature control of the reactor is by far the most important factor open to control.

The differential reactor is also useful because it offers an easy way of studying the further reactions of products such as C. Either feeding C alone, or mixtures of A, B and C, allows the determination of the kinetics of the over-reaction. This can be useful in that it allows comparison with results obtained in an integral reactor.

An integral reactor is one in which conversion is allowed to rise to higher levels. It may be operated isothermically or adiabatically, but both of these modes of operation may be expected to be approximations: it is remarkably difficult to ensure good heat transfer throughout a catalyst bed, or to ensure no heat leakage from or into a catalyst bed. Experimental results obtained in an integral reactor can, however, be of considerable use. Obviously they allow some measure of the optimal activity and selectivity that can be obtained with a catalyst and they can be used to give estimates of kinetics. In this connection, however, they are perhaps more of use in that they can be mathematically simulated using data obtained in a differential bed. As a result, they can act to check the accuracy of the rate data.

One of the main disadvantages of using tubular reactors is that it may be difficult to eliminate diffusion. Bulk diffusion (where the mass transfer across the boundary layer to the outside of the pellet is rate controlling) can be checked by varying the flow rate at constant contact time (ref. 2). Because the thickness of the boundary layer depends on flow rate, the conversion observed at the same contact time but different flow rates (and different weights of catalyst) will vary if bulk diffusion affects the results.

The classic tests for pore diffusion involve several procedures. The effect of pore diffusion on the reaction kinetics is to halve the activation energy and to change the order of reaction from m to $\frac{m+1}{2}$ (ref. 2). One good preliminary check is to measure the apparent activation energy over a range of temperatures. A decrease of ca. 50% in the value usually (but not always (ref. 12)) indicates that pore diffusion may be important. The second test involves grinding the catalyst (ref.3). The same weight (i.e. the same surface area) of catalyst should always produce the same conversion at the same flow rate unless diffusion effects are important (the effectiveness factor is less than one (ref. 2)). A plot of conversion versus particle size for the same contact time will soon show whether diffusion is important.

There is little that can be done to check the presence of surface diffusion, and it could be argued that little needs to be done. This results from the fact that surface diffusion (or mobile adsorption) is a function of the chemical nature of the catalyst and of the reactants/products: either they do adsorb (when surface diffusion may be important) or they do not. Altering the catalyst to enhance or retard surface diffusion is, as a result, liable to have a major effect on catalyst performance.

Tubular reactors are widely used for the measurement of kinetics and, on the whole, confidence is not misplaced. Other types of reactor can be more useful in some circumstances.

(b) 'Plug' reactors

This is used as a general term to indicate either that a plug of catalyst or a plug of reactants can be used. Most of these reactors can be used to measure kinetics, but the disadvantages to their use are significant.

The use of single pellet (refs. 5,6) or string of pellets (ref. 7) reactors has been discussed briefly above, and little more needs to be added. By-passing can be a serious problem and the advantages vis-a-vis tubular reactors may be quite small.

Plug flow reactors have been used to measure kinetics. These are reactors through which a plug of reactants is passed, and they operate chromatographically or non-chromatographically (refs. 9,10). The major difficulties are two. As the reactants pass through the bed, some separation usually occurs: this may be as a result of adsorption of the reactants to different extents, or as a result of mass transfer limitations. Indeed, the changes in shape of a plug of gas passing through a catalyst bed have been used as the basis of a measurement of diffusion coefficients (refs. 13,14). Whether or not chromatographic separation of the constituents occurs, the results have to be affected by adsorption, reaction and diffusion. As a result, it may be difficult to untangle the results sufficiently to obtain reliable kinetics.

Secondly, as stated above, the results may not be representative since the catalyst is not at equilibrium. During most reactions, the surface of a catalyst becomes modified in some way - for example, by the deposition of carbonaceous residues (ref. 15) - and this will affect the kinetics at the steady state, but will not affect the kinetics during the time that the pulse passes.

One reactor system which can be included in this section is an exception to the rule that 'plug' reactors are not very useful for kinetic measurements. This is the recycle reactor, which bears some similarities with the tubular reactor and some with the stirred reactor (ref. 16). The principle of the method is very simple. Since an integral reactor may give unreliable results because of temperature fluctuations, these problems could be avoided if high conversions were attained by recycling the reactants over the catalyst in such a way that only a small conversion occurred in each cycle. The catalyst could then have a chance to equilibrate temperature between each pulse.

In its simplest form, the recycle reactor is shown below. This arrangement (also known as a race track) is adjusted so that only a small conversion is obtained per pass.

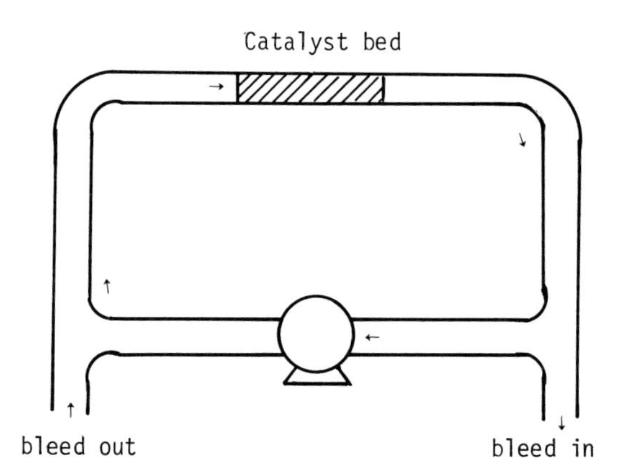

Fig. 6.3. A typical recycle flow reactor.

Although many different arrangements have been used (refs. 17,18) this schematic shows clearly the main disadvantage of the recycle reactor. This is that there is significant space in the system which must be kept hot but which is not filled with catalyst: as a result, homogeneous reactions are often significant. These can be reduced by proper design of the system (refs. 16,17,18) but, in defining what is required for the reactor, homogeneous reactions must be accepted as a possibility.

(c) Stirred reactors

It has been seen that temperature variations or mass transfer limitations can be problems with a tubular reactor system. In an attempt to avoid these problems various stirred tank reactors have been developed in which the gas is circulated through a static bed of catalyst or the catalyst is circulated through the gas.

The basis of the reactor is the assumption of good mixing. In conventional terms, a continuously stirred tank reactor (CSTR) is widely used on the same basis, and there are many similarities between the systems. A schematic diagram of an actual reactor (in which the catalyst is rotated) is shown in Fig. 6.4. This type of reactor, developed by Carberry (ref. 19) and by Brisk et al. (ref. 20) is not available commercially. A reactor in which gas is circulated through a static bed can be purchased from Union Carbide.

Treatment of the results follows that applied to the CSTR, and differs from the tubular reactor. If the system is perfectly mixed, the exit concentration will be the same as that existing at any point in the reactor. As a result, the plot shown in Fig. 6.2 will not apply, and the results must be analysed by setting up a mass balance across the reactor (ref. 20). Two problems are immediately apparent. For application of the theory the reactor must be well mixed and it is difficult to measure the true residence time in the reactor. Both of these potential difficulties can be resolved by measuring the residence time distribution in the reactor: this may be done by comparing the inlet and outlet shape of a pulse or step signal generated by an inert gas (ref. 3). Not only does this give the true residence time but, by comparison with predictions based on good mixing, it is possible to validate the assumption that mixing is complete. If this is so, then the temperature variation throughout the bed will also be minimal. It is useful to be able to check this and, in this respect, the static bed system offers experimental advantages over the rotating bed reactor.

Although the stirred reactor has many advantages, it is obvious that not all mass transfer limitations can be removed by its use. In the context of the catalyst, stirring increases the gas velocity over the surface and, as a result, would be expected to minimise bulk diffusion: pore diffusion, on the other hand, remains unaffected. This is less of a problem than it would appear, since pellets which are intended for use in the industrial reactor can be used in the basket, as well as ground pellets. This means that the time kinetics can be obtained (from the results on ground pellets) and the effectiveness factor of industrial pellets can be measured.

The other major problem inherent to the use of these reactors is the possibility of unwanted reactions. These originate from two sources: homogeneous reactions catalysed by the materials of construction of the reactor. It is often not appreciated that stainless steel has a high concentration of chromium oxides at

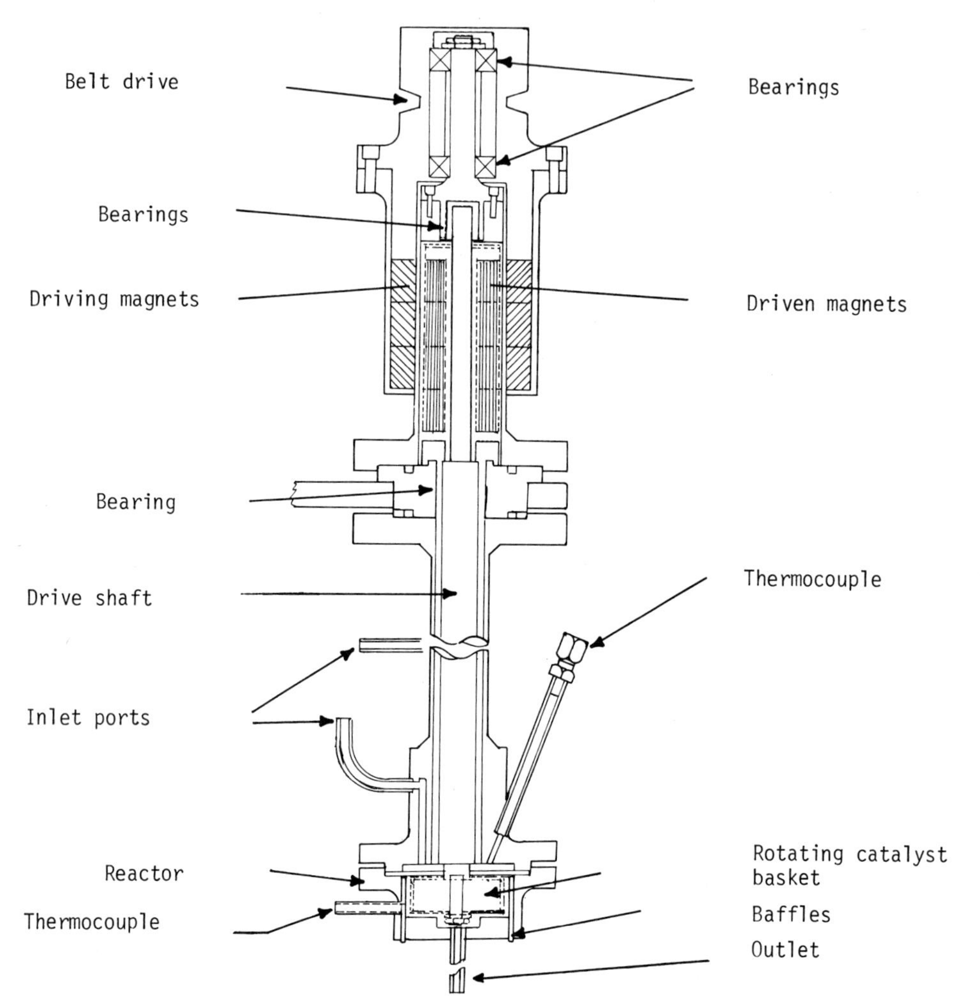

Fig. 6.4. A stirred tank reactor. The catalyst is maintained in a basket which can be rotated. Isolation of the reactor involves a magnetic drive.

the surface, and that these can be good catalysts (ref. 21). The effects can be minimised by plating the surfaces in aluminium (ref. 22), a material which shows minimal catalytic activity.

Any laboratory which is equipped to measure the kinetics of catalytic reactions should have both tubular reactor systems and stirred reactors. There are difficulties with both reactors, and the application of either depends on the reaction under study. Where mass transfer or thermal effects are expected to be significant, the stirred reactor offers many advantages. Other types of reactor can be useful in particular circumstances, but most of the basic measurements do not require their use.

IV. LONG-TERM CATALYST TESTING

As stated above, the life of a catalyst can be very significant in assessing a catalytic design. Assessing life of catalysts poses no new problems except those connected with the length of the test. The use of a side stream reactor may be included in the general running costs of a plant, but testing on a smaller scale (laboratory or pilot plant) can be expensive in terms of labour. As a result of this fact, considerable interest has been focused on computer (or micro processor) controlled testing rigs. There are many examples of these rigs now in operation: in all of them the basic principles are the same as those controlling the use of similar "manual" systems.

At the heart of the system lies a number of reactors which depend on the size of the operation. On a laboratory scale, anything up to 5-10 reactors are possible, while - in the pilot plant - only one reactor is usual. Gases are fed to these reactors, with one extra line going direct to the analysis system. The flow of gases is controlled by motorised valves and measured with flow meters capable of giving a signal to the computer. The temperature measurement and control of the reactors also involves the computer. Gases leaving the reactors may be analysed via a sample valve that can accept gas from each stream in turn. A combination of gas chromatography and mass spectrometry offers the optimal reliability, width of application and ease of analysis although, where a specific reaction is to be studied for very long times, specific gas meters may be advantageous. The output of the analysis line may be fed to temporary or permanent storage on the computer.

At the start of a test, the analysis line is calibrated and a programme of experimental conditions is fed to the computer. For a full analysis of kinetics and for life testing, details of flows, temperatures and times of sampling are used. The system is then set up in the following way:

(a) adjust pressure and temperature to conditions 1
(b) sample inlet and outlet from each reactor in turn
(c) repeat sampling

(d) when the products yields from each reactor are the same over a preset time interval (i.e. when (b) and (c) agree) continue testing for a preset time (life testing) or change to conditions 2 and repeat (kinetic study).

The computer reduces the manpower requirements, then, to a programmer, a scientist needed to set up the system and analyse the results and an assistant to watch over gas supplies etc. Since the same team can supervise one or more rigs containing several catalysts, a marked reduction in labour costs is observed.

Of course, no brief account of methods of experimental testing can cover all of the advantages and disadvantages of each reactor, and the interested reader is referred to other texts (refs. 2,3,4,23,24).

REFERENCES

1 S.J. Gregg and K.S.W. Sing, Adsorption, Surface Area and Porosity, Academic Press, New York, 1967.
2 J.M. Thomas and W.J. Thomas, Introduction to the Principles of Heterogeneous Catalysis, Academic Press, New York, 1967.
3 O. Levenspiel, Chemical Reaction Engineering, John Wiley, New York, 1962.
4 J.J. Carberry, Chemical and Catalytic Reaction Engineering, McGraw Hill, New York, 1976.
5 J.G. Firth and H.B. Holland, Trans. Farad. Soc., 65, 1121 (1969).
6 C.F. Cullis, T. Nevell and D.L. Trimm, Farad. Trans 1, 68, 1406 (1972).
7 L.L. Hegedus and E.F. Petersen, Chem. Eng. Sci., 28, 345 (1973).
8 T. Hattori and Y. Murakami, J. Catal., 12, 166 (1968).
9 M. Riassian, D.L. Trimm and P.M. Williams, J. Catal., 46, 82 (1977).
10 S.Z. Roginskii and A.L. Rozental, Kinet. Catal., 5, 86 (1964).
11 C.E. Schwartz and J.M. Smith, Ind. Eng. Chem., 45, 1209 (1953).
12 W. Lam and D.L. Trimm, In preparation (1979).
13 P.E. Eberley, Ind. Eng. Chem., 8, 25 (1969).
14 J. Corrie and D.L. Trimm, Chem. Eng. J., 4, 229 (1972).
15 D.L. Trimm, Catal. Rev. Sci. Eng., 16, 155 (1977).
16 G.P. Korneichuk and Yu I Pyatnitskii, Kinet. Catal., 3, 132 (1962).
17 D.W.T. Rippin, Ind. Eng. Chem. (Fund), 6, 488 (1967).
18 J.A. Mahoney, J. Catal., 32, 247 (1974).
19 J.J. Carberry, Ind. Eng. Chem., 56, 39 (1964).
20 M.L. Brisk, R.L. Day, M. Jones and J.B. Warren, Trans. Inst. Chem. Eng., 46, T3 (1968).
21 J.E. Germain, Catalytic Conversion of Hydrocarbons, Academice Press, New York, 1969.
22 P.H. Calderbank, Private communication.
23 F. Trifiro and I. Pasquon, Chimica e l'Industria, 944 (1967).
24 G.C. Chinchen, Paper read before the Fertilizer Society of London, 20 April 1978.

CHAPTER 7

SUMMARY OF SOME USEFUL GENERAL INFORMATION FOR CATALYST DESIGNERS

I. INTRODUCTION

Whatever else is required in the study of catalysis, it is essential that information be obtained from a variety of disciplines, and this necessity is reinforced in the context of the design of catalysts. Obviously the storehouse of knowledge that the designer has at his disposal increases with experience. Equally obviously, there are few designers who are completely self-sufficient. Indeed it is foolish to rely on experience if there is an expert available with whom a particular problem can be discussed. A ceramics expert will usually know much more about support materials than a catalyst designer, although he may know little about organic reaction mechanisms. The first general rule for a catalyst designer is, then, to use an expert if he is available.

The second general rule is that the basis of catalyst design involves a mixture of logic and common sense. In the chapters above, a general approach has been described, and this is illustrated in the following chapters. The reader may find that a given sequence of operations is not as logical to him as an alternative approach. Do not hesitate to adopt the alternative if you, yourself, feel more comfortable using it.

The third general rule is that, in catalyst design, there is every opportunity to illustrate the famous law. If something can go wrong, it probably will! As a result, the feedback between experimental testing and catalyst design is particularly important, and should be used to direct thinking and design.

The fourth general rule is perhaps the most important in undertaking a catalyst design. It is always advantageous to use general relationships reported in the literature where this is possible. These relationships may or may not have a sound theoretical basis, but if they can be used to predict the behaviour of catalysts in practice then they can be very useful indeed. As a result, it is always wise to have easy access to some listing of such information while working on a design.

The present chapter has been written in response to this need, in that it contains some relationships that I have found useful over the years. Note the use of the personal pronoun. Any catalyst designer has his own list of relationships which he has developed which may, or may not, coincide with those given below. Any relevant information can be useful, listed or not, and blank pages have been left in the text, where other useful relationships may be added by the reader.

Having listed these relationships, however, a word of warning is required. As with all general relationships, exceptions may be found. This is particularly true where activity patterns are based on only a limited number of studies. Oxidation or reduction activity patterns are usually well established. Hydration or reactions on sulphides are based on much less evidence, and may well be less reliable. An indication of reliability is given below, but it is necessary always to remember the famous law. No matter how reliable the correlation, there is always a chance that the reaction under study is the exception!

In one sense, the observations below cannot be classified since they may apply in a variety of circumstances. In an effort to systemise the information, some subdivision has been attempted. The reader is advised to use this only as a general guide.

II. METALS

1. Some properties of metals which could be of interest as catalysts are shown in Table 7.1.

2. Metals are mainly used for reactions involving the addition or removal of hydrogen, isomerisation and oxidation.

3. For reactions involving hydrogen, alone or with a hydrocarbon, the activity pattern may be broadly assessed as

Ru, Rh, Pd, Os, Ir, Pt > Fe, Co, Ni > Ta, W, Cr ∿ Cu.

4. Palladium is an unusual metal, often being both active and selective.

5. Activity can sometimes be related with percentage d-character of metals, as given in Table 7.2. This relationship should be treated with caution, as there are many exceptions.

6. Activity can sometimes be related with lattice parameters of metals, as shown in Figure 7.1. This relationship should be treated with caution, as there are some exceptions.

7. According to Dowden (ref. 4), the selectivity of metals in the saturation of multiple bonds with hydrogen is shown in Table 7.3, for the organic group in the left hand column reacting in the presence of the organic group along the top row. Note the particular importance of Pd.

8. Metals which are reasonably stable under oxidising and sulphiding conditions include

Rh Pd Ag
Ir Pt Au

TABLE 7.1

Data for the more common metals (ref. 1)

Metal	Atomic Weight	Crystal structure	Lattice parameters (nm)	Nearest Neighbour distance (nm)	Melting point (K)	Density (293 K) (10^3 kg m^{-3})
Al	26.98	f.c.c.	0.404	0.286	1033	2.70
Ba	137.34	b.c.c.	0.501	0.434	998	3.51
Be	9.01	h.c.p.	0.288;0.357	0.225	1551	1.85
Cd	112.40	h.c.p.	0.297;0.561	0.297	594	8.65
Ca	40.08	f.c.c.	0.557	0.343	1118	1.56
Cr	52.00	b.c.c.	0.289	0.249	2163	7.19
Co	48.93	f.c.c.	0.355	0.251	1768	8.90
Cu	63.54	f.c.c.	0.361	0.255	1356	8.96
Au	196.97	f.c.c.	0.407	0.288	1336	19.32
Hf	178.49	h.c.p.	0.320;0.508	0.316	2420	13.29
Ir	192.2	f.c.c.	0.383	0.271	2683	22.42
Fe	55.85	b.c.c.	0.286	0.248	1808	7.87
La	138.91	h.c.p.	0.372;0.606	0.371	1193	6.19
Pb	207.19	f.c.c.	0.349	0.349	601	11.35
Li	6.94	b.c.c.	0.350	0.303	452	0.534
Mg	24.31	h.c.p.	0.321;0.521	0.320	924	1.74
Mn	54.94	complex			1517	7.20
Mo	95.94	b.c.c.	0.314	0.272	2883	10.22
Ni	58.71	f.c.c.	0.352	0.249	1726	8.90
Os	190.2	h.c.p.	0.273;0.431	0.270	3273	22.57
Pd	106.4	f.c.c.	0.388	0.275	1825	12.02
Pt	195.09	f.c.c.	0.392	0.277	2042	21.45
K	39.10	b.c.c.	0.531	0.462	337	0.862
Re	186.2	h.c.p.	0.276;0.445	0.274	3453	21.02
Rh	102.91	f.c.c.	0.380	0.268	2239	12.41
Ru	101.07	h.c.p.	0.270;0.427	0.267	2523	12.41
Ag	197.87	f.c.c.	0.408	0.288	1234	10.50
Na	22.99	b.c.c.	0.428	0.371	371	0.971
Sr	87.62	f.c.c.	0.605	0.430	1042	2.54
Ta	180.95	b.c.c.	0.330	0.285	3269	16.60
Th	232.04	f.c.c.	0.508	0.360	2053	11.66
Sn	118.69	tetr. double b.c.c.	0.582;0.317	0.316	505	7.28
Ti	47.96	h.c.p.	0.295;0.468	0.293	1948	4.54
W	183.85	b.c.c.	0.316	0.274	3683	19.30
U	238.03	complex			1405	18.95
V	50.94	b.c.c.	0.302	0.263	2163	6.12
Zn	65.37	h.c.p.	0.266;0.494	0.266	692	7.12
Zr	91.22	h.c.p.	0.322;0.512	0.319	2125	6.53

TABLE 7.2

Percentage d-character in the metallic bond of transition elements (ref. 2)

IIIA	IVA	VA	VIA	VIIA	←	VIII	→	IB
Sc	Ti	V	Cr	Mn	Fe	Co	Ni	Cu
20	27	35	39	40.1	39.5	34.7	40	36
Y	Zr	Nb	Mo	Tc	Ru	Rh	Pd	Ag
19	31	39	43	46	50	50	46	36
La	Hf	Ta	W	re	Os	Ir	Pt	Au
19	29	39	43	46	49	49	44	

TABLE 7.3

Selectivity in hydrogenation (ref. 4)

compound hydrogenated \ in the presence of non-reacting	C ≡ C	C = C	⌬	C ≡ N	>C = O	RNO_2
C≡C	(*) Pd, Fe (Ni(**)S)	Pd, (Ni)	Ni, Pd, Pt	Pd	Pd	Pd
C=C		Pd,	Ni, Pd, Pt	Pd	Pd (conjugated bonds)	Pd
⌬			Ni, Pd, Pt			
C≡N			Co, Ni, Rh	Co(**)Ni,Pd, Pt, (mild condts)	Ni	
>C=O		Pd(**) N-base Pd(**)Fe^{2+}, Zn^{2+} Re, Ru, Cu, Zn (conjugated)	Ni, Cu Ru, Rh		Ru, Rh, Pd Pt	
RNO_2			Ni, Cu Ru, Rh	Pd, Pt	Pd	

(*) includes hydrogenation of acetylene to ethylene
(**) especially in the presence of ammonia

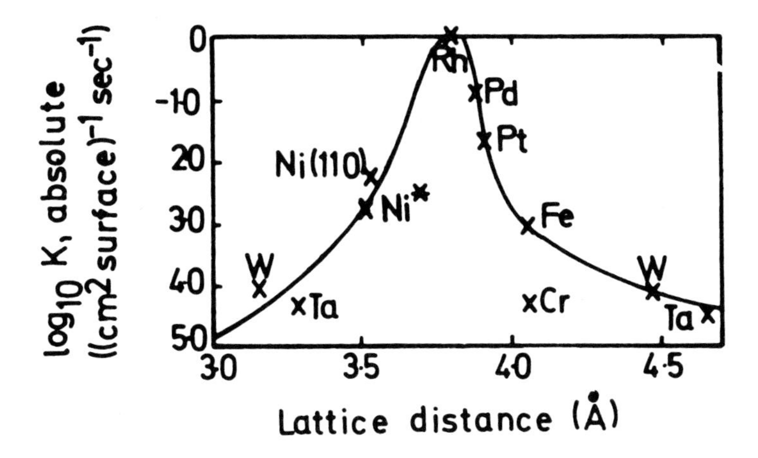

Fig. 7.1. Specific activity of thin metallic films as a function of lattice dimensions (ref. 3).

9. Catalyst design may be helped by reference to references 1 and 2.

III. OXIDES

1. Oxides of possible interest as catalysts are shown in Table 7.4 together with some of their properties.
2. Transitional metal oxides may catalyse oxidation and dehydrogenation reactions. Acid-base characteristics of oxides are summarised in section 7.5.
3. Simple oxides with several stable valency states are usually the most active catalysts (Table 7.5).
4. As a general rule, alkalies stabilise higher valency states and acids stabilise lower valency states.

For oxidation

5. Typical activity patterns are shown in Fig. 7.2. This is applicable to very many systems, although some deviations have been reported (ref. 6).
6. Activity and selectivity in catalytic oxidation are often inversely related.
7. Oxides containing metals capable of achieving d^0 or d^{10} electronic structures can be very selective oxidation catalysts (see Table 7.6).
8. Activity can also be related with the "strength of the oxygen bond" on the surface.
9. Morooka (ref. 7) relates activity to the heat of formation of a metal oxide divided by the number of oxygen atoms contained in the oxide (Fig. 7.3).

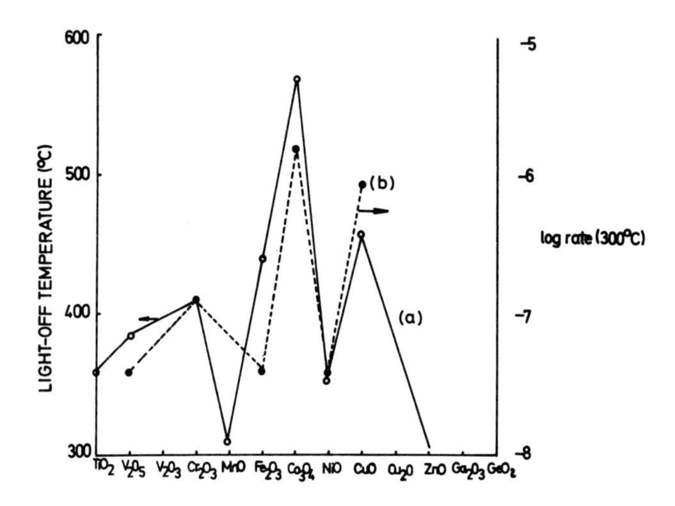

Fig. 7.2. Activity patterns of transition metal oxides for oxidation

-o-o- : Oxidation of ammonia (ref. 5)

--●-●---- : Oxidation of propylene (refs. 6,7)

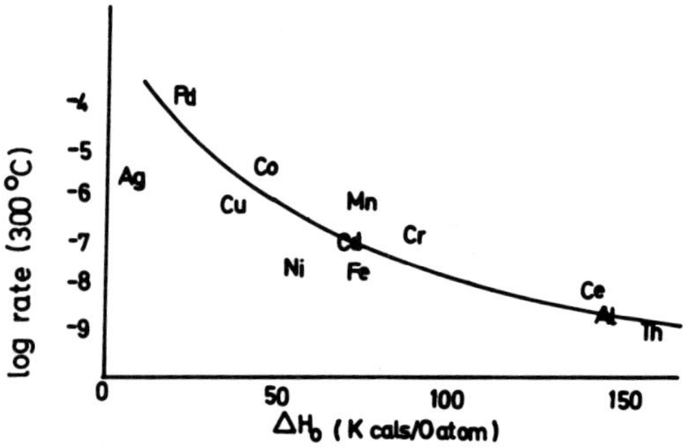

Fig. 7.3. Oxidation of propylene as a function of heats of formation per oxygen

Table 7.4

Some Properties of Non-Metallic Substances Used as Catalysts*

Substance	Structure (a)	Lattice type	M-V distance (Å)	(A)	M.P (b) (°C)	$-\Delta H_f^o$ (B) (kcal/mole)
AgO	mn.		2.03	p	d>100	6.3
Ag_2O	cb	Cuprite	2.05	p	d 300	7.31
Ag_2S	rh		2.87	n	tr 175, 825	7.6
$AlBr_3$	mn	$AlCl_3$	2.23		97.5	125.8
$AlCl_3$	mn	$AlCl_3$	2.15		190	166.8
AlN	hex	Wurtzite	1.87		>2200	131.4
α-Al_2O_3	hed	Corundum	1.85	n	2015	399.09
β-Al_2O_3	hex	Corundum				
γ-Al_2O_3	cb	Spinel	1.78	n	tr to α	384.8
α-AlOOH	rh	Diaspore	1.82	)		
γ-AlOOH	rh	Boehmite	1.87	)	- H_2O	
$Al(OH)_3$	am		1.85	)		304.2
Al_2S_3	hex	Wurtzite	2.31		s 1500	121.6
α-As	hed	α-As	2.80		817	0
As_4O_6	cb	Senarmontite	2.01		315	312.8
As_2O_5					s,d,500	
As_2S_3	mn		2.20			35
Au_2O_3				n	$(-O_2)$16	- 19.3
B	tet		1.94	p,n	2300	0
α-B_2O_3	hex		1.45	Ionic	460	302
$BaCl_2$	rh	$PbCl_2$	3.15		963	205.6
BaO	cb	NaCl	2.76	Ionic,p	1923	-133.4
BaO_2	tet	CaC_2	2.68	n	450	150.5
BeO	hex	Wurtzite	1.65		2530	146.0
Bi	hed	α-As	3.10		271.3	0

*Column identifications:

A - Prominent type of conductivity at low temperature

B - from element at 25 °C: crystalline state

a - The following abbreviations are used: cb - cubic, hex - hexagonal, mn - monoclinic, rh - rhombic, hed - rhombohedral, tet - tetragonal, am - amorphous

b - s = sublimes, d = decomposes

continued on p. 138

Table 7.4 (cont'd.)

Substance	Structure (a)	Lattice type	M-V distance (Å)	(A)	M.P.(b) (^oC)	$-\Delta H_f^o$ (B) (kcal/mole)
α-Bi_2O_3	mn	Valentinite	2.38		820	137.9
α-Bi_2O_3	cb	Mg_3P_2	2.45	p	tr 704	
Bi_2S_3	rh	Antimonite	2.60	n	d 685	43.8
C	cb	Diamond	1.54	p	> 3550	0
C	hex	Graphite	1.42	n		0
$CaCl_2$	rh	Deform.rutile	2.73	Ionic	772	190
CaF_2	cb	Fluorite	2.38	Ionic	1360	290.3
CaO	cb	NaCl	2.40	p,n	2580	151.9
$Ca(OH)_2$	rh	CdI_2	2.36		$-H_2O$,580	235.8
CdO	cb		2.35	n	< 1462	60.86
α-CdS	hex	Wurtzite	2.52	n	1750	34.5
β-CdS	cb	Sphalerite	2.53	n		
CeO_2	cb	Fluorite	2.34	n	ca2600	233
$CoCl_2$	rh	$CdCl_2$	2.62		724	77.8
CoO	cb	NaCl	2.12	p	1935	57.2
Co_2O_3	cb	Spinel	1.75	n	d 895	
Co_3O_4	cb	Spinel	1.92	p		210
$Co(OH)_2$	rh	CdI_2	2.10		d	131.2
$Co(OH)_3$	hex		2.13			176.6
CoS	hex		2.56		> 1120	20.2
$CrCl_3$	rh		2.43		824	134.6
CrN	cb		2.07	n	d 1700	29.8
CrO	cb	NaCl	2.24	p		
CrO_2	tet	Rutile	1.90	n	300	
CrO_3	rh		1.76	n		
Cr_2O_3	rh	Corundum	2.01	p.n	196	269.7
CrS	hex	NiAs	2.62	n	1550	
Cr_2S_3	rh	Corundum	2.42			
αCsCl	cb	CsCl	3.57		646	103.5
Cs_2O	rh	Anti-$CdCl_2$	2.86		d 400	75.9
CuBr	cb	Sphalerite	2.46	n,p	492	25.1
$CuBr_2$	mn		2.76		498	33.8
CuCl	cb	Sphalerite	2.34	n,p	430	32.5
$CuCl_2$	mn		2.64		620	49.2
CuI	cb	Sphalerite	2.62		605	16.2

Table 7.4 (cont'd.)

Substance	Structure (a)	Lattice type	M-V distance (Å)	(A)	M.P.(b) (^oC)	$-\Delta H_f^o$ (B) (kcal/mole)
CuO	mn	Tenorite	1.95	n,p	1326	37.1
Cu_2O	cb	Cuprite	1.84		1235	39.84
$Cu(OH)_2$	rh	CdI_2	2.00		- H_2O	-107.2
CuS	hex	Covellite	2.54		tr 103	11.6
Cu_2S	cb	Fluorite	2.24	p	1100	19.0
Dy_2O_3	cb	Mn_2O_3	2.28		2340	
Er_2O_3	cb	Mn_2O_3	2.25			
Eu_2O_3	cb	Mn_2O_3	2.32			96.8
$FeCl_3$	rh	$CrCl_3$	2.46		d 315	0.9
Fe_2N	rh				d 200	63.7
FeO	cb	NaCl	2.16	p	1420	96.5
α-Fe_2O_3	rh	Corundum	1.91	n	1565	
β-Fe_2O_3	cb	Mn_2O_3	2.01			
γ-Fe_2O_3	cb	Spinel	1.88	n		
Fe_3O_4	cb	Spinel	1.93	n	d 1538	267
α-FeOOH	rh	Lepidocrocite	1.96			
γ-FeOOH	rh	Goethite	1.92			
$Fe(OH)_2$	rh	CdI_2			d	135.8
$Fe(OH)_3$	cb		2.14			197
FeS	hex	NiAs	2.57	p	1196	22.72
FeS_2	cb	Pyrite	2.26	p.n		42.52
FeS_2	rh	Marcasite			tr 450	36.88
$FeSi_2$	tet	$FeSi_2$	2.31	n		
GaN	hex	Wurtzite	1.92		s 800	25
Ga_2O_3	rh	Corundum	1.95		1900	258
Ga_2S_3	cb	Sphalerite	2.22	n	965	
Gd_2O_3	cb	Mn_2O_3	2.31			
Ge	cb	Diamond	2.45	p.n	937.4	
GeO_2	tet	Rutile	1.87		1086	128.3
HfO_2	cb	Fluorite	2.21	n,p	2812	271.5
$HgCl_2$	rh	$HgCl_2$	2.25		276	55
HgO	rh		2.01	n	d 500	21.68
HgO	hex		2.03	n		
HgS	cb	Sphalerite	2.53	n	583.5	13.9
In_2O_3	cb	Mn_2O_3	2.15	n		222.5

continued on p. 140

Table 7.4 (cont'd.)

Substance	Structure (a)	Lattice type	M-V distance (Å)	(A)	M.P.(b) (^oC)	$-\Delta H_f^o$ (B) (kcal/mole)
KBr	cb	NaCl	3.29		730	93.7
KCl	cb	NaCl	3.15	Ionic	776	104.2
KF	cb	NaCl	2.67		846	277.0
KI	cb	NaCl	3.53		686	78.3
K_2O	cb	Fluorite	2.80		d 350	86.4
KOH	cb	NaCl	2.89		360	101.8
La_2O_3	rh	La_2O_3	2.42	n	2315	458
$La(OH)_3$	hex	UCl_3	2.46		d	
Li_2O	cb	Fluorite	2.00		>1700	142.4
LiOH	tetrag	PbO	1.97		450	116.5
$MgBr_2$	rh	CdI_2	2.69		700	123.7
$MgCl_2$	rh	$ChCl_2$	2.54	Ionic	708	153.4
MgF_2	tet	Rutile	2.05	Ionic	1266	263.5
MgO	cb	NaCl	2.10	n,p,Ionic	2800	143.8
$Mg(OH)_2$	rh	CdI_2	2.09		$-H_2O$,350	221
MnO	cb	NaCl	2.22	p	5	92.0
β-MnO_2	tet	Rutile	1.87	n	535(-O)	124.5
γ-MnO_2	rh	Goethite	1.84	n	1080(-O)	
Mn_2O_3	cb	Mn_2O_3	2.02			232.1
Mn_3O_4	cb	Spinel	2.05		1705	331.4
$Mn(OH)_2$	am		1.87		d	165.8
$Mn(OH)_3$	am				d	212.2
α-MnS	cb	NaCl	2.61		d	48.8
β-MnS	cb	Sphalerite	2.42		d	47.6
γ-MnS	hex	Wurtzite	2.41			
Mo_2C	hex	Mo_2C	2.01		2687	- 4.3
MoO_2						130
MoO_3	mn	Deform.rutile	2.00	p		
MoO_3	rh	MoO_3	1.75	n	795	180.3
MoS_2	hex	Molybdenite	2.35	p	1185	55.5
MoS_3					d1100	
NaBr	cb	NaCl	2.98			86
NaCl	cb	NaCl	2.81	Ionic		98.2
NaF	cb	NaCl	2.31			136.0
NaI	cb	NaCl	3.24			68.8
Na_2O	cb	Fluorite	2.41			99.4
NaOH	rh	NaCl	2.51			175.2

Table 7.4 (cont'd.)

Substance	Structure	Lattice type	M-V distance (Å)	(A)	M.P.(b) (°C)	$-\Delta H_f^o$ (B) (kcal/mole)
$NiBr_2$	rh	$CdCl_2$	2.74		963	54.2
Ni_3C						- 11.0
$NiCl_2$	rh	$CdCl_2$	2.43		1001	75.5
NiO	cb	NaCl	2.09	p	1990	58.4
Ni_2O_3	rh	Corundum	1.80	n		
$Ni(OH)_2$	rh	CdI_2	2.09		d 230	162.1
NiS	hex	NiAs	2.56	p	797	213.0
NiSe	hex	NiAs	2.72			
$PbBr_2$	rh	$PbCl_2$	3.01		373	66.2
$PbCl_2$	rh	$PbCl_2$	2.86		501	85.9
PbI_2	rh	CdI_2	3.08		402	41.85
PbO,red	tet	PbO	2.30	n,p	888	52.40
PbO,yellow	rh	PbO	2.24	p		52.07
β-PbO_2	tet	Rutile	2.16	n	d 290	66.12
Pb_3O_4			2.18		d 500	175.6
Pbs	cb	NaCl	2.96	n,p	1114	22.5
PdO	tet		2.2		870	20.1
PtO					d 550	
PtO_2			1.9	n	450	
Pt_3O_4	cb		2.19		d	
PtS					d	20.8
ReO_2	mn	Deform.rutile	1.99		d1000	
ReS_2	hex	Molybdenite	2.34			44.3
Rh_2O_3	rh	Corundum	2.01		d1100	68.3
RuO_2	tet	Rutile	2.00		d	52.5
Sb	rh	α-As	2.87		630.5	0
Sb_2O_3	cb	Senarmonite	2.22	p	656	
Sb_2O_4	rh				$(-O_2)$930	214
Sb_2O_5	cb			n	$(-O_2)$380	234.4
Sb_2S_3	rh	Antimonite	2.39	p.n	550	43.5
Sc_2O_3	cb	Mn_2O_3	2.10	p.n		
α-Se	rh	a-Se	2.32	p	217	0
Si	cb	Diamond	2.35	n,p	1410	0
αSiO_2	cb	α-quartz	1.61			205.4
β-SiO_2	hex	β-quartz	1.59			204.8
SiO_2	hex	β-tridymite	1.54		1713	205.0
β-SiO_2	tet	α-cristobalite	1.59			

continued on p. 142

Table 7.4 (cont'd.)

Substance	Structure (a)	Lattice type	M-V distance (Å)	(A)	M.P.(b) (^oC)	$-\Delta H_f^o$(B) (kcal/mole)
β-SiO_2	cb	β-cristobalite	1.58			
SiS_2					s 1090	34.7
Sm_3O_3	cb	Mn_2O_3	2.22			
γ-Sn	cb	Diamond	2.80	n,p	2260	
$SnCl_2$	rh	$PbCl_2$	2.78		246	83.6
SnO	tet	PbO	2.21	p.n	d 1080	68.4
SnO_2	tet	Rutile	2.06	n	1127	138.8
SnS	rh	SnS	2.62	p	882	18.6
SnS_2	rh	CdI_2	2.55		d 600	
SrO	cb	NaCl	3.57	p,ionic	2430	141.1
SrS	cb	NaCl	3.00		> 2000	108.1
TaC	cb	NaCl	2.23		3880	
TaO_2	tet	Rutile	2.02			
Ta_2O_8	rh	U_3O_8	2.00	n	1800	
ThO_2	cb	Fluorite	2.42	n		292
TiC	cb	NaCl	2.16	n	3140	54
$TiCl_2$	rh	CdI_2	2.52		s, H_2	114
α-$TiCl_3$	rh	AsI_3	2.50			165
γ-$TiCl_3$	rh	BiI_3	2.50		d 440	73
Tin	cb	NaCl	2.12	p	2930	
TiO	cb	NaCl	2.12	p	1750	
TiO_2	tet	Anatase	1.95			
TiO_2	rh	Brucite	1.87		1825	
TiO_2	tet	Rutile	1.94	n	1830-50	218
Ti_2O_3	rh	Corundum	2.03	p	d 2130	367
TiS	hex	NiAs				
TiS_2	rh	CdI_2	2.42			
Tl_2O_3	cb	Mn_2O_3	2.26	n		
Tl_2S	rh	CdI_2	3.23			20.8
UO_2	cb	Fluorite	2.37	p	2500	270
UO_3	cb	ReO_3	2.08	n		302
U_3O_8	rh	U_3O_8	2.06;2.31	n	d 1300	898
VC	cb	NaCl	2.08		2810	
VCl_3	rh	BiI_3	2.14		d	137
VN	cb	NaCl	2.14	p	2320	41
VO	cb	NaCl	2.05		ignites	

Table 7.4 (cont'd.)

Substance	Structure (a)	Lattice type	M-V distance (Å)	(A)	M.P. (b) (^oC)	$-\Delta H_f^o$ (B) (kcal/mole)
V_2O_4	tet	Rutile	1.76-2.05		1967	52(gas)
V_2O_3	rh	Corundum	2.05		1970	344
V_2O_5					690	373
V_2O_8	rh	V_2O_8	1.54-2.81	n	290	
VS	hex	NiAs	2.66		d	
WC	hex		2.18		2870	9.1
W_2C	rh		2.06		2860	
WO_2	tet	Rutile	2.00	n	1500-1600 (N_2)	136.3
WO_3	rh	WO_3	1.83	n	1473	200.8
WS_2	hex	Molybdenite	2.48		d 1250	46.8
Y_2O_3	cb	Mn_2O_3	2.27	n	2410	
$ZnCl_2$	rh	$CdCl_2$	2.31		283	99.4
ZnO	hex	Wurtzite	1.99	n	1975	83.2
$Zn(OH)_2$	rh	CdI_2	1.95		d 125	153.5
ZnS	hex	Wurtzite	2.34	n	1850	48.5
ZnS	cb	Sphalerite	2.36	n	tr1020	45.3
ZrC	cb	NaCl	2.34		3540	45
ZrN	cb	NaCl	2.32	p	2980	82.2
ZrO_2		Fluorite	2.20	n	2715	
ZrO_2	mn	Baddeleyite	2.04			258.2
ZrS_2	rh	CdI_2	2.58		∿ 1550	

TABLE 7.5

Elements showing easily accessible variable valency states

IIA	IIIA	IVA	VA	VIA	VIIA	←	VIII	→	IB	IIB	IIIB	IVB	VB	VIB	VIIB
Be											B	C	N	O	F
Mg											Al	Si	P	S	Cl
Ca	Sc	Ti	V	Cr	Mn	Fe	Co	Ni	Cu	Zn	Ga	Ge	As	Se	Br
Sr	Y	Zr	Nb	Mo	Tc	Ru	Rh	Pd	Ag	Cd	In	Sn	Sb	Te	I
Ba	La	Hf	Ta	W	Re	Os	Ir	Pt	Au	Hg	Tl	Pb	Bi	Po	At
Ra	Ac	Th	Pa	U											

Table 7.6 Elements capable of forming d^0 or d^{10} cations in their usual oxides

IA	IIA	IIIA	IVA	VA	VIA	VIIA	←	VIII	→	IB	IIB	IIIB	IVB	VB	VIB	VIIB	VIIIB
Li	Be											B	C	N	O	F	Ne
Na	Mg											Al	Si	P	S	Cl	Ar
K	Ca	Sc	Ti	V	Cr	Mn	Fe	Co	Ni	Cu	Zn	Ga	Ge	As	Se	Br	Kr
Rb	Sr	Y	Zr	Nb	Mo	Tc	Ru	Rh	Pd	Ag	Cd	In	Sn	Sb	Te	I	Xe
Cs	Ba	La	Hf	Ta	W	Re	Os	Ir	Pt	Au	Hg	Tl	Pb	Bi	Po	At	Rn
Fr	Ra	Ac	Th	Pa	U												

——— elements capable of forming d^0 or d^{10} cations in their usual oxides

– – – – – oxides unselective in oxidation

× × × × × oxides selective in oxidation

10. Germain (ref. 6) notes some inconsistencies in this approach and points out - quite correctly - that the enthalpy of formation of higher and lower metal oxides per oxygen atom may be different. He summarised the literature as shown in Table 7.7.

11. There are grounds to suggest that activity should be related to ease of removal of oxygen from the surface. In this case, the equilibrium pressure of oxygen released from oxides heated under high vacuum should be indicative (Table 7.7).

12. Alternatively, the accessibility of oxygen could be best measured by the ease of isotopic exchange of oxygen. The limited data available is summarised in Table 7.7.

13. In complex oxides, oxygen mobility seems to be associated with layer crystal structures, as - for example - in bismuth molybdates (ref. 12).

14. Many forms of oxygen adsorption are possible, and some of these are shown in Table 7.8.

For dehydrogenation

15. Dowden (ref. 13) has produced the activity pattern shown in Fig. 7.4.

16. Catalyst design may be helped by reference to references 4,6,13 and 14.

IV. SULPHIDES

1. Elements with sulphides often used as catalysts under reducing conditions are shown in Tables 7.4 and 7.9.

2. Sulphides are usually used as hydrotreating catalysts when deactivation of the catalyst may be expected.

3. Typical activity patterns for reactions catalysed by sulphides are shown in Figs. 7.5 and 7.6.

4. When information is not available on sulphides, the catalysts may be treated as analogous to oxides.

5. In dealing with sulphide catalysts, particular attention should be paid to the pore size distribution in the final catalyst. Preferred porosity will vary from system to system, but - as a general rule - a combination of micropores and of pores of diameter of ca. 10 nm is a good starting point. See, however, ref. 18.

6. Catalyst design is greatly assisted by reference to reference 19.

V. ACID-BASE CHARACTERISTICS

1. Acid-base properties of simple oxides are shown in Table 7.10.

2. The relative acidities of mixed oxides are shown in Table 7.11.

3. The increase of acidity for metal ions in complexes, based on the reaction

TABLE 7.7

Some properties of transition metal oxides: measures of oxygen "accessibility"

Oxide	ΔH_f/O atom kcal mole^{-1}		ΔH_f/O atom (lower oxide) kcal mole^{-1}		ΔH_{Des} kcal mole^{-1}	E_A kcal mole^{-1}
	(8)	(9)	(9)	(10)	(11)	(11)
TiO_2	109	112.7	88	69	59	
Ti_2O_3	-	120.9	-	-		
V_2O_5	87.4	74.4	30	29	43	46
V_2O_4	95.6	85.5	-	-		
CrO_2	71.2	69.6	8.4	15.3		
CrO_2O_3	91	90	-	-	26	42
MnO_2	61.5	62.2	19	16.9	20	22
Mn_2O_3	74.3	76.5	25.4	-		
Fe_2O_3	66.1	65.4	55	-	34	33
Fe_3O_4	64.3	63.2	-	-		
Co_3O_4	49.1	54	45	38.4	16	16
CoO	57.5	57.1	-	-		
NiO	58.4	57.5	57.5	58.4	19	24
CuO	38.5	37.1	34	34.3	18	26
Cu_2O	42.5	40	40	-		
ZnO	83.5	83.2	83.2	83.1	54	40

ΔH_f: heat of formation of oxide ΔH_{Des}: heat of desorption of oxygen

E_A: activation energy of isotopic exchange of oxygen

Table 7.8 Different forms of adsorbed oxygen

IA	IIA	IIIA	IVA	VA	VIA	VIIA ←	VIII	→		IB	IIB	IIIB	IVB	VB	VIB	VIIB	VIIIB
Li	Be											B	C	N	O	F	Ne
Na	Mg											Al	Si	P	S	Cl	Ar
K	Ca	Sc	Ti	V	Cr	Mn	Fe	Co	Ni	Cu	Zn	Ga	Ge	As	Se	Br	Kr
Rb	Sr	Y	Zr	Nb	Mo	Tc	Ru	Rh	Pd	Ag	Cd	In	Sn	Sb	Te	I	Xe
Cs	Ba	La	Hf	Ta	W	Re	Os	Ir	Pt	Au	Hg	Tl	Pb	Bi	Po	At	Rn
Fr	Ra	Ac	Th	Pa	U												

——— elements capable of forming peroxides (O_2^{2-})

- - - - elements capable of forming superoxides (O_2^-)

—·—· elements capable of forming peroxy acids (-OOH)

Table 7.9 Elements with sulphides commonly met in reducing conditions

IA	IIA	IIIA	IVA	VA	VIA	VIA ←	VIII	→		IB	IIB	IIIB	IVB	VB	VIB	VIIB
Li	Be											B	C	N	O	F
Na	Mg											Al	Si	P	S	Cl
K	Ca	Sc	Ti	V	Cr	Mn	Fe	Co	Ni	Cu	Zn	Ga	Ge	As	Se	Br
Rb	Sr	Y	Zr	Nb	Mo	Tc	Ru	Rh	Pd	Ag	Cd	In	Sn	Sb	Te	I
Cs	Ba	La	Hf	Ta	W	Re	Os	Ir	Pt	Au	Hg	Tl	Pb	Bi	Po	At
Fr	Ra	Ac	Th	Pa	U											

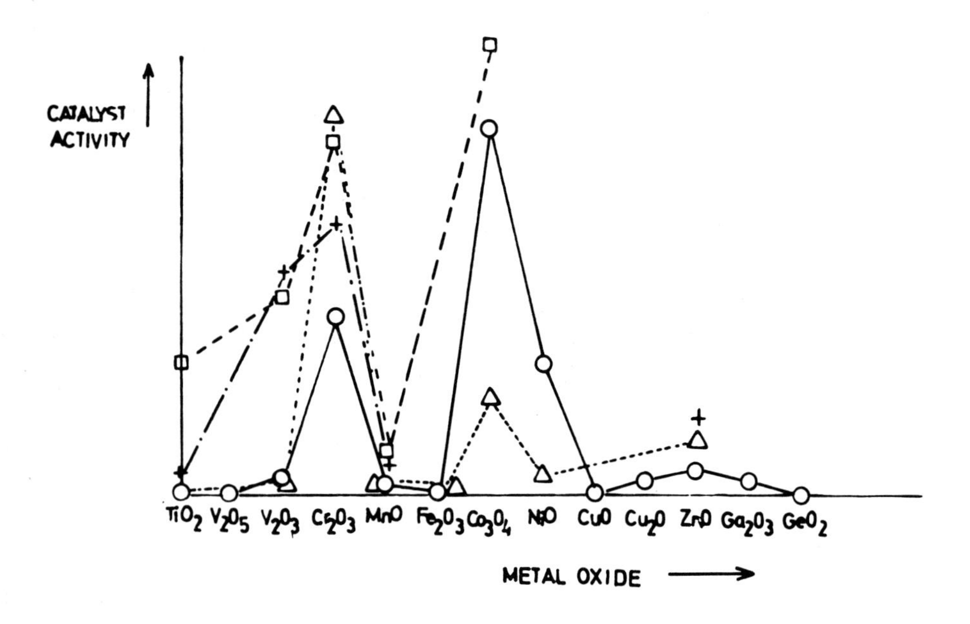

Fig. 7.4. Activity patterns for oxides (refs. 13,15).

: cyclohexene disproportionation (200-450°C)

: propylene dehydrogenation (550°C)

: H_2/D_2 exchange (80°C)

: ethylene hydrogenation (-120-400°C)

Reproduced, with permission, from reference 13.

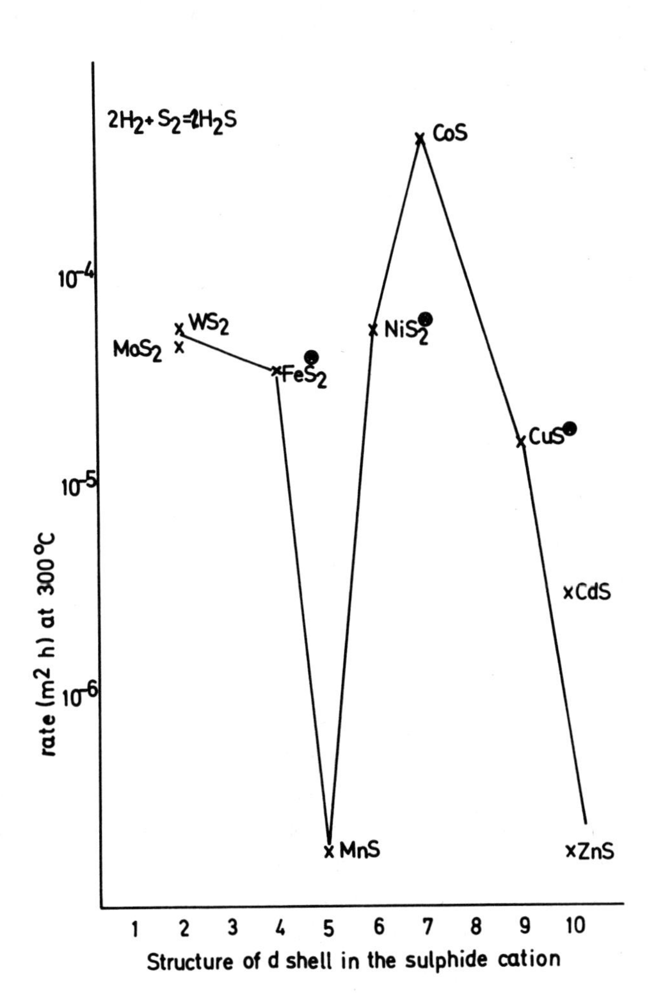

Fig. 7.5. Activity patterns for metal sulphides
(i) $H_2 + S_2 = H_2S$ (ref. 16)
● structure of sulphide uncertain

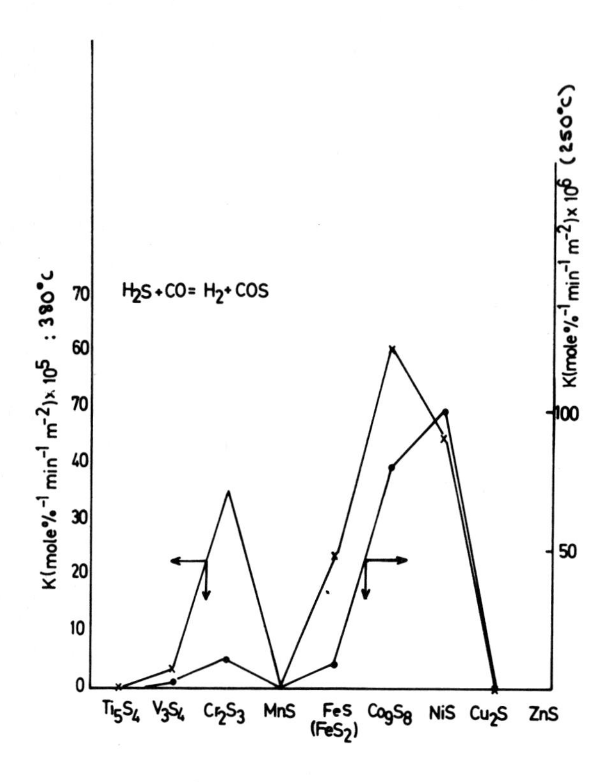

Fig. 7.6. Activity patterns for metal sulphides. (ii) $H_2S + CO = H_2 + COS$ (ref.17). Reproduced, with permission, from reference 17.

Table 7.10 Acid-base properties of oxides

IA	IIA	IIIA	IVA	VA	VIA	VIIA ←	VIII	→		IB	IIB	IIIB	IVB	VB	VIB	VIIB	VIIIB
Li	Be											B	C	N	O	F	Ne
Na	Mg											Al	Si	P	S	Cl	Ar
K	Ca	Sc	Ti	V	Cr	Mn	Fe	Co	Ni	Cu	Zn	Ga	Ge	As	Se	Br	Kr
Rb	Sr	Y	Zr	Nb	Mo	Tc	Ru	Rh	Pd	Ag	Cd	In	Sn	Sb	Te	I	Xe
Cs	Ba	La	Hf	Ta	W	Re	Os	Ir	Pt	Au	Hg	Tl	Pb	Bi	Po	At	Rn
Fr	Ra	Ac	Th	Pa	U												

——— Acidic oxides

– – – – Basic oxides

TABLE 7.11 Relative acidities.

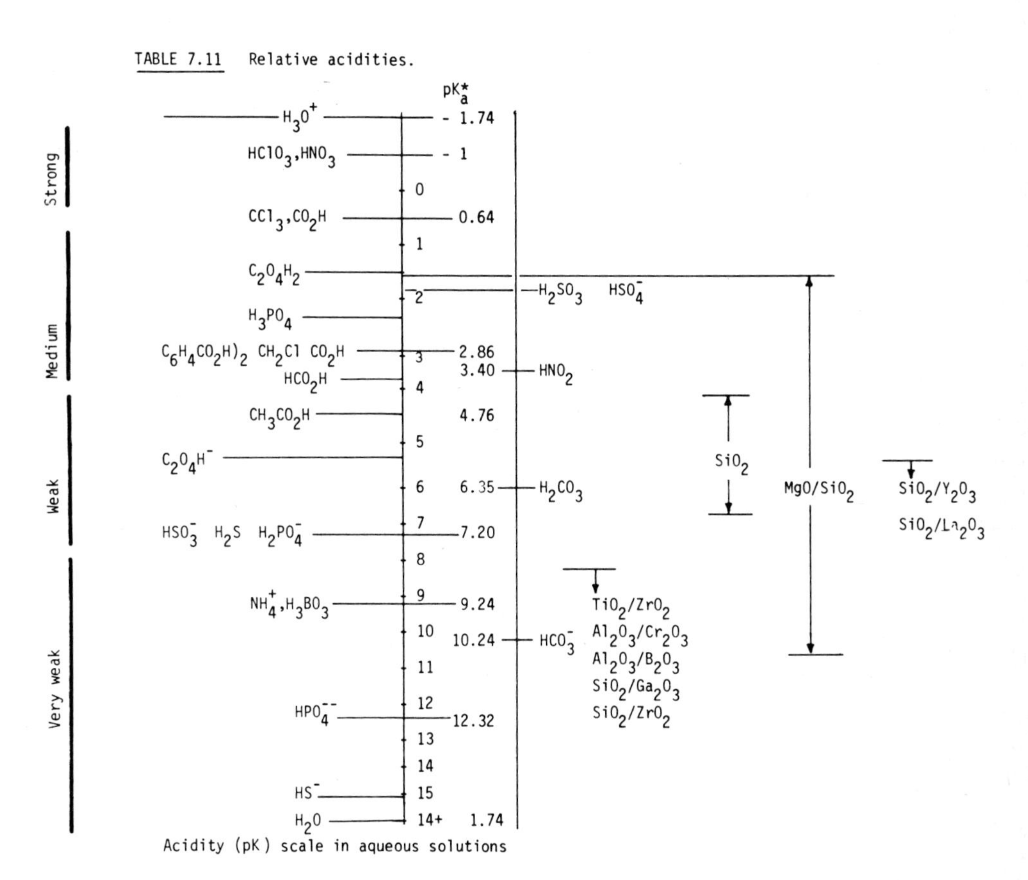

$$M(H_2O)^{n+} \rightleftharpoons M(OH)(H_2O)^{(n-1)+} + H^+$$

is (ref. 14)

$Tl < Ba^{2+} < C^{2+} < Mg^{2+} < Ni^{2+} < Fe^{2+} < Zn^{2+} < Cd^{2+} < Cu^{2+} < Pb^{2+} < Ac^{3+} < Sc^{3+} < Hg^{2+} < In^{3+} < Cr^{3+} < Fe^{3+}$

4. Metal ions are stronger acids in tetrahedral complexes than in octahedral complexes, because of the reduced charge shielding.

5. Acidity in metals increases as (ref. 15)

(W-Mo) > Rh > Ni > Co > Fe

6. Acidity favours carbon formation: basicity inhibits carbon formation.

VI. SUPPORTS

1. Common supports are listed according to their acidity and melting point in Table 7.12.

2. Typical properties of common supports are listed in Table 7.13.

3. Elements with oxides reasonably stable under reducing conditions are listed in Table 7.14.

4. The conversion between the various forms of alumina is represented in Fig. 7.7 (ref. 20).

5. In general, the presence of water vapour at high temperatures means potential trouble.

6. In general the melting point of oxides is greater than that of sulphides which is greater than that of halides.

7. Catalyst design is greatly helped by reference to references 1 and 2.

REFERENCES

1 J.R. Anderson, Structure of Metal Catalysts, Academic Press, London, 1975.
2 G.C. Bond, Catalysis by Metals, Academic Press, London, 1962.
3 O. Beeck, Disc. Farad. Soc., 8, 118 (1950).
4 D.A. Dowden, La Chimica e L'Industria, 55, 639 (1973).
5 D.L. Trimm and J.A. Busby, Chem. Eng. J., 13, 149 (1977).
6 J.E. Germain, Intra Science Report, 6, 101 (1972).
7 Y. Morooka and A. Ozaki, J. Catal., 5, 116 (1966).
8 F.R. Bichowsky and F.D. Rossini, The Thermochemistry of Chemical Substances, Reinhold, New York, 1936.
9 C.J. Smithels, Metals Reference Book, Butterworths, London, 1967.
10 K. Klier, J. Catal., 8, 14 (1967).
11 G.K. Boreskov, Adv. in Catal., 15, Academic Press, New York, 1964.
12 A.W. Sleight and W.J. Linn, Annals N.Y., Acad. Sci., 272, 22 (1976).
13 D.A. Dowden, Catal. Rev. Sci. Eng., 5, 1 (1972).

14 O.V. Krylov, Catalysis by Non Metals, Academic Press, New York, 1970.
15 D.A. Dowden, Proc. IV Inter. Congr. on Catalysis, Moscow, 1968, p.201.
16 V.A. Zazhigolov, S.V. Gerei and M.Ya. Rubanik, Kinet. Catal., 16, 837 (1975).
17 K. Fukuda, M. Dokida, T. Kameyana and Y. Kotera, J. Catal., 49, 379 (1977).
18 Symposium on effect of pore size on catalytic behaviour, Am. Chem. Soc. (Div. Petrol. Chem.), Miami Beach, Sept. 10-15 (1978).
19 O. Weisser and S. Landa, Sulphide Catalysts: their properties and applications, Pergamon, Oxford and Viewig, Braunschweig, 1973.
20 W.H. Gitzen, Alumina as a Ceramic Material, Am. Cer. Soc., Columbus, Ohio, 1970.
21 Catalyst Handbook, Wolfe Scientific Texts, London, 1970.

Table 7.12 Possible support materials (and melting points in ^oC).

Bases	Amphoters	Neutral	Acids
MgO (2800)	Al_2O_3 (2015)	$MgAl_2O_4$ (2135)	SiO_2 (1713)
CaO (1975)		$CaAl_2O_4$ (1600)	$SiO_2 \cdot Al_2O_3$
ZnO (1975)	TiO_2 (1825)	$Ca_3Al_2O_4$ (d 1535)	Zeolites
MnO (1600)	ThO_2 (3050)		Al phosphates
	Ce_2O_3 (1692)		
	CeO_2 (2600)	$MgSiO_2$ (1910)	Carbon
	Cr_2O_3 (2435)	Ca_2SiO_4 (2130)	
		$CaTiO_3$ (1975)	
		$CaZnO_3$ (2550)	
		$MgSiO_3$ (d 1557)	
		Ca_2SiO_3 (1540)	
		Carbon	

d = decomposes

Table 7.13 Typical properties of supports

Support	Preparation/notes		Typical surface area $m^2\ g^{-1}$	Typical pore diameter
High surface area silica	amorphous: prepared as a silica gel		200-800	2-5 nm
Low surface area silica	powdered glass		0.1 - 0.6	2-60 μm
Alumina (γ)	see table		150-400	variant
Alumina (α)	see table		0.1 - 5	0.5 - 2 μm
Magnesia	slit shaped pores		∿ 200	∿ 2 nm
Thoria	slightly radioactive:	prepared as gel	∿ 80	1 - 2 nm
		heat gel to 770 K	∿ 20	
		heat gel to 1270 K	∿ 1.5	
Titania	Anatase		40-80	
	Rutile		up to 200	
	Calcine at > 1050 K		10	
Zirconia	hydrogel		150-300	
Chromia	As gel, heated to 370 K		80-350	< 2 nm
	heated to > 770 K in air		10-30	
Activated carbon	A variety of pores with approximately three maxima		up to "1000"	< 2 nm, 10-20 nm and > 500 nm
Graphite			1-5	-
Carbon molecular, sieves	Slit shaped pores		up to "1000"	0.4 - 0.6 nm
SiC, mullite, zircon, $CaAl_2O_4$			0.1 - 0.3	10-90 μm
Zeolites	Acidic		"500-700"	0.4 - 1 nm
Silica-alumina	Acidic		200-700	3.5 nm

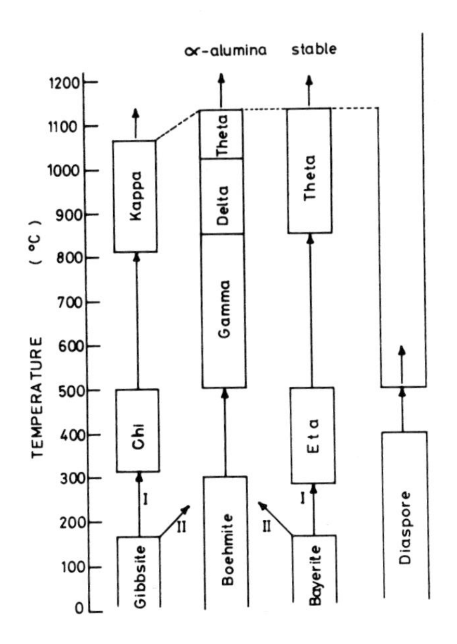

Fig. 7.7. Interconversion of aluminas and aluminium hydroxides (ref. 20).
path I - favoured by dry small crystal sizes
path II - favoured by wet coarse crystals.

Table 7.14 Elements with oxides reasonably stable under reducing conditions

IA	IIA	IIIA	IVA	VA	VIA	VIIA	←	VIII	→	IB	IIB	IIIB	IVB	VB	VIB	VIIB
Li	Be											B	C	N	O	F
Na	Mg											Al	Si	P	S	Cl
K	Ca	Sc	Ti	V	Cr	Mn	Fe	Co	Ni	Cu	Zn	Ga	Ge	As	Se	Br
Rb	Sr	Y	Zr	Nb	Mo	Tc	Ru	Rh	Pd	Ag	Cd	In	Sn	Sb	Te	I
Cs	Ba	La	Hf	Ta	W	Re	Os	Ir	Pt	Au	Hg	Tl	Pb	Bi	Po	At
Fr	Ra	Ac	Th	Pa	U											
		Rare Earths														

Space left for addition of general information useful to individual catalyst designers.

Space left for addition of general information useful to individual catalyst designers.

Space left for addition of general information useful to individual catalyst designers.

PART 2.

SPECIFIC EXAMPLES OF CATALYST DESIGN

INTRODUCTION

The second part of this book is concerned with examples of catalyst design carried out over the last ten years. As stated in the introduction, a deliberate choice has been made to present the designs as they were carried out, with comments being added as to the strengths and weaknesses of the design. Most of the designs were carried out by students working for an M.Sc. degree, and I should record my gratitude for their hard work and their enthusiasm.

Several general observations may be made. The designs were concerned with catalysts for new reactions and with catalysts for well-known reactions. In most cases, the design of the primary catalyst component has been completed but less attention has been paid to secondary components. This reflects the circumstances under which the designs were carried out, in that experimental testing was usually not possible. Depending on the person/people involved, the designs are based either on a simple or a complex description of reactions which may occur in the system.

The first example to be dealt with is a catalyst for the conversion of olefins to aromatics, since this system has often been used as an illustrative example in Part 1 of this book. In Chapter 8, the design is presented as a whole.

Attention is then focused on catalyst designs based on a full description of the reactions that are possible. Catalyst designs are presented for steam reforming, the oxidation of butenes, the hydrogenation of benzaldehyde, the hydrogenation of methyl formate and the hydrogenation of acetylene.

Finally, several catalyst designs are presented which are based on a less complete listing of possible reactions. These include catalysts for the manufacture of terpenes, for methanation and for the reduction of nitrogen oxides.

As discussed in Part 1, many of the designs involve postulation of reactions occurring on a surface. For ease of presentation, metal centres are assumed to be in an octahedral configuration, even though there is no scientific reason why the metal should not adopt some other configuration.

CHAPTER 8

THE DESIGN OF A CATALYST FOR THE CONVERSION OF OLEFINS TO AROMATICS

I. DESCRIPTION OF THE IDEA

One trend that emerged in the early nineteen-seventies was the desire to reduce pollution from motor vehicles. It was soon realised that this would require some form of catalytic cleaning of the exhaust (ref. 1). This posed problems not only in effecting the cleaning, but also in the fact that suitable catalysts could be poisoned by compounds (such as lead or phosphorus) that were emitted with the exhaust gas. Lead, in particular, was a problem, because it was added to gasoline (as lead tetra-alkyl) in order to increase the octane rating of the fuel. If it was necessary to remove lead, then the octane rating would have to be increased in some other way, of which the most obvious route was to increase the percentage of aromatics and alkylaromatics in the fuel.

At the same time, a perennial problem arose in connection with the manufacture of terephthalic acid, an important intermediate in the plastics industry. The acid was made from p-xylene (ref. 2), which, in turn, was produced together with the less valuable o- and m-xylenes (ref. 2). A process that could produce p-xylene selectively would be of considerable interest.

As a result, attention was focused on a general reaction

$$2CH_3-\overset{\displaystyle R}{\overset{|}{C}}=CH_2 \longrightarrow \text{R-C}_6\text{H}_4\text{-R (p-disubstituted benzene ring)}$$

where R could either be H or CH_3. Regrettably, insufficient attention was paid to the economics of the reaction, particularly with respect to the future, since it is now probably more profitable to turn benzene into propylene! Nonetheless, the general reaction above was chosen as the subject of a design and, over the years, the design of the primary and secondary components has been tested experimentally (refs. 3,4,5,6).

Initial checks of the thermodynamics of the system showed that the conversion of propylene to benzene and of isobutene to p-xylene was possible. A literature search revealed no information on the specific reactions, although relevant papers have since appeared (refs. 7-11). The closest reaction reported in the literature involved the polymerisation/cyclisation of olefins over acidic (ref. 12) or basic (ref. 13) catalysts, but these reactions favoured branched-chain polymers. There was also some evidence to suggest that olefin dimerisation in the presence of oxygen could occur (refs. 14,15): cobalt and nickel oxides (refs. 14,15) or platinum on alumina (ref. 16) were reported to be slightly active.

Using the conversion of propylene to benzene, the desired reactions were then written out without reference to the surface:

$$2CH_2 = CH\text{-}CH_3 \xrightarrow{1} CH_2 = CH\text{-}CH_2\text{-}CH_2\text{-}CH = CH_2 + H_2 \quad (1)$$

$$\text{benzene} + H_2 \xleftarrow{3} \text{cyclohexene (from step 2)} + H_2$$

Linear dimerisation (step 1), cyclisation (step 2) and dehydrogenation (all steps) were seen to be involved. Of these, dehydrogenation can be carried out on metals (ref. 16), on acidic metal oxides (ref. 17) and on metal oxides in the presence of oxygen (ref. 14). Polymerisation and cyclisation are known to occur over acidic metal oxides (refs. 12,18) and may involve a free radical mechanism. The use of metallic catalysts is known to be complicated by carbon formation (ref. 19), and free radical reactions are usually non-selective; as a result, both of these reactions were kept in reserve.

The difficulty with the use of acidic or basic catalysts lies mainly in the fact that dimerisation/polymerisation involves the production of branched-chain compounds (refs. 10,18). Simple experimental testing showed that this was the case with both propylene and hexene, the yields of desired products being very small. As a result, entry to the design of the primary constituents was made without significant background knowledge.

II. DESIGN OF PRIMARY CONSTITUENTS

To simplify the discussion for this design, attention is focused on the line of argument that eventually proved successful. Other reaction mechanisms were proposed, and designs based on these mechanisms gave, at best, small yields of desired products.

The reactions listed above were considered in turn. Although little could be found concerning reaction 1, inspection of the literature showed that a somewhat analogous reaction existed, involving the oxidation of olefins to unsaturated aldehydes or to diolefins (ref. 20). The reaction mechanism involves pi-allylic intermediates produced by the oxidative removal of hydrogen

$$CH_3\text{-}CH = CH_2 \xrightarrow{O_2} HO^- + [CH_2\text{-}CH\text{-}CH_2] \xrightarrow{O_2} CH_2 = CH\cdot CHO \qquad (4)$$

It was soon realised that an analogous mechanism could be suggested for reaction (1)

$$2CH_3\text{-}CH = CH_2 \xrightarrow{O_2} [CH_2\text{-}CH\text{-}CH_2 + CH_2\text{-}CH\text{-}CH_2] \xrightarrow{O_2} CH_2 = CH\text{-}CH_2\text{-}CH_2\text{-}CH = CH_2 \qquad (5)$$

The essential difference between the reactions revolves around the production of two molecules of intermediate in the latter case, which should be adsorbed adjacent to one another. The production of one molecule should lead to acrolein, as in reaction 4.

Extending the analogy, it was known that the function of the intermediate on the surface involved lattice oxygen and charge transfer (ref. 21):

$$CH_3\text{-}CH = CH_2 + O^{2-}M^{n+}O^{2-} \longrightarrow [CH_2\text{-}CH^-\text{-}CH_2] \longrightarrow [CH_2\text{-}\overset{\delta+}{CH}\text{-}CH_2]$$
$$\downarrow \qquad\qquad \downarrow$$
$$HO^-\text{-}M^{n+}\text{-}O^{2-} \qquad HO^-\text{-}M^{(n-1)+}\text{-}O^{2-} \qquad (6)$$

Now, from Chapters 2 and 3, we know that strong pi adsorption of olefins occurs on metal ions with the electronic structure d^o, d^1, d^2, d^3, d^8, d^9 or d^{10}; potential catalysts, then, should have this structure. In addition, however, the product (hexadiene) is not required (at this stage) to react further and, as a result, the catalyst should not be able to pi adsorb hexadiene. Desorption would be favoured if the catalyst - after reaction - did not have any of the electronic structures above.

In view of the necessity to have two intermediate molecules adsorbed adjacent to each other, production of the two intermediates on one ion centre was taken as a working assumption: as shown later, this is not strictly necessary. With this assumption, the metal ion must, then, be capable of accepting two electrons (see sequence 6), the higher valency involving d^o, d^1, d^2, d^3, d^8, d^9 or d^{10} and the lower valency involving none of these. Inspection of the periodic table revealed that Sn^{2+}/Sn^{4+}, Tl^+/Tl^{3+}, Pb^{2+}/Pb^{4+}, Bi^{3+}/Bi^{5+}, In^{1+}/In^{3+}, Cd^o/Cd^{2+} and Hg^o/Hg^{2+} were potential catalysts on these grounds. Neglecting mercury, because of the difficulty in handling Hg^{o}, and bismuth and lead (in the first instance) because they are not very thermally stable, acts to reduce this list. In addition, since the lower valency states tend to become more stable as Groups III B, IV B and V B are descended (ref. 22), and the electron attraction increases with the stability of the lower valency state, thallium would appear to be the obvious first candidate.

Similar arguments can be applied directly to the design of the primary components for reactions 2 and 3, as can easily be seen if the reactions are written on the surface:

$$CH_2 = CH.CH_2CH_2.CH = CH_2 \longrightarrow O^{2-} \; C \; M^{n+} \; C \; O^{2-}$$

$$OH^- \; C \rightarrow M^{n+} \leftarrow C \; OH^- \longrightarrow M^{(n-2)} + 2OH^-$$

$$\longrightarrow \text{(cyclohexene)} + M^{(n-2)+} \xrightarrow{OMO} \text{(benzene)}$$

III. EXPERIMENTAL TESTING

Preliminary testing of the catalysts suggested above was then carried out. The results have been reported in detail (refs. 3,4), but a summary of the findings is useful in the present discussion. Thallium oxides were found to be superb dimerisation catalysts (ref. 3) but thallous oxide rapidly sublimed out of the reactor. This was an obvious fault in the design, and should have been checked before testing.

Indium oxide was an effective catalyst for both dimerisation and cyclisation, a typical product-time curve being shown in Fig. 8.1. The selectivity of the heterogeneous reaction was found to increase with temperature (ref. 4), but a homogeneous reaction became significant above ca. 480°C. Acrolein and hexadiene were important initial products, the production of benzene increasing at longer contact times (fig. 8.1). Carbon dioxide was a major product.

The basis of the design was also supported by the finding that bismuth, tin (refs. 7,8), cadmium and lead (ref. 9) oxides catalyse the dehydrodimerisation of olefins. Indeed, as discussed below, bismuth-based catalysts are the most effective for the conversion of olefins to aromatics by this route.

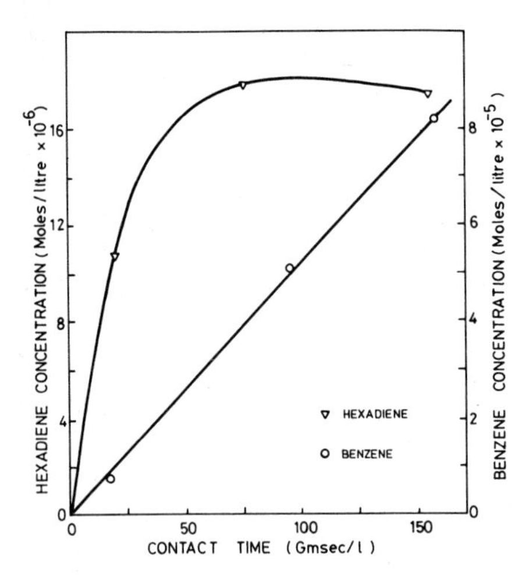

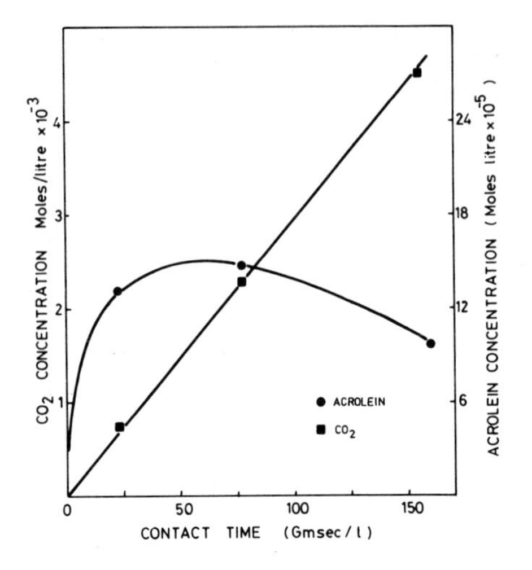

Fig. 8.1. Oxidation of propylene over indium oxide. Propylene = oxygen = 8.9×10^{-3} moles/litre.

IV. DESIGN OF SECONDARY COMPONENTS

Improvements to the catalyst must depend upon the results of experimental testing. As can be seen from Fig. 8.1, the main problems are the overoxidation to carbon dioxide, and the production of acrolein.

Since there was no obvious use of major importance for the catalyst, "mechanisatic" design of secondary components was not undertaken, but attention was focused on the more simple approach (see Chapter 4).

Whatever else is true, the production of carbon dioxide requires considerably more oxygen than any other product. Assuming that most of this oxygen is adsorbed on the catalyst, any additive that decreases adsorption should decrease carbon dioxide production. Oxygen could be adsorbed as a peroxy entity (ref. 23) and, as a result, additives such as calcium or barium (which would favour peroxide formation) should be eliminated. Oxygen is often adsorbed as a negatively charged species (ref. 23), and the addition of an electronegative species to the catalyst should decrease the adsorption of oxygen. Such an effect has been reported, with bismuth phosphate being a more selective catalyst than bismuth oxide (ref. 7). Tests of the catalysts discussed above have been carried out (refs. 5,6), and these are considered later, together with the tests of other additives.

The production of acrolein can be expected if two allylic species are <u>not</u> adsorbed adjacent to each other. At this stage in the design it was realised that an error had been made in the design of the primary constituents. It had been assumed that the two allylic species would be adsorbed on the same site and, as a result, that desired catalytic ions would have valency requirements differing by two. This is not necessarily true, however. Considering indium as an example, In^{1+}

and In^{3+} have been identified, but it is possible that In^{2+} exists - albeit only for a short time. As a result

$$
\begin{array}{cccc}
H_3C\text{-}CH{=}CH_2 & CH_2{=}CH\text{-}CH_3 & CH_2\text{-}\overset{\ominus}{C}H\text{-}CH_2 & CH_2\text{-}\overset{\ominus}{C}H\text{-}CH_2 \\
\downarrow & \downarrow & \cdots\cdots\downarrow & \downarrow\cdots\cdots \\
O^{2-} - In^{3+} & In^{3+} - O^{2-} \longrightarrow & HO^{-} - In^{3+} & In^{3+} - OH^{-} \\
 & & \downarrow & \\
 & & CH_2\text{-}CH\text{-}CH_2 & CH_2\text{-}CH\text{-}CH_2 \\
 & & \cdots\cdots\downarrow & \downarrow\cdots\cdots \\
 & & HO^{-} - In^{2+} & In^{2+} - OH^{-} \\
 & & \downarrow & \\
 & & CH_2{=}CH{\cdot}CHO & CH_2{=}CH\ CHO \\
 & & + & \\
 & & HO^{-} - In^{2+} & In^{2+} - OH^{-} \\
 & & \downarrow \text{fast} & \\
 & & HO^{-} - In^{1+} \;+ & In^{3+} - OH^{-}
\end{array}
$$

Since indium oxide is an n-type semiconductor, transfer of an electron between two In^{2+} species could be expected to be fast - even if the two sites are some distance apart. As a result, the valency requirements of the system are met, but acrolein - rather than hexadiene - is produced.

The only way that this suggestion could be checked would be to prepare a very dilute solid solution of indium oxide (Chapter 4), but it was obvious at this stage of the proceedings that the catalyst would not be of tremendous importance. As a result, other approaches were adopted.

The production of acrolein may also be influenced by the location of the charge transferred to the catalyst. Peacock et al. (ref. 21) have argued that location of the transferred electron at the adsorption site could lead to strong bonding and to an increased chance of complete oxidation. If the electron can be moved away from the active site, weaker bonding will result and partial oxidation will be favoured. Extending these arguments to the oxidation of olefins to aromatics, localisation of the electron at the adsorption centre would favour a cationic adsorbed intermediate, which would react with anionic oxygen to produce acrolein. In addition, cationic intermediates would be reluctant to react with each other to produce dimers. If, on the other hand, the adsorbed intermediates were more radical-like, the dimerisation would be easier and reaction with oxygen less favoured. As a result, it should be desirable to transfer the electron away from the adsorption site.

This could be done in two ways. The location of the electron will depend, to some extent, on the acidity of the surface, with a basic "adsorption site" favouring transfer of the electron. As a result, doping with alkali should favour dimerisation. The addition of 5 at.% of Na_2O and Li_2O to stannic oxide has been reported to raise the selectivity of the oxidation of propylene to benzene (ref. 7), even though stannic oxide is not a particularly good catalyst for the reaction. Results of the addition of alkali to indium- and bismuth-based catalysts are reported below.

The second alternative is to dope the catalyst with an additive that facilitates electron transfer. One such additive is bismuth oxide (ref. 7) and it is interesting to note that bismuth-based catalysts are active and selective (refs. 7-11). It would be instructive, however, to dope indium oxide with bismuth oxides; this has been tested experimentally, and is described below.

It is convenient to consider possible effects of the support at this stage because, in fact, little attempt was made to design the support. It was realised, however, that diffusion limitations would favour over-oxidation and, as a result, a support having wide pores would be most suitable.

Experimental testing of these concepts was carried out using two catalysts. As above, indium oxide was known to be a suitable primary component. In the period between design and secondary component design and testing, Seiyama et al. (ref. 7) had reported that bismuth-based catalysts were more selective than indium oxides. As a result, experimental tests were also carried out with a bismuth oxide-based catalyst.

Experiments were carried out with the oxidation of isobutene to methacrolein and p-xylene over indium oxide (ref. 5). A maximum in the yield of desired products was obtained at a loading of ca. 6 wt.% oxide and, as would be suggested from the above, the selectivity to methacrolein was higher with higher acidity supports. This trend was confirmed by doping the catalyst with oxides of phosphorus, antimony and arsenic: the higher the acidity, the higher the selectivity to methacrolein.

Experimental testing of the reverse effect was more complex, however, since doping with Na_2O or Li_2O markedly increased the yield of methacrolein rather than, as expected, increasing the yield of the diene and p-xylene. This was in contradiction to the results over stannic oxide (ref. 7), and is difficult to explain.

Doping the catalyst with bismuth oxide markedly reduced the selectivity of the oxidation to carbon dioxide, reduced slightly the selectivity to the diene but increased markedly the selectivity to methacrolein. This probably results, however, from the preparation of the doped catalysts, in that several samples of the same batch of supported indium oxide were doped with different additives. No attempt was made to get the bismuth oxide to dissolve in the indium oxide (ref. 24) and, as a result, the catalyst should be considered as involving bismuth oxide supported on indium oxide. In these terms it is not surprising that the selectivity to methacrolein increased.

Similar testing was carried out with a bismuth oxide-based catalyst, where the addition of phosphorus oxides was again found to be very beneficial (ref. 6). In this case, the design of the secondary components was extended towards the mechanistic approach, by measuring the kinetics of the reactions. Although Langmuir-Hinshelwood models are notoriously unreliable in the extension from kinetics to mechanism, it was found that the best fit with the experimental observations was obtained with a surprising model. For dimethylhexadiene, the results supported a mechanism involving one or two isobutene molecules reacting with monatomic oxygen. For methacrolein, a model based on one adsorbed isobutene did not give good predictions: instead, the reaction model was found to involve two molecules reacting with diatomic oxygen. This model was so opposed to that used for the basis of the secondary design, and so different from the obvious route to methacrolein, that the design was abandoned in some confusion.

It is interesting to note that experimental testing of all doped catalysts revealed that their activity was less than the undoped catalyst (refs. 5,6), although the selectivity changed markedly. This is a function of the preparation procedure which, in order to obtain meaningful comparisons, was carried out using standard original catalysts rather than adopted to optimise the use of each dopant. Were the catalyst to be used commercially, it is obvious that considerable attention would have to be paid to preparation.

V. COMMENTS

This is an interesting design, largely because the paperwork and the experimental testing have been completed for most stages of the design.

In the design of the primary constituents, no attempt was made to use the geometric approach, and this is a fault. Heats of adsorption were considered, but no data was available. It was certainly a mistake not to test all the suggestions of the primary design in more detail, since this led to the neglect of the (suggested) more active and more selective bismuth oxide-based catalysts.

Little attention was paid to the design of the support, with the exception that the necessity of removing heat from the catalyst and of having wide pores was recognised. Indeed, an acidic support (which, from the design, was known to favour unwanted side reations) was tested; this reflects the lack of attention to this aspect (Chapter 5).

Perhaps, however, the most serious fault was the lack of definition at the start of the design. The economic analysis of the reaction was totally inadequate and, as a result, the design reverted from an interesting project to an intellectual exercise.

REFERENCES

1 J. Wei, Adv. Chem. Ser., 148, 1 (1975).

2 A.L. Waddams. Chemicals from Petroleum. 4th edn., Wiley, London, 1978.
3 L.A. Doerr and D.L. Trimm, J. Catal., 23, 49 (1971).
4 L.A. Doerr and D.L. Trimm, J. Catal., 26, 1 (1972).
5 M.R. Goldwasser and D.L. Trimm, Acta Chem. Scand., B32, 286 (1978).
6 M.R. Goldwasser and D.L. Trimm, J. Appl. Chem., (in press).
7 T. Seiyama, M. Egashira, T. Sakamoto and I. Aso, J. Catal., 24, 76 (1972).
8 M. Egashira, H. Sumie, T. Sakamoto and T. Seiyama, Kogyo Kagaku Zasshi, 73, 860 (1960).
9 Dutch Pat. 66-04526.
10 F. Solymosi and F. Bozso, Proc. Vi Intern. Congr. on Catalys, London, p.365, 1976.
11 M. Akimoto, E. Echigoya, M. Okada and Y. Tomatsu, Proc. VI INtern. Congr. on Catalysis, London, 872, 1976.
12 H. Charcosset, A. Revillon and A. Guyot, J. Catal., 8, 326, 334 (1967).
13 A.W. Shaw, C.W. Bittner, W.V. Bush and G. Hovzurah, J. Org. Chem., 30, 3286 (1965).
14 R.G. Schultz, J.M. Schuck and B.S. Wildi, J. Catal., 6, 385 (1966).
15 S.J. Thomson and G. Webb, Heterogeneous Catalysis, Oliver and Boyd, Edinburgh, p.18, 1962.
16 M. Jouy and J.C. Balaceanu, Actes II Congres Int. Catalyse, Technip, Paris, 645, 1960.
17 Ch. Marcilly and B. Delmon, J. Catal., 24, 336 (1972).
18 K. Ziegler, Agnew. Chem., 72, 829 (1952).
19 D.L. Trimm, Catal. Rev. Sci. Eng., 16, 155 (1977).
20 C.R. Adams and T.J. Jennings, J. Catal., 2, 63 (1963); 3, 549 (1964).
21 J.M. Peacock, A.J. Parker, P.G. Ashmore and J.A. Hockey, J. Catal., 15, 373, 379, 387, 398 (1969).
22 P.A. Batist, A.H.W.M. der Kinderis, Y. Leeuwenburgh, F.A.M.G. Metz and G.C.A. Schuit, J. Catal., 12, 45 (1968)
23 D.A. Dowden, Chem. Eng. Prog. Symp. Ser., 63 (73), 90 (1967).
24 A.W. Sleight and W.J. Linn, Ann. N.Y. Acad. Sci., 272, 22 (1976).

CHAPTER 9

THE DESIGN OF A STEAM REFORMING CATALYST

I. INTRODUCTION

It is convenient now to turn to the first example of the type of design favoured by Dowden (refs. 1,2,3), in which attention is focused more on a design on paper. This is not to say that experimental testing is considered to be unnecessary - far from it - but rather to say that greater emphasis is placed on the identification (and design to accommodate) of both desired and undesired reactions before experimental testing is initiated. Several such designs have been reported and will be discussed below, but one illustrative design was completed for the steam reforming of hydrocarbons. The reaction is well known (refs. 1,5) and is used industrially for the production of methane or of hydrogen. Nickel-based catalysts are widely used since, although they are not the most active catalysts (ref. 4), they offer best value for money.

The essential difference in the approach is reflected in the detail in which the reaction network is presented. The target transformations

$$CnH_{2n+2} + nH_2O \longrightarrow nCO + (2n+1)H_2$$

$$CnH_{2n+2} + 2nH_2O \longrightarrow nCO_2 + (3n+1)H_2$$

were considered in terms of the characteristic chemistry of butane as in Table 9.1. Although this list is extensive, it will be seen later that it is far from complete and, as a deliberate simplification, some important reactions were ignored. Thermodynamic data for the reactions were based on methods of group increments (ref. 6) and on values quoted by Rossini et al. (ref. 7).

As written, the reaction scheme is seen to involve several different types of reaction:

Cracking	$C_4H_{10} \longrightarrow C_3H_6 + CH_4$
Dehydrogenation	$C_2H_{10} \longrightarrow C_4H_8 + H_2$
Hydration	$C_4H_8 + H_2O \longrightarrow C_6H_9OH$
Steam cracking	$C_4H_8 + H_2O \longrightarrow C_2H_5.CHO + CH_4$
Decarboxylation	$C_2H_5CHO \longrightarrow CO + C_2H_6$

Water gas shift $CO + H_2O \rightleftharpoons CO_2 + H_2$

Polymerisation $2C_2H_4 \longrightarrow C_4H_8$

The critical reactions in this scheme (steam cracking and hydration) are not favoured thermodynamically (Table 9.1), and a catalyst should be used to improve the kinetics of the reactions.

II. DESIGN OF PRIMARY COMPONENTS

Largely as a result of the information available in the literature, the initial approach was based on activity patterns. Both metals and metal oxides were considered. Thus, for cracking the order of activity has been shown to be:

Metals $W \sim Mo > Rh > Ni > Cr > Fe > CO$

Metal oxides $NiO_2 > MoO_2 > V_2O_5 > Cr_2O_3$

This activity pattern would seem far from complete, since many metal oxides (and combinations of metal oxides) are known to catalyse cracking. However, two reactions must be avoided - the polymerisation of olefins and the formation of carbon. Note that this latter possibility was deliberately excluded from Table 9.1 and, as will be discussed later, this is a very important reaction. Both of these reactions are favoured by acidity, which decreases (Chapter 2) in the order

$SiO_2/Al_2O_3 > SiO_2/MgO > SiO_2 >> \gamma Al_2O_3 > TiO_2 > ZrO_2 > MgAl_2O_4 > UO_2 > CaO \sim MgO$

Similarly, for dehydrogenation, the patterns of activity are:

Metals Precious metals $> Ni > Co > W \sim Cr >> Fe$

Oxides also catalyse dehydrogenation, and - as shown in Chapters 2 and 3 - oxides of transition metals with configuration d^1 to d^4 and d^6 are known to be active. However, it is also true that steam reforming leads to hydrogen and, under the conditions of reaction, most transition oxides will be reduced to metals. It is also true that combinations such as Cr_2O_3/Al_2O_3 can catalyse dehydrogenation (ref. 8): in these cases, however, the acidity of the catalyst will lead to polymerisation and to carbon formation. As a result, there are no obvious oxidic catalysts suitable for dehydrogenation in the context of steam reforming.

Little information is available concerning the dissociation of steam, but Dowden suggests that the reaction over oxides can be represented by

$$MO_x + H_2O \longrightarrow MO_{x+1} + H_2$$

TABLE 9.1

Steam reforming

Reaction	Type	ΔG (900 K) (kcal $mole^{-1}$)
Target transformation		
$C_4H_{10}+4H_2O \; = \; 4CO+9H_2$	steam reforming	- 48.2
$C_4H_{10}+3H_2O \; = \; 4CO_2+13H_2$	steam reforming	- 53.9
Primitive processes		
$C_4H_{10} \; = \; C_4H_8+H_2$	dehydrogenation	+ 1.9
$C_4H_8 \; = \; C_4H_6+H_2$	dehydrogenation	+ 2.1
$C_4H_{10} \; = \; C_3H_6+CH_4$	demethanation	- 13.2
$C_4H_8 \; = \; 2C_2H_4$	cracking	- 3.5
Cross interaction		
$C_4H_8+H_2O \; = \; C_4H_9OH$	hydration	+ 21.5
$CH_3-CH=C=CH_2-H_2O \; = \; C_2H_5-CO-CH_3$	hydration	- 4.3
$CH_3-C{=}CH+H_2O \; = \; CH_3-CO-CH_3$	hydration	- 4.8
$CH_3-CH_2-CH=CH_2+H_2O \; = \; C_3H_8-CH_2O$	steam cracking	+ 7.75
$CH_3-CH_2-CH=CH_2+H_2O \; = \; C_2H_5\cdot CHO+CH_4$	steam cracking	- 2.9
$CH_3-CH=C=CH_2+H_2O \; = \; CH_2O+CH_2=CH-CH_3$	steam cracking	- 3.0
Primitive processes of intermediates		
$C_4H_9OH \; = \; C_3H_7-CHO+H_2$	dehydrogenation	- 28.8
$C_3H_7-CHO \; = \; C_3H_8+CO$	decarbonylation	- 28.8
$CH_3-CO-CH_3 \; = \; CH_4+CH_2=C=O$	cracking	- 8.0
$CH_2=C=O+H_2O \; = \; CH_3-COOH$ (SQ)	hydration	- 4.9
$CH_3-COOH \; = \; CH_4+CO_2$	decarboxylation	- 29.7
'Equilibration' reactions		
$CH_4+H_2O \; = \; CO+3H_2$	methane reforming	- 0.5
$CO+H_2O \; = \; CO_2+H_2$	water gas shift	- 1.4
Undesired reactions		
$C_4H_{10}+H_2 \; = \; C_3H_8+CH_4$		- 14.3
$2C_2H_4 \; = \; C_4H_8$		+ 3.5

If the non-stoichiometry of the oxide is important, then oxides that maintain a state between MO_x and MO_{x+1} under steam reforming conditions may be used

$UO_4/U_3O_8 > MoO_2/MoO_3 > WO_2/WO_3 > Pr_2O_3/Pr_4O_{10} > Cr_2O_3/CrO_2 > CrO_2/Cr_2O_3$

This is an interesting agreement, but it must be pointed out that the scission of water into H and OH is more probable, and the reaction between hydrocarbon fragments and OH is just as feasible as with O. In this case, the basicity of oxides may be important, and the reverse of the order of acidity given above is directive.

For metals, an activity pattern based on the ability of metals to split hydroxyl groups in alcohols (ref. 9) was used:

Precious metals > Ni >> W > Fe > Ag

Dowden also cheated a little in the design by using the activity pattern for methane steam reforming

$$CH_4 + H_2O \longrightarrow CO + 3H_2$$

This has been reported to be:

Ru > W > Rh > Ni > Co > Os > Pt > Fe > Mo > Pd > Ag

In one sense this is a little unfair, in that the design of the catalyst was being carried out in the knowledge that activity patterns for naphtha steam reforming were available (ref. 4). These, as would be expected, are very similar to methane steam reforming. Since the design was being carried out either as an exercise or to confirm that no obvious catalyst can be suggested other to those in use then reference to the activity pattern for methane reforming is close to assuming the answer before it is known!

Of the reactions in Table 9.1, hydrogenolysis is undesired, with an activity pattern (ref. 10):

Ni > precious metals > other metals

In addition, if the design is extended to consider the undesired carbon formation reaction, the activity pattern is:

Fe > Ni > Co > precious metals

Dowden then turned his attention to the support, which he selected on two bases. The thermodynamics of hydrogen production favour high temperature reactions and, as a result, the support must be thermally stable. Arranging the oxides in the order of their melting points gives:

$ThO_2 > ZrO_2 > CrO_2 > MgO > CaO > BaO > UO_2 > Al_2O_3 > Cr_2O_3 > MoO_2 > WO_2$

Excluded from this list was silica, for the unstated reason that silica volatilizes slowly in steam. Also unstated was the fact that phase transfer (and hence sintering) is markedly accelerated in alumina in the presence of steam.

Supporting oxides and promotors are needed to maintain a good distribution of active metal and, in this respect, oxides which form reducible solid solutions with oxides of the active metals are particularly effective. The order of decreasing ease of reaction with other oxides is (ref. 11):

$MoO_3 \sim WO_3 > NiO_2 > TiO_2 > Al_2O_3 > MgO > CaO$

The combination of these activity patterns was then used to arrive at the following suggestions:

Catalyst: Precious metals > Ni > Co > Fe
Support: $MgAl_2O_4 > Ca(Sr)\text{-}\beta Al_2O_3 > \alpha Al_2O_3 > CrO_2$ (stabilised)
Promotor: $UO_2 > Pr_2O_3(CeO_2) > MoO_2 > CaO > MgO > Al_2O_3$

A more complete consideration of the design would, in fact, bring out the importance of MgO as a possible support. This oxide meets many of the requirements above and also acts to catalyse the gasification of carbon by steam (ref. 4).

Accepting the suggestions of the activity patterns, Dowden then moved to consider reactions on the surface. For simplicity, he viewed the active sites as a collection of octahedrally coordinated ions, and wrote - for the first stages of the reaction - the reaction scheme shown on the next page.

No attempt was made to extend these mechanisms, nor to base aspects of the design upon the mechanisms. Instead, Dowden turned to kinetic considerations, in which experimental results were used to assess the importance of the differing reactions. The findings of these support the suggestion of the activity patterns that nickel supported on alumina or magnesia is an active catalyst for steam reforming.

III. COMMENTS

Although this design is valuable in that it shows the advantages to be gained by considering a wider range of reactions, it is not a good design. There are several reasons for this.

What is clearly evident is that activity patterns for the component reactions (or analogous reactions) can be very important to a design. Analysis of the above shows, in fact, that they can - on their own - predict active catalysts.

In a deliberate attempt to simplify and to clarify, Dowden did not consider carbon formation reactions. In practice, this should not be excluded, since carbon formation is the major undesirable reaction in steam reforming, and the prevention of carbon deposition (and the gasification of deposits) is a major factor influencing steam reforming catalysts.

Comparing the design with the full procedure discussed in Chapter 2, it is obvious that many of the lines of approach which are possible have not been attempted. Steam reforming in a complicated reaction because of the number of reactions that are possible (ref. 5) and lack of space probably precluded a more complete analysis of the reactions. The design does serve, however, to illustrate well the approach of using a more complete list of reactions and the benefits that can accrue from the application of activity patterns. To extend the concept, it is better to turn to another design completed by Dowden for the oxidation of butene.

REFERENCES

1 D.A. Dowden, C.R. Schnell and G.T. Walker, Proc. IV Intern. Congr. on Catalysis, Moscow, 201, 1968.
2 D.A. Dowden, Chem. Eng. Prog. Symp. Ser., No. 73, Vol. 63, 90 (1967).
3 D.A. Dowden, Chim. Ind. (Milan), 55, 639 (1973).
4 J. Rostrup Nielsen, Steam Reforming Catalysts, Teknisk Forlag, Copenhagen, 1975.
5 Catalyst Handbook, Wolfe Scientific Texts, London, 1970.

6 G.J. Janz, Estimation of Thermodynamic Properties of Organic Compounds, Academic Press, New York, 1958.
7 F.D. Rossini, K.S. Pitzel, R.J. Arnett, R.M. Brann and G.C. Pimenkel, Selected Values of Physical and Thermodynamic Properties of Hydrocarbons and Related Compounds, Carnegie Press, 1953.
8 O.V. Krylov, Catalysis by Non-Metals, Academic Press, New York, 1970.
9 J.R. Anderson and C.K. Kemball, Trans Faraday Soc., 51, 966 (1955).
10 G.C. Bond, Catalysis by Metals, Academic Press, New York, 1962.
11 G. Tamman and W. Rosenthal, Z. Anorg. Allg. Chem., 156, 20 (1926).

CHAPTER 10

THE DESIGN OF A CATALYST FOR THE CONVERSION OF BUTENES TO MALEIC ANHYDRIDE

I. INTRODUCTION

In many ways the design of a catalyst for the oxidation of butenes to maleic anhydrides offers a more complete example of the approach adopted by Dowden. First reported in 1967 (ref. 1), the design illustrates the genesis of the general mechanism and the way that this may be translated to the surface; less attention is paid to the physical requirements of the system with respect to mass and heat transfer.

The general mechanism was based on reactions that had been recognised in the gas phase. Homogeneous oxidation reactions are very complex (ref. 2), but if the type of reaction relevant to the catalytic reaction is recognised, possible reactions include:

1. Stepwise hydrogenation: peroxide route.

Adapting the mechanism suggested by Blundell and Skirrow (ref. 3), butene-2 can be expected to react as follows

$$CH_3\text{-}CH\text{=}CH\text{-}CH_3 \xrightarrow{O_2} CH_3\text{-}CH\text{=}CH\text{-}CH_2\cdot + HO_2\cdot \quad (1)$$

$$CH_3\text{-}CH\text{=}CH\text{-}CH_2\cdot + O_2 \longrightarrow CH_3\text{-}CH\text{=}CH\text{-}CH_2OO\cdot \quad (2)$$

$$CH_3\text{-}CH\text{=}CH_2\text{-}OO\cdot + RH \longrightarrow CH_3\text{-}CH\text{=}CH_2OOH + R \quad (3)$$

$$CH_3\text{-}CH\text{=}CH_2OOH \longrightarrow CH_3\text{-}CH\text{=}CH\cdot CHO + H_2O \quad (4)$$

$$CH_3\text{-}CH\text{=}CH_2OOH \longrightarrow \begin{array}{c} HC\text{=}CH \\ | \quad | \\ H_2C \quad CH_2 \\ \diagdown \; \diagup \\ O \end{array} + H_2O \quad (5)$$

This reaction scheme is unlikely, for several reasons. Initial hydrogen removal would be more likely to involve the tertiary rather than the primary hydrogen. Peroxides are known to be chain initiators and, under the conditions necessary for reaction 1 to occur, decomposition to two radicals would be more common than dehydration (reaction 4). Cyclisation involving the primary hydrogen (reaction 5) is improbable in the presence of a suitably placed tertiary hydrogen.

Nonetheless, reactions of this type are possible, and could be favoured in the presence of a catalyst.

2. Dehydrogenation to butadiene: diperoxide route

Dehydrogenation of butene to butadiene opens another possible route to maleic anhydride:

$$CH_3\text{-}CH_2\text{-}CH\text{=}CH_2 \xrightarrow{-2H} CH_2\text{=}CH\text{-}CH\text{=}CH_2 \text{ or } CH_2\text{-}CH\text{-}CH\text{-}CH_2 \quad (6)$$

$$CH_2\text{-}CH\text{-}CH\text{-}CH_2 + O_2 \longrightarrow CH_2\text{=}CH\text{-}CH\text{-}CH_2OO\cdot \quad (7)$$

$$\searrow \cdot H_2C\text{-}CH\text{=}CH\text{-}CH_2OO\cdot \quad (8)$$

$$\cdot H_2C\text{-}CH\text{=}CH\text{-}CH_2OO + O_2 \longrightarrow \cdot OOH_2C\text{-}CH\text{=}CH\text{-}CH_2OO\cdot \quad (9)$$

$$\cdot OOH_2C\text{-}CH\text{=}CH\text{-}CH_2OO\cdot + 2RH \longrightarrow HOOH_2C\text{-}CH\text{=}CH\text{-}CH_2OOH + 2R\cdot \quad (10)$$

$$HOOH_2C\text{-}CH\text{=}CH\text{-}CH_2OOH \xrightarrow{-H_2O} OHC\text{-}CH\text{=}CH\text{-}CHO \quad (11)$$

$$OHC\text{-}CH\text{=}CH\text{-}CHO \xrightarrow[O_2]{-H_2O} \text{maleic anhydride (HC=CH, C(=O)-O-C(=O) ring)} \quad (12)$$

Side reactions could either involve $CH_2\text{=}CH\text{-}CH\text{-}CH_2OO\cdot$ or a complex polymerisation (ref.4). This reaction scheme is slightly more plausible in the gas phase. Diperoxides are known (refs. 2,5) and have been suggested as intermediates (reactions 10/11). Even if the reactions represented in eqns. (11) and (12) are not real, peroxides and hydroperoxides are known to give acids (refs. 2,5), which could dehydrate to give maleic anhydride.

3. Dehydrogenation to butadiene: cyclic peroxide route.

Formation of butadiene could lead to a monoperoxide, which could also lead to maleic anhydride:

$$CH_2\text{=}CH\text{-}CH\text{=}CH_2 \;(\text{cis}) + O_2 \longrightarrow \text{cyclic } CH_2\text{-}O\text{-}O\text{-}CH_2\text{-}CH\text{=}CH \text{ (ring)} \quad (13)$$

$$\begin{array}{c} O-O \\ CH_2 \quad\quad CH_2 \\ CH{=}CH \end{array} \longrightarrow \begin{array}{l} HC{-}CH_2OH \\ \Vert \\ HC{-}CHO \end{array} \xrightarrow[-H_2O]{O_2} \begin{array}{l} \quad\;\; O \\ \quad\;\; \Vert \\ HC - C \\ \Vert \quad\quad\;\; O \\ HC - C \\ \quad\;\; \Vert \\ \quad\;\; O \end{array} \tag{14}$$

This route is quite feasible, since intramolecular peroxide formation across a double bond is also possible (ref. 5), leading to epoxides and carbonyls.

Dowden rejects the possibility of dimerisation of butenes followed by ring closure and a series of oxidative decarbonylations to give maleic anhydride. This is almost certainly not a major route to the anhydride, although a Diels-Alder type condensation to give carbonaceous residues may be an unwanted side-reaction:

$$\begin{array}{l} CH_2 \\ /\!/ \\ CH \\ | \\ CH \\ \backslash\!\backslash \\ CH_2 \end{array} + \begin{array}{l} CH_3 \\ / \\ CH \\ \Vert \\ CH \\ \backslash \\ CH_3 \end{array} \longrightarrow \begin{array}{l} \quad CH_2 \quad\; CH_3 \\ CH \quad\;\; CH \\ \Vert \quad\quad\;\; | \\ CH \quad\;\; CH \\ \quad CH_2 \quad\; CH_3 \end{array} \longrightarrow \text{carbon} \tag{15}$$

Dowden also noted that many of these reactions involved the oxidation of male-aldehyde, and he set up reaction schemes for this too.

4. Oxidation of maleic aldehyde

Oxidation of the aldehyde was suggested to involve two possible routes. Homolytic oxidation via an acyl radical led to

$$\begin{array}{l} CH\cdot CHO \\ \Vert \\ CH\cdot CHO \end{array} + RH \longrightarrow \begin{array}{l} CH\cdot \dot{C}{=}O \\ \Vert \\ CHCHO \end{array} + R\cdot \tag{16}$$

$$\begin{array}{l} CHC{=}O \\ \Vert \\ CH\cdot CHO \end{array} + O_2 \longrightarrow \begin{array}{l} \quad\quad\; O \\ \quad\;\; /\!/ \\ CH{-}C \\ \Vert \quad\quad \diagdown OO\cdot \\ CH\cdot CHO \end{array} + RH \longrightarrow \begin{array}{l} \quad\quad\; O \\ \quad\;\; /\!/ \\ CHC \\ \Vert \quad\quad \diagdown OOH \\ CH\cdot CHO \end{array} + R\cdot \tag{17}$$

$$\begin{array}{l} \quad\quad\; O \\ \quad\;\; /\!/ \\ CH\cdot C \\ \Vert \quad\quad \diagdown OOH \\ CH\cdot CHO \end{array} \longrightarrow \begin{array}{l} \quad\quad\quad O \\ \quad\quad\;\; /\!/ \\ CH - C \\ \Vert \quad\quad\quad\; O \\ CH - C \\ \quad\quad\;\; \backslash\!\backslash \\ \quad\quad\quad O \end{array} + H_2O \tag{18}$$

In acidic solutions, oxidation of aldehydes is often preceded by enolisation (ref. 7), and an alternative mechanism was suggested

$$\begin{array}{l} HC-CHO \\ \| \\ HC-CHO \end{array} \xrightarrow{H^+} \begin{array}{l} HC-CH-OH^+ \\ \| \\ HC-CHO \end{array} \qquad (19)$$

$$\begin{array}{l} CH\cdot CH-OH^+ \\ \| \\ CH-CHO \end{array} + OMO^- \longrightarrow \begin{array}{l} HC(OH)-CHOMO \\ \| \\ HC-CHO \end{array} \qquad (20)$$

$$\longrightarrow \begin{array}{l} HC - C(=O)OH \\ \| \\ HC - CHO \end{array} + OMO^- + H^+ \qquad (21)$$

$$\begin{array}{l} CH-C(=O)OH \\ \| \\ CH-CHO \end{array} \longrightarrow \text{repeat to give anhydride} \qquad (22)$$

It will be seen later that the catalyst is required to be acidic for other reasons, and the corresponding reaction in basic media was noted but not listed.

General inspection of these reaction schemes revealed that they had much in common. Isomerisation of olefins was an obvious possibility, as was stabilisation of the double bond in the centre of the molecule (schemes 1, 2, 3 and 4) and the dehydration of hydroperoxides (schemes 1, 2 and 4). On the assumption that production of maleic anhydride would be greatest for the most rapid route from butene to maleic anhydride, schemes 3 and 4 would appear attractive, but this situation could well change when the reactions on a surface are considered.

II. DESIGN OF PRIMARY COMPONENTS

Dowden chose to enter the design by considering possible adsorbed states. Both sigma and pi-adsorbed forms of butene and butadiene were discussed.

For butene adsorbed on the surface of a metal oxide lattice, possible adsorbed forms include (refs. 8,9,10,11):

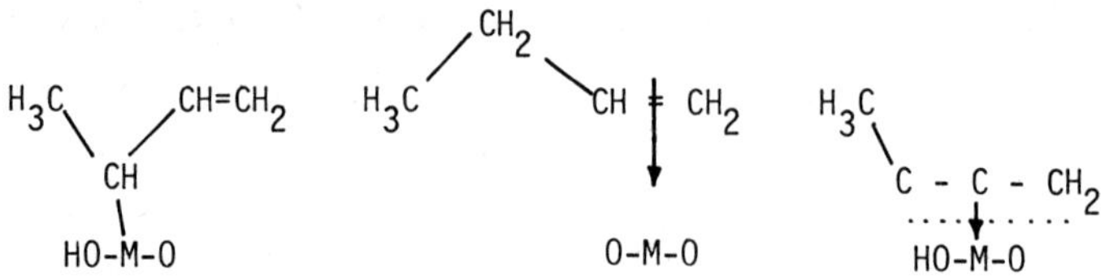

H_3C—CH(—$CH=CH_2$), CH ↓ HO-M-O

sigma

H_3C—CH_2—CH ≠ CH_2 ↓ O-M-O

pi-bonded butene-1

H_3C—C - C - CH_2 ↓ HO-M-O

pi-bonded allyl species

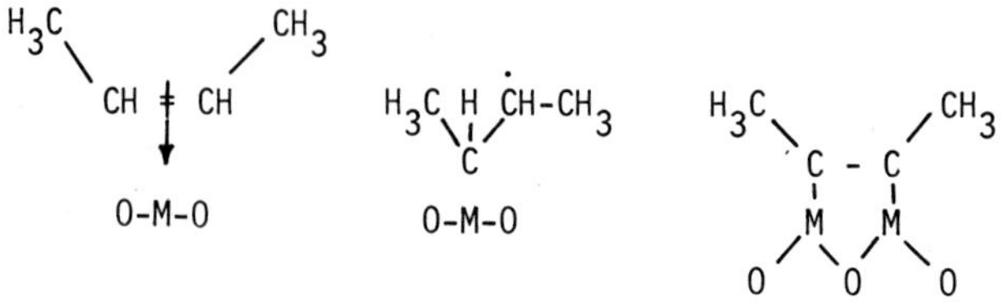

H_3C—CH ≠ CH—CH_3 ↓ O-M-O

pi-bonded butene-2

H_3C H $\dot{C}H$-CH_3 C O-M-O

split double bond mono sigma bond

H_3C—C - C—CH_3 M M O O O

split double bond: di sigma bond

Loss of a second hydrogen from a sigma or a pi-bonded allyl species could lead to pi-adsorbed butadiene

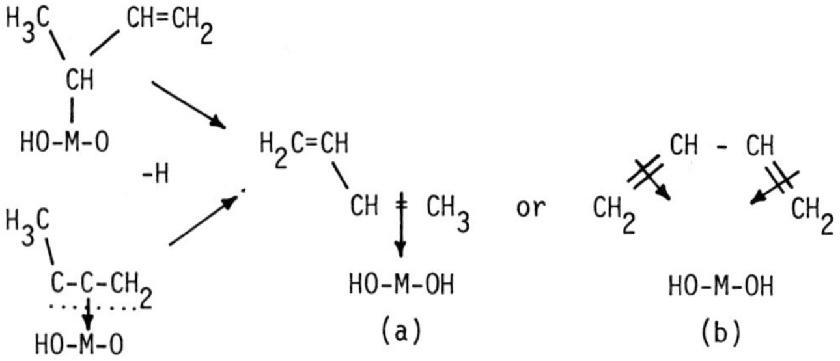

H_3C—CH(—$CH=CH_2$) | HO-M-O

-H

H_3C—C-C-CH_2 ↓ HO-M-O

→ $H_2C=CH$—CH ≠ CH_3 ↓ HO-M-OH (a) or CH - CH, CH_2 ↘ ↙ CH_2 HO-M-OH (b)

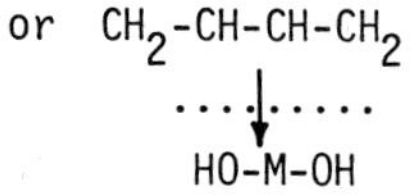

or CH_2-CH-CH-CH_2 ↓ HO-M-OH (c)

As Dowden points out, the confirmation of the adsorbed butadiene is open to question. A "mono-adsorbed" species (as above) is probably very similar to a mono-olefin. For a "di-adsorbed" species (such as (b)), an edge site may be appropriate ((d) below)

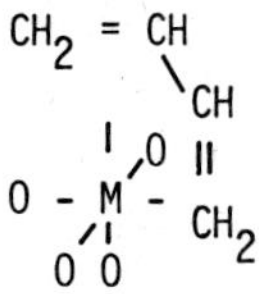

CH_2 = CH—CH ‖ CH_2; O - M - (O, O O)

(d)

The d_{z^2} bond will be stronger than the $d_{x^2-y^2}$ bond. Only one pi-bond would be possible, as the butadiene lies in the xy plane of the central atom.

Dowden argued that, in order to maintain the central double bond, emphasis should be placed on pi-bonded intermediates. Strong bonding requires the overlap of a filled bonding pi orbital of the olefin with an empty sigma d_{z^2} orbital of the metal, together with back donation from occupied pi d_{xz} d_{yz} orbitals of the metal into the empty antibonding pi orbitals of the olefin. These symmetry and electronic requirements imply the following metal ions (at least in the square pyramidal coordination)

D orbitals only	d^1	d^2	d^3	
	Ti(3)	Ti(2)	(Ti(1)	
	V (4)	V (3)	V (2)	
	Cr(5)	Cr(4)	Cr(3)	
		Mn(5)	Mn(4)	(etc.)

D and P orbitals	d^8	d^9	d^{10}	
	Fe(0)			
	Co(1)	Co(0)		
	Ni(2)	Ni(1)	Ni(0)	
		Cu(2)	Cu(1)	
			Zn(2)	(etc.)

Valency states which would not be expected to be present in appreciable quantities in a low partial pressure of oxygen (Ti(3), Ti(2), Ti(1), Fe(0), Co(1), Co(0), Ni(0)) or in a high pressure of oxygen (Mn(5)) were then eliminated, as were those states energetically unfavourable to the formation of pi complexes (Zn(2)).

The results of the full consideration of possible valency states are summarised in Table 10.1. Also presented are the results for butadiene, which has similar requirements to butene, at least with respect to the electronic requirements of pi bonding.

Moving on to consider the adsorption of oxygen, the following diatomic adsorbates were considered

```
 .
 O                            ..
 |                         /  O:
 O        O ‡ O    :O:        O - O          O -
 |          ↓        ↓        |   |          |
O-M-O     O-M-O   O-M-O       M   M          M
'radical'  pi adsorbed        split double bond
  (e)       (f)     (g)       (h)            (i)
```

TABLE 10.1

Possible chemisorbed states (square pyramidal coordination)

d^x =	1	2	3	4	5	6	7	8	9	10	S=1	2
Accessible	Ti(3)											
states	V(4)	V(3)	V(2)		Fe(3)	Fe(2)		Pt(3)	Cu(2)	Cu(1)	Zn(1)	
	Cr	Cr	Cr(3)	Cr(2)		Co(3)	Co(2)	Pd(2)		Sn(4)		Sn(2)
	Mo( )	Mo( )										
	W	W		Mn(2)	Mn(2)		Ni(3)	Ni(2)		Sb(5)		Sb(3)
Chemisorbed states												
Butene												
pi-bond	+	+	+					+	+	+		
sigma bond				+	+	+	+	+			+	+
Butadiene												
monodentate	+	+	+					+	+	+		
bidentate		+								+		
Oxygen												
pi-bond	+	+	+					+	+	+	+	+
radical			+	+	+	+	+	+				
donor lone pair atoms	+	+	+	+	+	+	+	+	+	+	+	+

Adsorption in the radical form requires that an orbital containing a single electron should project from the surface cation, while the pi-bond structure requires the excited form of the oxygen molecule (ref. 12). Structures (g) and (h) are similar to the corresponding forms of the olefin and require d^0 to d^3 or configurations around d^{10}.

Monoatomic adsorption requires two adjacent sites and, if the mechanism proposed does not require this, dissociation can be slowed down by dilution of the active species with a non-oxygen adsorbing species: this generally requires a species involving only closed shells. The active sites can be metal or oxide: there are no geometric barriers to the dissociation of pi-adsorbed complexes to give two peroxide anions. The coordination around the cation would then be similar to the peroxy ions of chromium or vanadium $(VO_8)^{3-}$ (ref. 13).

Although Dowden did not use the fact, there is evidence that a general sequence of oxygen adsorption can be suggested (refs. 14, 15):

$$O_2 + e \longrightarrow O_2^- \quad (23)$$

$$O_2 + e \longrightarrow 2O^- \quad (24)$$

$$O^- + e \longrightarrow O^{2-} \quad (25)$$

where e is an electron. Increasing temperature favours O^{2-} at the expense of O_2^-. It has also been reported that, for Ti, V and Mo ions at least, oxygen adsorbs as O_2^- on a tetrahedrally coordinated ion, while no such adsorption was observed on square pyramidal or octahedral ions (ref. 16). For a monolayer catalyst, tetrahedral coordination is favoured by silica supports and octahedral coordination is favoured by alumina.

Dowden did not consider the adsorption of maleic anhydride as such although, as discussed below, he did consider the reaction route as a function of surface geometry.

The conclusions arrived at concerning the possibility of various adsorbed states are summarised in Table 10.1. There are, of course, many assumptions that have been used in the derivation of this table, but adsorbed states that are desirable for the reaction in question centre around two groups, (d^2-d^3) and (d^9-d^{10}). These are, respectively, the n- and p-type of semi-conducting oxides, of which n-type are generally known to favour selective oxidation while p-type favour total oxidation. What is not shown in this table are all of the non-transition metal oxides, which are known to favour dehydrogenation (e.g. Bi (2-3)). In general, oxides of this type (SnO_2, Sb_2O_5, Bi_2O_5, TiO_2 etc.) can be expected to be more selective than p-type transition metal oxides.

Moving to the reaction sequence itself, Dowden noted that the activity pattern for dehydrogenation (Chapter 3) shows that oxides of transition metals with configuration d^1 to d^4 and d^6 are preferred. He also noted that the reducible n-type oxides of non-transitional metals are known to have moderate or high dehydrogenation activity. As a result, the activity patterns indicate that dehydrogenation should be efficient on the following oxides:

Transitional	(d^1)	Ti(3), V(4), Cr, Mo, W(5)
	(d^2)	V(3), Cr, Mo, W(4)
	(d^3)	Cr(3), V(2)
	(d^4)	Cr(2), Mn(3)
	(d^6)	(strong field) Co(3)
Non-transitional		Zn(1), Sn(2), Bi(2), Bi(3)

Dehydrogenation can be expected to produce water rather than hydrogen and, by analogy with the oxidation of olefins (ref. 17), should be the rate-determining step.

For the oxygen insertion reaction, the situation was complicated by the fact that little is known about the reaction. Realistically, Dowden assumed that both monatomic and diatomic oxygen species were present, and adopted the old idea that activity increases but selectivity decreases with the strength of the metal-oxygen bond. Carbon dioxide formation would be favoured by high concentrations of loosely held oxygen on the surface.

Dowden then made a jump in the design to consider the geometries of the reactions on vanadium pentoxide, a catalyst known to be effective for the desired reaction (ref. 18). This was done in order to illustrate the geometric requirements of the dehydrogenation and oxygen addition reactions. A complete account of comparisons of bond lengths and angles in the reacting molecules and in the catalyst was not given, but it was stated that dehydrogenation and oxygen addition via dihydroperoxides were both feasible on V_2O_5 or $V_{12}O_{26}$ lattice planes. An example of the approach was given, but this has been dealt with more clearly in other designs. The point was made, however, that the active centre could involve five ions in the surface: two to maintain two peroxy adsorbates, one to pi-adsorb butene which, from geometric reasons, also covered two other sites. No special requirements were envisaged for the oxidation of malealdehyde, since this should occur easily once a proton is available for enolisation.

Geometric arguments were not used to choose between the various alternatives presented in Table 10.1, although this could have been done. Of the ions suggested, vanadium was chosen as the primary component of most interest, mainly on the grounds of Table 10.1 and of experience (refs. 18,19). In fact, however, there are problems with a catalyst based on vanadia. First, it is too active and non-selective, a fact which can be partially overcome by using a supported catalyst of very low area. Second, mechanistic studies suggest that V(4) is the lowest valency of vanadium that persists during oxidation (refs. 15,16,17). From the design, it is seen that maximal concentrations of V(4), V(3) and, if possible, V(2) are desired. Improvements should, then, result from attempts to increase the concentration of lower valencies.

III. DESIGN OF SECONDARY COMPONENTS

In this design, the distinction between primary and secondary components is less clear, since many of the suggested secondary components are, in fact, present in high concentrations. It is advantageous, however, to stay with the general approach.

The first argument involved the fact that the production of carbon oxides requires a high concentration of loosely bound oxygen. This will be favoured if there is a high surface concentration of ions that can adsorb oxygen and, as a result, ways were sought to reduce the total oxidation.

One route suggested was to replace vanadium or molybdenum oxides by niobium or tantalum oxides, on the grounds that they would be more selective and less active. This was rejected on the grounds of cost and efficiency. What was not done was to turn to the plot of activity versus heat of formation (Chapter 3), which would have directed attention at chromium and (possibly) cerium oxides (ref. 20).

Secondly, it was suggested that dilution of the catalyst could be beneficial. According to the model, dehydrogenation requires three cations but, since mobile adsorption is known to be possible (refs. 8-11), reaction can also occur on two or, less probably, one cation. As a result, if the catalyst is diluted with an oxide that has dehydrogenation capacity and some capacity for combining with oxygen, selectivity should be improved. To favour the oxidation of malealdehyde, acidity is also required.

The oxides of boron, titanium, zirconium, phosphorus and sulphur are acidic and have the capacity to form peracid compounds. Oxides of niobium, molybdenum, tungsten and uranium are weakly acidic (in their higher valency states), form peracids and have dehydrogenation ability. Oxides of tin, lead, antimony, bismuth, selenium and tellurium have dehydrogenation ability and a capacity to take up and lose oxygen.

At this point it is convenient to remember that it is desired to maximise the concentration of lower valencies of vanadium. Of the commonly available oxides, molybdenum and tungsten trioxides (with valencies of six) would be suitable additives, but the latter does not form solid solutions with vanadia. As a result, a solid solution of MoO_3 in V_2O_5 (25 mole % solubility) should induce the formation of V(4).

At the same time, however, the acidity of the catalyst would not be expected to rise appreciably. Were it possible to form a solid solution of vanadia in an acid oxide, reduction of V(5) to V(4) could be stabilised by the acidity of the matrix, thereby giving an active (high V(4)) and acidic (more selective) catalyst. The obvious additive is sulphur dioxide, as used in the oxidation of naphthalene (ref. 21). However, sulphates can only be formed by reaction with a basic salt (e.g. to form an alkali metal sulphate), and the resulting catalyst may contain a high proportion of lower valency vanadium and yet not be acidic enough.

Boric oxide is another alternative, but it dissolves only a small amount of V_2O_5, even at 1200^oC. Phosphorus oxide dissolves more vanadia, even at moderate temperatures, and it would seem to offer the best compromise involving more acidity and more lower valency states of vanadium.

IV. SUMMARY

To summarise, then, the design suggests the following catalysts for the oxidation of butene to maleic anhydride. Important factors considered include electronic requirements of adsorption and reaction (e), geometric factors (g), dilution of the active phase (d) and acidity (a).

Primary components (e,g):	V(2), V(3), V(4), Cr(2), Cr(3), Cr(4), Cr(5), Mo(4), Mo(5), W(4), W(5), SnO_2, SbO_5, Bi_2O_5, TiO_2 etc.
Secondary components (d,a):	P, B, Mo oxides, Possibly Se, Te oxides.
Support (d):	low surface area, wide pores.

V. COMMENTS

This is a good, well balanced design of the chemistry of the catalyst. No detailed attempt has been made to design the support nor to consider heat and mass transfer in the system. Perhaps too little attention has been paid to heats of adsorption and to activity patterns for oxidation reactions although, as the author stated (ref. 1), little is known concerning the oxygen insertion reaction.

In an attempt to clarify the procedure, several important steps have been summarised by the author. It is important that these should be recognised. They include:

1. Probable adsorption states have been derived only for the square pyramidal coordination.
2. Full details of the geometric approach have not been given.
3. The mechanism of the reaction was never written down, although it was assumed that the butadiene route was important.

4. The author did consider all the plausible mechanisms but, for clarity and brevity, discussed only one of these.
5. Experimental testing was carried out but was not discussed in detail.

Comparing the design with that carried out for the conversion of olefins to aromatics, it is seen that much more detail is used in describing the idea, and this will be extended in the next design to be considered. Comparing the present design with the steam reforming catalyst design is like comparing a Rolls Royce to a Model T. Ford. In the present design, most of the approaches open to the designer have been used, while, for steam reforming, the choice of catalyst was based, essentially, on activity patterns.

In extending the description of the idea to its logical conclusion, considerable complexity can be introduced. This emerges from consideration of one other design reported by Dowden (refs. 1,22): the hydrogenation of benzaldehyde to benzyl alcohol.

REFERENCES

1 D.A. Dowden, Chem. Eng. Prog. Symp. Ser., No. 73 vol 63, 90 (1967).
2 J.L. Franklin, Ann. Rev. Phys. Chem., 18, 261 (1967).
3 A. Blundell and G. Skirrow, Proc. Roy. Soc. (London), A284, 331 (1958).
4 E.G.E. Hawkins, Organic Peroxides, Spon, London, 1961.
5 C.F. Cullis, A. Fish, M. Saeed and D.L. Trimm, Proc. Roy. Soc. (London), A289, 402 (1966).
6 C.F. Cullis, J.A. Garcia Dominques, D. Kiraly and D.L. Trimm, Proc. Roy. Soc. (London), A291, 235 (1966)
7 W.A. Waters, Mechanisms of Oxidations of Organic Compounds, Methuen, London, 1964.
8 F.C. Gault, J.J. Rooney and C.K. Kemball, J. Catal., 1, 255 (1962).
9 J.L. Garnett and W.A. Sollich-Baumgartner, Aust. J. Chem., 14, 481 (1961).
10 J.L. Garnett and W.A. Sollich-Baumgartner, Adv. Catal., 16, 95 (1966)
11 Z. Knor, Adv. Catal., 22, 51 (1972).
12 F.A. Cotton and G. Wilkinson, Advanced Inorganic Chemistry, Interscience, London, 1962.
13 Ph.A. Batist, B.C. Lippens and G.C.A. Schuit, J. Catal., 5, 55 (1966).
14 M. Ya. Kon, V.A. Schvets and V.B. Kazanskii, Kinet. Catal., 13, 65 (1972).
15 V.A. Shvets, V.M. Vorotynstev and V.B. Kazanskii, Kinet. Catal., 10, 287 (1969).
16 V.A. Shvets and V.B. Kazanskii, J. Catal., 25, 123 (1972).
17 C.R. Adams, Ind. Eng. Chem., 61, (6), 31 (1969).
18 C.F. Cullis, Ind. Eng. Chem., 59, 18 (1967).
19 B.L. Moldavskii and Yu.D. Kernos, Kinet. Catal., 1, 242 (1960).
20 Y. Moro-Oka, Y. Morikawa and A. Ozaki, J. Catal., 7, 23 (1967).
21 P. Mars and J.G.H. Maessen, J. Catal., 10, 1 (1968).
22 D.A. Dowden, Chim. Ind. (Milan), 55, 639 (1973).

CHAPTER 11

THE CONVERSION OF BENZALDEHYDE TO BENZYL ALCOHOL

The hydrogenation of benzaldehyde to benzyl alcohol was chosen by Dowden as a straightforward example of catalyst design in which minimisation of parasitic reactions was important. The problem was considered in 1967 (ref. 1) and in 1973 (ref. 2), and it is interesting and instructive to compare the two approaches.

I. DESCRIPTION OF THE IDEA

In the earlier design, the description of the idea was fairly straightforward, the target reaction

$$C_6H_5CHO + H_2 \rightleftharpoons C_6H_5CH_2OH \qquad (1)$$

being listed together with

(a) primitive reactions

$$C_6H_5CHO \longrightarrow C_6H_6 + CO \qquad (2)$$

(b) self- and cross-interactions

$$2\,C_6H_5CHO \rightleftharpoons C_6H_5CH(OH)-C(=O)C_6H_5 \qquad (3)$$

$$C_6H_5CHO + 3H_2 \longrightarrow C_6H_{11}CHO \qquad (4)$$

(c) sequential reactions

$$C_6H_5CH_2OH \rightleftharpoons C_6H_5CH_2-O-CH_2C_6H_5 + H_2O \qquad (5)$$

$$C_6H_5CHO + xH_2 \rightleftharpoons C_6H_5CH_3 + H_2O \quad (6)$$

$$\rightleftharpoons C_6H_{11}CH_2OH \quad (7)$$

and

(d) cross-sequential reactions

$$C_6H_5CH_3 + 3H_2 \rightarrow C_6H_{11}CH_3 \quad (8)$$

$$CO + H_2O \rightleftharpoons CO_2 + H_2 \quad (9)$$

$$CO + 3H_2 \longrightarrow CH_4 + H_2O \quad (10)$$

etc.

The thermodynamics of the target reaction was then considered, together with the physical state of reactants and products. The reaction

$$C_6H_5CHO + H_2(g) \rightleftharpoons C_6H_5CH_2OH \quad (1)$$

has a heat of reaction of 17 kcal/mole, while the free energy is ca. -9 kcal/mole. Since benzaldehyde (m.p. = -26^oC, b.p. = 178^oC) is to be converted to benzyl alcohol (m.p. -15.3^oC, b.p. = 205^oC), the reaction could be easier to handle above 205^oC. The equilibrium is such that, at 300^oC, little benzaldehyde will be present but, because of the reaction exothermicity, the reactor should be operated at low temperatures.

As would be expected, solid products could be obtained if condensation reactions are important (see below). Thus, $C_6H_5COOCH_2C_6H_5$ melts at 328^oC and $C_6H_5CH(OH)COC_6H_5$ melts at 344^oC. The importance of such products will obviously depend upon the selectivity of the reaction.

Most of the undesired reactions are also favoured thermodynamically up to 300^oC, e.g.:

$$C_6H_5CHO \rightleftharpoons C_6H_6 + CO \qquad \Delta G^o_{298\ K} = -2\ \text{kcal/mole} \qquad \Delta G^o_{373\ K} = -4.4\ \text{kcal/mole}$$

$$C_6H_5CHO \rightleftarrows C_6H_6 + CO \qquad \Delta G^o_{298\,K} = -2 \text{ kcal/mole} \qquad \Delta G^o_{373\,K} = 4.4 \text{ kcal/mole}$$

The formation of polymers and of carbon can also be expected to become more important as the temperature increases. As a result, there is no simple thermodynamic route to increased selectivity, and control of kinetics by the careful selection of a catalyst is seen to be of prime importance.

The design reported in 1973 was based on a much more complete description of the idea:

The target transformation

$C_6H_5.CHO + H_2 \rightleftarrows C_6H_5CH_2OH$ (hydrogenation) (1)

The characteristic chemistry

(a) Primitive processes

$C_6H_5CHO \rightleftarrows C_6H_5 + CO$ (decarbonylation) (2)

(b) Self-interactions

$2C_6H_5CHO \rightleftarrows C_6H_5COOCH_2C_6H_5$ (condensation)

$2C_6H_5CHO \rightleftarrows C_6H_5CH(OH)C(=O)C_6H_5$ (condensation) (3)

(c) Cross-interactions

$C_6H_5CHO + H_2 \rightleftarrows C_6H_5CH_2OH$ (hydrogenation) (1)

$C_6H_5CHO + H_2 \rightleftarrows C_6H_7CHO$ (hydrogenation) (11)

(d) Derived primitive processes

$C_6H_7CHO \rightleftarrows C_6H_8 + CO$ (decarbonylation) (11)

(e) Derived self-interactions

$2C_6H_6 \rightleftarrows (C_6H_5)_2 + H_2$ (dehydrogenation) (13)

$$2CO \rightleftarrows CO_2 + C \qquad \text{(disproportionation-oxygen transfer)} \tag{14}$$

$$2C_6H_5CH_2OH \rightleftarrows (C_6H_5CH_2)_2O + H_2O \qquad \text{(dehydration)} \tag{5}$$

(f) Sequential reactions

$$C_6H_8 + H_2 \rightleftarrows C_6H_{10} \qquad \text{(hydrogenation)} \tag{15}$$

(g) Sequential cross-interactions

$$C_6H_6 + H_2 \rightleftarrows C_6H_8 \qquad \text{(hydrogenation)} \tag{16}$$

$$CO + H_2 \rightleftarrows HCHO \qquad \text{(hydrogenation)} \tag{17}$$

$$C_6H_5CH(OH)CC_6H_5 + H_2 \rightleftarrows C_6H_5CH(OH)CH(OH)C_6H_5 \qquad \text{(hydrogenation)} \tag{18}$$

$$\rightleftarrows C_6H_5CH_2\underset{\underset{O}{\|}}{C}C_6H_5 + H_2O \qquad \text{(hydrogenolysis)} \tag{19}$$

$$C_6H_5CH_2OH + H_2 \rightleftarrows C_6H_5CH_3 + H_2O \qquad \text{(hydrogenation)} \tag{20}$$

$$C_6H_7CHO + H_2 \rightleftarrows C_6H_9CHO \tag{21}$$

$$\text{and so on to} \rightleftarrows C_6H_{11}CHO \tag{22}$$

$$C_6H_7CHO + H_2 \rightleftarrows C_6H_9CH_2OH \tag{23}$$

$$\text{and so on to} \rightleftarrows C_6H_{11}CH_2OH \tag{24}$$

(h) Interjacent primitive processes

$$C_6H_5CH(OH)CH(OH)C_6H_5 \rightleftarrows C_6H_5\overbrace{CH\text{-}CH}^{O}C_6H_5 + H_2O \qquad \text{(dehydration)} \tag{25}$$

$$C_6H_5CH_2C\cdot C_6H_5 \rightleftarrows C_6H_5CH_2C_6H_5 + CO \qquad \text{(decarbonylation)} \tag{26}$$

(i) Interjacent self-interactions

$$2C_6H_{11}CH_2OH \rightleftarrows (C_6H_{11}CH_2)_2O + H_2O \qquad \text{(dehydration)} \tag{27}$$

(j) Interjacent cross-interactions

$CO + H_2O \rightleftarrows CO_2 + H_2$		(dehydrogenation-oxygen exchange)	(9)
$CO_2 + H_2 \rightleftarrows HCOOH$		(hydrogenation)	(28)
$C_6H_{10} + H_2 \rightleftarrows C_6H_{12}$		(hydrogenation)	(29)
$HCHO + H_2 \rightleftarrows CH_2OH$		(hydrogenation)	(30)
$CH_3OH + H_2 \rightleftarrows CH_4 + H_2O$		(hydrogenolysis)	(31)
$C_6H_5CH_2C(=O)C_6H_5 + H_2 \rightleftarrows C_6H_5CH_2CH(OH)C_6H_5$		(hydrogenation)	(32)
$C_6H_5CH_3 + H_2 \rightleftarrows C_6H_7CH_3$		(hydrogenation)	(33)
and so on to $\rightleftarrows C_6H_{11}CH_3$			(8)
$C_6H_{11}CH_2OH + H_2 \rightleftarrows C_6H_{11}CH_3$		(hydrogenation)	(34)
Cyclic dienes $\rightarrow$ polymers		(polymerisation)	(35)

Hydrogenations

$C_6H_5CHO + H_2 \rightleftarrows C_6H_5CH_2OH$ (1)

$C_6H_5CHO + H_2 \rightleftarrows C_6H_7CHO$	(11)	$C_6H_5CHO + 3H_2 \rightleftarrows C_6H_{11}CHO$*	(4)
$C_6H_7CHO + H_2 \rightleftarrows C_6H_9CHO$	(21)		
$C_6H_9CHO + H_2 \rightleftarrows C_6H_{11}CHO$	(22)		
$C_6H_5CH_2OH + H_2 \rightleftarrows C_6H_7CH_2OH$		$C_6H_5CH_2OH + 3H_2 \rightleftarrows C_6H_{11}CH_2OH$	(36)
$C_6H_7CH_2OH + H_2 \rightleftarrows C_6H_9CH_2OH$	(23)		
$C_6H_9CH_2OH + H_2 \rightleftarrows C_6H_{11}CH_2OH$	(24)		
$C_6H_6 + H_2 \rightleftarrows C_6H_8$	(20)	$C_6H_6 + 3H_2 \rightleftarrows C_6H_{12}$*	(37)
$C_6H_8 + H_2 \rightleftarrows C_6H_{10}$	(19)		
$C_6H_{10} + H_2 \rightleftarrows C_6H_{12}$	(29)		

$CO + H_2 \rightleftarrows HCHO$ (17)
$HCHO + H_2 \rightleftarrows CH_3OH$ (30)
} $CO + 2H_2 \rightleftarrows CH_3OH^*$ (38)

$C_6H_5CH(OHCOC_6H_5 + H_2 = C_6H_5CH(OH)CH(OH)C_6H_5^*$ (18)

$C_6H_5CH_3 + H_2 \rightleftarrows C_6H_7CH_3$ (33)
$C_6H_7CH_3 + H_2 \rightleftarrows C_6H_9CH_3$ (39)
$C_6H_9CH_3 + H_2 \rightleftarrows C_6H_{11}CH_3$ (40)
} $C_6H_5CH_3 + 3H_2 \rightleftarrows C_6H_{11}CH_3^*$ (8)

Hydrogenolysis*

$C_6H_5CH(OH)C(=O)C_6H_5 + H_2 \rightleftarrows C_6H_5CH_2C(=O)C_6H_5^* + H_2O$ (18)

$C_6H_5CH_2OH + H_2 \rightleftarrows C_6H_5CH_3 + H_2O^*$ (6)

$CH_3OH + H_2 \rightleftarrows CH_4 + H_2O^*$ (31)

$C_6H_{11}CH_2OH + H_2 \rightleftarrows C_6H_{11}CH_3^*$ (34)

$C_6H_5CHO \rightleftarrows C_6H_6 + CO^*$ (12)

$C_6H_7\ CHO \rightleftarrows C_6H_8 + CO$ (41)

$C_6H_{11}CHO \rightleftarrows C_6H_{12} + CO^*$ (42)

$C_6H_5CH_2CC_6H_5 \rightleftarrows C_6H_5CH_2C_6H_5 + CO^*$ (26)

Condensation*

$2C_6H_5CHO \rightleftarrows C_6H_5CH(OH)C(=O)C_6H_5^*$ (3)

$2C_6H_5CHO \rightleftarrows C_6H_5COOCH_2C_6H_5$ (43)

Dehydration*

$C_6H_5CH(OH)CH(OH)C_6H_5 \rightleftarrows C_6H_5CH\text{-}CHC_6H_5$ (epoxide, O bridging the two CH) $+ H_2O^*$ (25)

$2C_6H_5CH_2OH \rightleftarrows (C_6H_5CH_2)_2 + H_2O^*$ (5)

$2C_6H_{11}CH_2OH \rightleftarrows (C_6H_{11}CH_2)_2O + H_2O^*$ (27)

Miscellaneous*

$2CO \rightleftarrows CO_2 + C$ (disproportionation)* (14)

$2C_6H_6 \rightleftarrows (C_6H_5)_2 + H_2$ (dehydrogenation) (13)

$CO + H_2O \rightleftarrows HCOOH$ (hydration) (44)

There are many other consecutive reactions of minor importance which can be written down, but in the light of the overall process these will be seen to be unimportant. Alternatively, reactions can be written down in terms of the basic reactions involved in the overall process.

$HCOOH \rightleftarrows CO_2 + H_2$ (dehydrogenation) (45)

cyclic dienes $\rightleftarrows$ polymers (polymerisation) (35)

Note that a given reaction appears in more than one place, and that a reaction which is recognised as - for example - a cross-sequential reaction, need not necessarily be recognised in terms of the chemistry to be expected. In other words, it is more desirable to duplicate a reaction at this stage than to ignore a reaction.

Combining the most important of these reactions gave the basic scheme for the overall reaction as shown in Fig. 11.1.

Dowden first used activity patterns and selectivity patterns to recognise that rhodium, nickel, palladium, platinum and copper catalysts would be active, and that nickel or copper could be active, selective and cost-effective. Copper has the advantage that it does not catalyse the saturation of aromatics rings or the disproportionation/hydrogenation of carbon monoxide. Because the inherent activity is low, the catalyst should be used at higher temperatures, where parasitic reactions will be favoured and sintering more rapid (melting point = 1083^oC). This means that particular attention must be paid to support and promotion.

Nickel is more active, and can be used at lower temperatures where the parasitic reactions are slower. Nickel is, in any case, more resistant to sintering than copper (melting point of nickel = 1453^oC) although hydrogen is know to favour sintering (ref. 3).

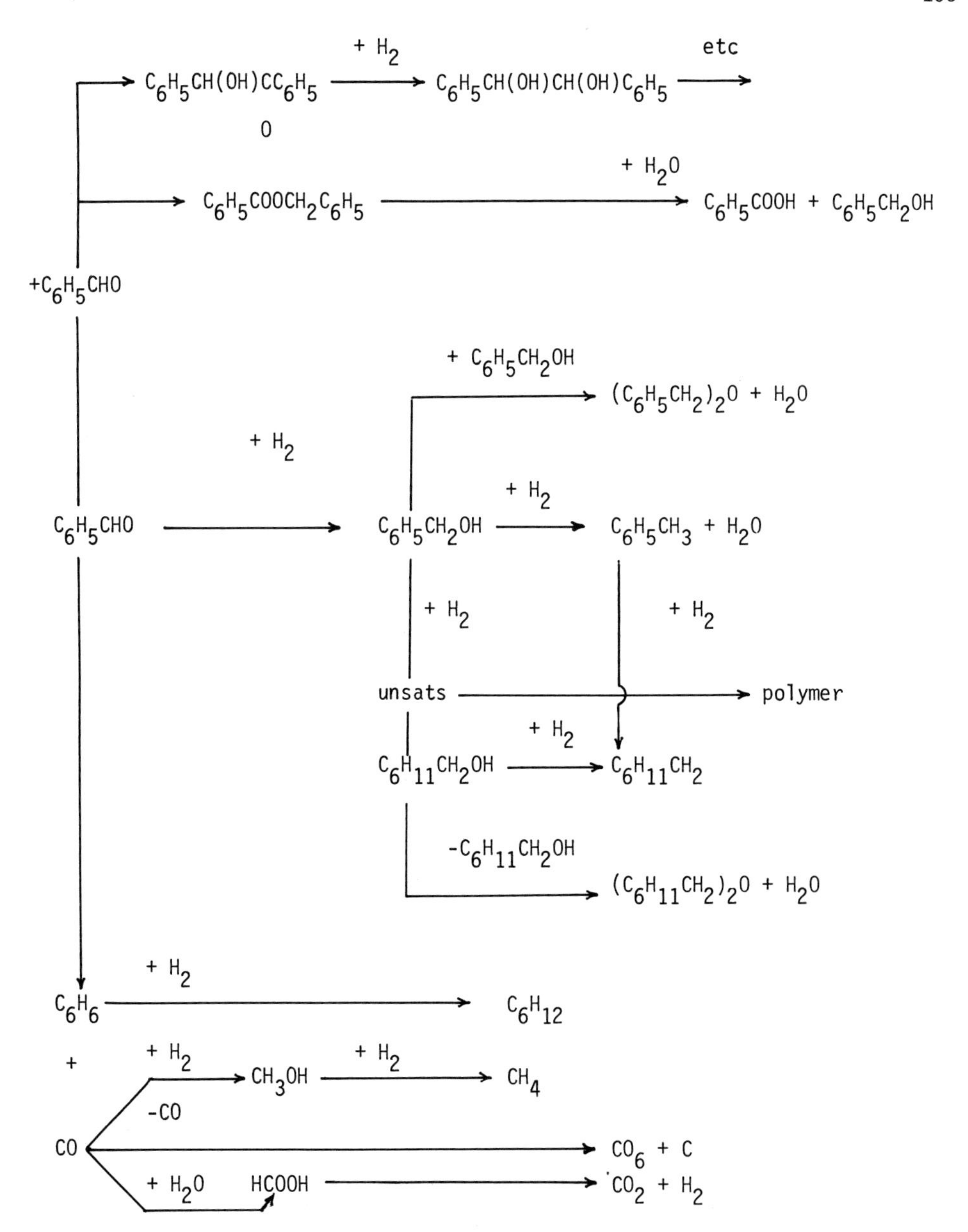

Fig. 11.1. Basic scheme for the overall reaction benzaldehyde to benzyl alcohol.

Dowden then considered possible kinetics on the two catalysts. The aromatic ring is relatively resistant to hydrogenation but, once the first double bond has been hydrogenated, subsequent complete hydrogenation would be easier. Hydrogenolysis may also be slow but may be assisted by polar factors (see later). Polymerisation would certainly be favoured by acidic oxides. Reflecting the activities and selectivities of the two catalysts, the case of hydrogenation would then be expected to be:

Nickel: aldehyde > ring saturation > hydrogenolysis
Copper: aldehyde > hydrogenolysis > ring saturation

At this point in the design of the primary components, Dowden effectively finalised the choice to copper or nickel, although he did continue to consider chemisorbed complexes and the reaction scheme on the surface. Chemisorbed complexes considered included:

Hydrogen: H ··· H weakly bonded to two M (weak); H–M (strong)

Benzaldehyde: $(H)(C_6H_5)C{=}O{-}M$; $M{\leftarrow}O{=}C(H)(C_6H_5)$; $M{-}O{-}C(M)(H)(C_6H_5)$

Benzyl alcohol: $H{-}O(\rightarrow M){-}C(H)(H){-}C_6H_5$; $(H)(H)C(O{-}M){-}C_6H_5$; $H{-}C(H)(M){-}C_6H_5 + H{-}O{-}M$

or, on polar solids

$$C_6H_5CH_2OH \longrightarrow HO^- - - - {}^+CH_2C_6H_5$$

$$C_6H_5CH_2OH \longrightarrow C_6H_5CH_2OH_2^+$$

$$C_6H_5CH_2OH + H^+ \longrightarrow C_6H_5CH_2OH_2^+$$

$$C_6H_5CH_2OH + H^+ \longrightarrow C_6H_5CH_2^+ + H_2O$$

Benzene $\underset{M}{\overset{C_6H_6}{|}}$ or $\underset{M}{\overset{C_6H_5}{|}}$ $\underset{M}{\overset{H}{|}}$

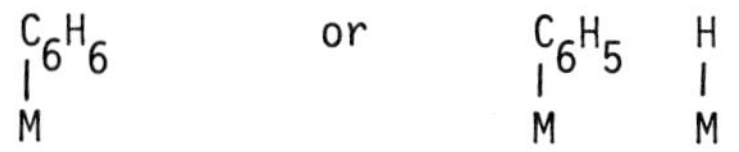

Carbon monoxide

$\begin{matrix} O \\ | \\ C \\ | \\ M \end{matrix}$ $\begin{matrix} O \\ \| \\ C \\ / \ \backslash \\ M \quad M \end{matrix}$

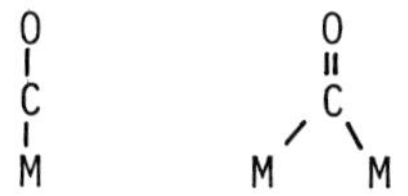

M represents a metal atom on the surface. Phenyl groups were taken to lie with the rings flat and as close to the surface as possible. Chemisorption on copper can be expected to be less strong than on nickel, carbon monoxide, for example, not being adsorbed. Polar adsorbed species can be expected to predominate only on ionic surfaces such as the support.

In considering the reactions involving these species on the surface, Dowden made some simplifying illustrations in the presentation. For clarity, the reaction centre was assumed to involve octahedrally coordinated metal atoms although, as he stated, this is probably neither correct nor the best active centre: it is just a convenient way of envisaging the reaction. For similar reasons, adsorbed hydrogen atoms were used, even though it is known that both monoatomic and diatomic species may be present.

With these points in mind, the reactions were considered in turn:

Hydrogenation

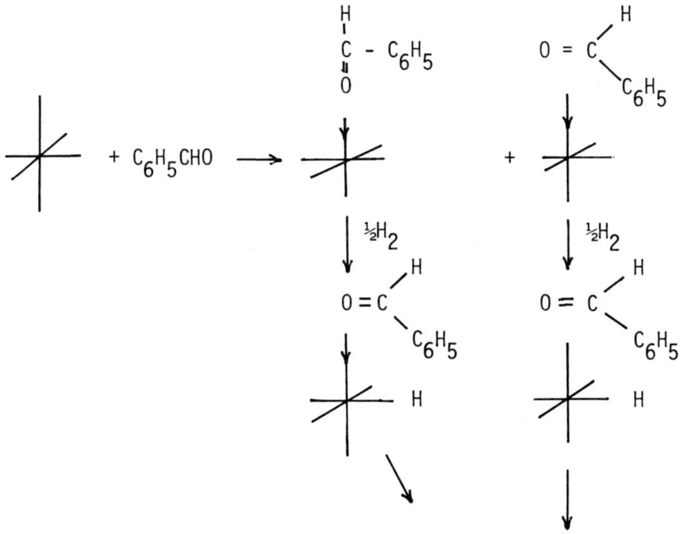

H
C - C_6H_5
O

O = C (H)(C_6H_5)

+ C_6H_5CHO → +

½H_2 ½H_2

O = C (H)(C_6H_5) O = C (H)(C_6H_5)

H H

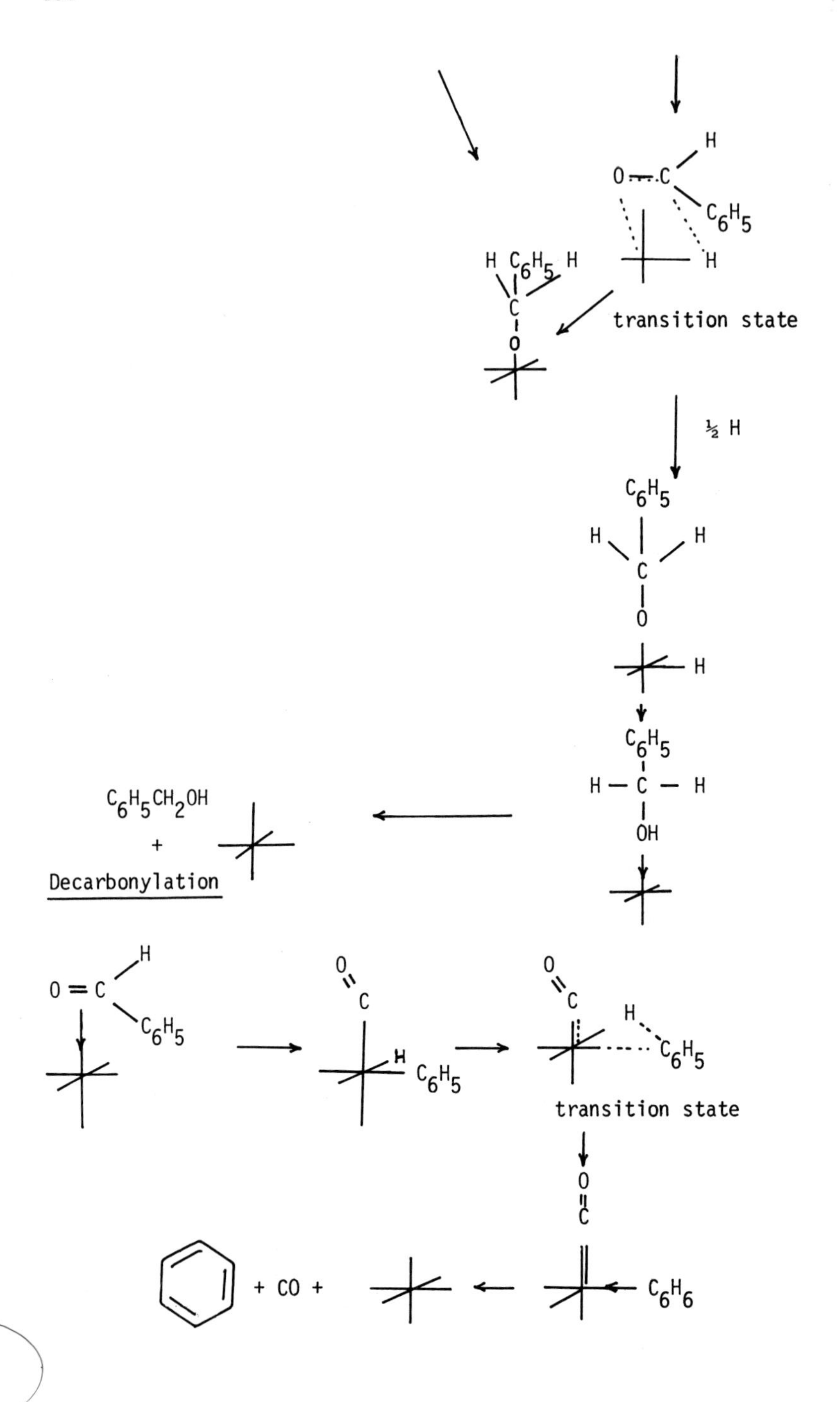
transition state
½ H
C6H5CH2OH
+
Decarbonylation
transition state
+ CO +
C6H6

Hydrogenolysis

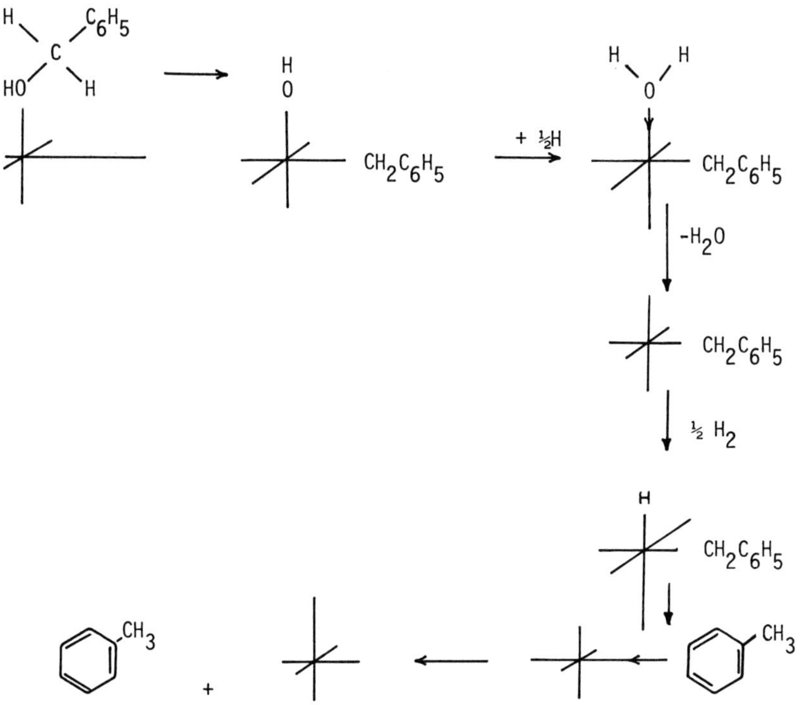

Similar reactions may be written to involve more than one metal atom (ref. 1).

Now these reactions all refer to a metal atom adsorption centre, and it is known that ionic adsorption is also possible, given an appropriate catalyst. As a result, a protonic mechanism may also be suggested:

$$C_6H_5CH_2OH + H^+ \rightleftarrows C_6H_5CH_2OH_2^+$$

$$C_6H_5CH_2OH_2^+ \rightleftarrows C_6H_5CH_2^+ + H_2O$$

$$C_6H_5CH_2^+ + C_6H_5CH_2OH \rightleftarrows C_6H_5CH_3 + C_6H_5CHO + H^+$$

$$C_6H_5CH_2^+ + C_6H_5CH_2OH \rightleftarrows (C_6H_5CH_2)_2O + H^+$$

These reactions are in equilibria dictated by reaction conditions but may be avoided if the catalyst is not acidic. If, on the other hand, the catalysts are too basic, then condensation reactions will be favoured. A neutral support or promotor is indicated.

II. DESIGN OF SUPPORT AND PROMOTOR

Since the design of secondary components carried out by Dowden was more closely connected with promotion by reduction of sintering, it is convenient to discuss this together with the support.

Choice of support was limited to the amphoteric and weakly basic metal oxides, for the reasons given above. Of these (Al_2O_3, SiO_2, TiO_2, Cr_2O_3, MnO, ZnO and spinels), titania may be eliminated because it is not readily available as a high surface area support and manganese oxide was avoided because it has only small hydrogenation-dehydrogenation activity. Silica tends to be available with small pores, and was avoided because of possible diffusion limitations. If alumina or chromia is used as a support, careful neutralisation of acidity should improve selectivity, particularly as a result of the fact that dehydration will be minimised.

At higher temperatures (> ca. 400°C) copper will react with zinc oxide to form brasses

$$Cu + ZnO + H_2 \rightleftharpoons \alpha CuZn + H_2O$$

which are less active than copper. On the other hand, it is necessary to minimise copper sintering, and this can best be done by using a spacer. Traditionally, this is done either by co-deposition of non-interactive compounds, or by isolating the active metal in a matrix of metal/metal oxide (alloys or solid solutions). Given the above, this would indicate that copper and alumina could be co-deposited on a support, or that copper chromite could be used as the active phase. This may react as copper in chromia or it may have an inherent activity of its own.

Thus the catalyst should consist of:

(a) Cu metal on alkali-doped alumina
(b) Cu metal on alkali-doped chromia
(c) Cu in chromia/copper chromite on alumina: alkali-doped
(d) Cu metal co-deposited with alumina on an alumina support: alkali-doped
(e) Ni metal on alkali-doped alumina
(f) Ni metal on any suitable support.

Because of the smaller inherent activity of copper and of its propensity for sintering, maximal copper surface area should be sought during preparation, and careful control of temperature should be maintained during operation.

III. COMMENTS

Although this design offers a good example of the development of the description of the idea, it suffers from the disadvantage that the real choice of catalyst was made on the basis of activity patterns. Despite the fact that possible adsorbed species were considered, as were reactions on the surface, this approach was not used in design. This arose primarily because the designer was beguiled by the knowledge that copper and nickel were active catalysts for the reaction. It was unfortunate, in particular, that the fact that some oxides do possess hydrogenation activity was not considered, although, in fact, the acidity of these oxides would tend to promote unwanted side reactions.

As with most designs reported by Dowden, little attention was paid to physical factors affecting the catalyst. More detailed attention to heat transfer is required as, indeed, was pointed out by Dowden (refs. 1,2).

Given the fact that the primary component choice rested primarily on activity patterns, design of secondary components - apart from in terms of structural promotion - is difficult. Subsequent to the 1973 design, information has been reported that would certainly influence secondary component design. Thus, for example, carbon formation on nickel has been established to require an ensemble of several sites (refs. 4,5), while the hydrogenation of benzaldehyde requires, as written, only one or two sites. Dilution of nickel by copper could lead to an active catalyst in which side reactions were minimised.

REFERENCES

1 D.A. Dowden, Chem. Eng. Prog. Symp. Ser., No. 73, Vol. 63, 90 (1967).
2 D.A. Dowden, Chim. Ind. (Milan), 55, 639 (1973).
3 M. Moayeri and D.L. Trimm, J. Chem. Soc. Faraday Trans., 1, 73, 1245 (1977).
4 G.A. Martin, M. Princet and J.A. Dalmon, J. Catal., 53, 321 (1978).
5 G.A. Martin, C.R. Acad. Sci., Ser. C, 284, 479 (1977).

CHAPTER 12

THE DESIGN OF A CATALYST FOR THE PRODUCTION OF METHANOL FROM METHYL FORMATE

I. THE IDEA

The decision to design a catalyst for the hydrogenolysis of methyl formate

$$HCOOCH_3 + 2H_2 = 2CH_3OH$$

was made in the context of developing an alternative route for the production of methanol. Work carried out in the early part of this century showed that it was feasible to carbonylate methanol to give methyl formate, and subsequent hydrogenolysis resulted in the gain of a methanol molecule over the total reaction. Although the direct combination of carbon monoxide and hydrogen to give methanol proved more viable, it seems possible that the two stage process could be carried out at lower temperatures. With recent increased interest in the conservation of energy, attention was focused once again on the two stage process.

Apart from very early references, little has been published on the hydrogenolysis of methyl formate, although more is available on the hydrogenolysis of esters in general. Most of this work has been carried out in connection with the production of long chain aliphatic alcohols, and a "copper chromite" catalyst is favoured by many workers. The exact nature of the catalyst under reaction conditions is open to considerable question, since most workers have been concerned much more with the mechanism of the reaction to be catalysed than with the nature or role of the catalyst.

II. THE DESCRIPTION OF THE IDEA

The students concerned with the design chose to describe the possible reaction mechanisms in detail, adopting the approach and nomenclature favoured by Dowden (ref. 1).

The stoichiometric statement

A. The target transformation

$$H\overset{O}{\overset{\|}{C}}OCH_3 + 2H_2 = 2CH_3OH \qquad \text{hydrogenolysis}$$

B. The characteristic chemistry

B.1.1. Primitive Processes

a. $HCOOCH_3 \rightleftarrows HOCH_3 + CO$ decarbonylation

b. $\rightleftarrows CH_4 + CO_2$ decarboxylation

c. $\rightleftarrows 2C + 2H_2O$ carbon formation

B.1.2. Self-interactions

a. $2H\overset{\overset{O}{\|}}{C}OCH_3 \rightleftarrows CH_3O\overset{\overset{O}{\|}}{C}\overset{\overset{O}{\|}}{C}OCH_3 + H_2$ dehydrogenation

b. $\rightleftarrows H\overset{\overset{O}{\|}}{C}OCH_2CH_2O\overset{\overset{O}{\|}}{C}H + H_2$ dehydrogenation

c. $\rightleftarrows CH_3O\overset{\overset{O}{\|}}{C}CH_2O\overset{\overset{O}{\|}}{C}H + H_2$ dehydrogenation

B.1.3. Cross-interactions

a. $H\overset{\overset{O}{\|}}{C}OCH_3 + H_2 \rightleftarrows H_2CO + HOCH_3$ hydrogenolysis

b. $\rightleftarrows H\overset{\overset{OH}{|}}{C}OCH_3$ hydrogenation

c. $\rightleftarrows H\overset{\overset{O}{\|}}{C}OH + CH_4$ hydrogenolysis

B.2.1. Derived primitive processes

a. $CH_3OH \rightleftarrows CH_2O + H_2$ dehydrogenation

b. $CH_3O\overset{\overset{O}{\|}}{C}\overset{\overset{O}{\|}}{C}OCH_3 \rightleftarrows CH_3\overset{\overset{O}{\|}}{C}OCH_3 + CO_2$ decarboxylation

c. $CH_3O\overset{\overset{O}{\|}}{C}\overset{\overset{O}{\|}}{C}OCH_3 \rightleftarrows CH_3O\overset{\overset{O}{\|}}{C}OCH_3 + CO$ decarbonylation

d. $H\overset{\overset{O}{\|}}{C}OCH_2CH_2O\overset{\overset{O}{\|}}{C}H \rightleftarrows H\overset{\overset{O}{\|}}{C}OCH_2CH_3 + CO_2$ decarboxylation

e. $\rightleftarrows H\overset{\overset{O}{\|}}{C}OCH_2CH_2OH + CO$ decarbonylation

f. $H_2C(OH)OCH_3 \rightleftarrows H_2CO + CH_3OH$ — disproportionation

g. $H_2C(OH)OCH_3 \rightleftarrows HCOOH + CH_4$ — disproportionation

h. $HC(=O)OH \rightleftarrows CO + H_2O$ — decarbonylation

B.2.2. Derived self-interactions

a. $2CO \rightleftarrows C + CO_2$ — disproportionation

b. $2HOCH_3 \rightleftarrows CH_3OCH_3 + H_2O$ — dehydration

c. $2CH_2O \rightleftarrows HC(=O)OCH_3$ — condensation

d. $\rightleftarrows HOCH_2CHO$ — condensation

e. $2HCOOH \rightleftarrows OCHOCHO + H_2O$ — dehydration

f. $2H_2C(OH)OCH_3 \rightleftarrows CH_3OCH_2OCH_2OCH_3 + H_2O$ — dehydration

B.2.3. Derived cross-interactions

a. $CH_3OH + H_2 \rightleftarrows CH_4 + H_2O$ — hydrogenolysis

b. $CO + H_2 \rightleftarrows CH_2O$ — hydrogenation

c. $\rightleftarrows H_2O + C$ — hydrogenolysis

d. $CO_2 + H_2 \rightleftarrows HCOOH$ — hydrogenation

e. $\rightleftarrows H_2O + CO$ — hydrogenolysis

f. $HCOOH + H_2 \rightleftarrows H_2CO + H_2O$ — hydrogenolysis

g. $H_2C(OH)OCH_3 + H_2 \rightleftarrows CH_3OCH_3 + H_2O$ — hydrogenolysis

h. $\rightleftarrows\ 2CH_3OH$ — hydrogenolysis

i. $\rightleftarrows\ H_2C(OH)_2 + CH_4$ — hydrogenolysis

j. $H\overset{O}{\overset{\|}{C}}OCH_2CH_2O\overset{O}{\overset{\|}{C}}H + H_2 \rightleftarrows H\overset{O}{\overset{\|}{C}}OCH_2CH_2CCH_2OH$ — hydrogenation

k. $HOCH_3 + H\overset{O}{\overset{\|}{C}}OH \rightleftarrows H\overset{O}{\overset{\|}{C}}OCH_3 + H_2O$ — dehydration

l. $H_2CO + H_2O \rightleftarrows H_2C(OH)_2$ — hydration

m. $H_2CO + HOCH_3 \rightleftarrows HOCH_2CH_2OCH_3$ — condensation

n. $HOCH_2OCH_3 + H_2O \rightleftarrows H_2C(OH)_2 + CH_3OH$ — hydrolysis

o. $CH_3O\overset{O}{\overset{\|}{C}}\overset{O}{\overset{\|}{C}}OCH_3 + H_2O \rightleftarrows CH_3O\overset{O}{\overset{\|}{C}}OH + H\overset{O}{\overset{\|}{C}}OCH_3$ — hydrolysis

p. $H_2CO + H_2\overset{OH}{\overset{|}{C}}OCH_3 \rightleftarrows HOCH_2OCH_2OCH_3$ — condensation

q. $H_2CO + HCOOH \rightleftarrows H\overset{O}{\overset{\|}{C}}OCH_2OH$ — condensation

r. $HOCH_2OCH_3 + HCOOH \rightleftarrows H\overset{O}{\overset{\|}{C}}OCH_2OCH_3 + H_2O$ — dehydration

s. $H\overset{O}{\overset{\|}{C}}OCH_3 + HCOOH \rightleftarrows HOCH_2OCH_3 + CO_2$ — decarboxylation

t. $H\overset{O}{\overset{\|}{C}}OCH_2CH_2OCH + H\overset{O}{\overset{\|}{C}}OOH \rightleftarrows HOCH_2OCH_2CH_2O\overset{O}{\overset{\|}{C}}H + CO_2$ — decarboxylation

B.3.1. Sequential primitive processes

a. $CH_3O\overset{O}{\overset{\|}{C}}OCH_3 \rightleftarrows CH_3OCH_3 + CO_2$ — decarboxylation

b. $CH_3OCOOH \rightleftarrows CH_3OH + CO$ — decarboxylation

c. $H\overset{O}{\overset{\|}{C}}OCH_2CH_3 \rightleftarrows HOCH_2CH_3 + CO$ — decarbonylation

d. $H\overset{O}{\overset{\|}{C}}OCH_2CH_2OH \rightleftarrows HOCH_2CH_2OH + CO$ — decarbonylation

e. $H\overset{\overset{O}{\|}}{C}OCH_2OH \rightleftarrows HOCH_2OH + CO$ decarbonylation

f. $H\overset{\overset{O}{\|}}{C}O\overset{\overset{O}{\|}}{C}H \rightleftarrows H_2CO + CO$ decarbonylation

g. $H\overset{\overset{O}{\|}}{C}O\overset{\overset{O}{\|}}{C}H \rightleftarrows HCOOH + CO$ decarbonylation

h. $H\overset{\overset{C}{\|}}{C}OCH_2CH_2OCH_2OH \rightleftarrows HOCH_2CH_2OCH_2OH + CO$ decarbonylation

i. $H\overset{\overset{O}{\|}}{C}OCH_2OCH_3 \rightleftarrows HOCH_2OCH_3 + CO$ decarbonylation

B.3.2. Sequential self-interactions

a. $2CH_2(OH)_2 \rightleftarrows HOCH_2OCH_2OH + H_2O$ dehydration

b. $2H\overset{\overset{O}{\|}}{C}OCH_2CH_2OH \rightleftarrows (H\overset{\overset{O}{\|}}{C}OCH_2CH_2)_2O + H_2O$ dehydration

c. $2H\overset{\overset{O}{\|}}{C}CH_2OH \rightleftarrows H_2\overset{\overset{O}{\|}}{C}CH_2OCH_2\overset{\overset{O}{\|}}{C}H + H_2O$ dehydration

d. $2H\overset{\overset{O}{\|}}{C}OCH_2CH \rightleftarrows H\overset{\overset{O}{\|}}{C}OCH_2OCH_2O\overset{\overset{O}{\|}}{C}H + H_2O$ dehydration

e. $2HOCH_2OCH_3 \rightleftarrows CH_3OCH_2OCH_2OCH_3 + H_2O$ dehydration

f. $2H\overset{\overset{O}{\|}}{C}OCH_2CH_2OCH_2OH \rightleftarrows (H\overset{\overset{O}{\|}}{C}OCH_2CH_2OCH_2)_2O + H_2O$ dehydration

g. $2HOCH_2CCH_2OCH_3 \rightleftarrows (CH_3OCH_3CCH_2)_2O + H_2O$ dehydration

B.3.3. Sequential cross-interactions

Only a small fraction of the most significant possible interactions will be listed here.

a. $CH_3\overset{O}{\overset{\|}{C}}OCH_3 + H_2O \rightleftarrows CH_3CCOH + HOCH_3$ hydrolysis

b. $CH_3O\overset{O}{\overset{\|}{C}}OCH_2 + H_2O \rightleftarrows CH_3O\overset{O}{\overset{\|}{C}}OH + HOCH_3$ hydrolysis

c. $CH_3OCH_3 + H_2 \rightleftarrows CH_4 + HOCH_3$ hydrogenolysis

d. $HOCH_2OCH_3 + H_2 \rightleftarrows CH_3OCH_3 + H_2O$ hydrogenolysis

e. $\rightleftarrows 2CH_3OH$ hydrogenolysis

f. $\rightleftarrows HOCH_2OH + CH_4$ hydrogenolysis

g. $CH_3OCH_2OCH_3 + H_2 \rightleftarrows CH_4 + HOCH_2OCH_3$ hydrogenolysis

$\rightleftarrows CH_3OH + CH_3OCH_3$ hydrogenolysis

The next stage in the description is to rearrange these reactions in terms of reaction type. Where this was available, thermodynamic data (see later) is also given.

Tabulation of reaction types

A. Hydrogenation

Reaction	ΔH^o_{500K}	$\log K_{P500K}$
1. $H\overset{O}{\overset{\|}{C}}OCH_3 + H_2 \rightleftarrows H_2\overset{OH}{\overset{\mid}{C}}OCH_3$		
2. $CO + H_2 \rightleftarrows CH_2O$	-2.20	-5.29
3. $CO + H_2 \rightleftarrows HCOOH$	2.75	-6.65
4. $H\overset{O}{\overset{\|}{C}}OCH_2CH_2O\overset{O}{\overset{\|}{C}}H + H_2 \rightleftarrows H\overset{O}{\overset{\|}{C}}OCH_2CH_2OCH_2OH$		

B. Hydrogenolysis

Reaction	ΔH^o_{500K}	$\log K_{P500K}$
1. $H\overset{O}{\overset{\|}{C}}OCH_3 + H_2 \rightleftarrows H_2CO + HOCH_3$	6.89	-2.14
2. $\rightleftarrows H\overset{O}{\overset{\|}{C}}OH + CH_4$	-25.55	10.90
3. $CH_3OH + H_2 \rightleftarrows CH_4 + H_2O$	-27.88	12.29

	Reactants		Products	ΔH^{o}_{500}	$\log K_{p500K}$
4.	$CO + H_2$	$\rightleftarrows$	$H_2O + C$	35.80	-18.37
5.	$CO_2 + H_2$	$\rightleftarrows$	$H_2O + CO$	9.51	- 2.11
6.	$HCOOH + H_2$	$\rightleftarrows$	$H_2CO + H_2O$	4.56	- 0.76
7.	$H_2\overset{OH}{\overset{\mid}{C}}OCH_3 + H_2$	$\rightleftarrows$	$CH_3OCH_3 + H_2O$		
8.		$\rightleftarrows$	$2HOCH_3$	-14.31	
9.		$\rightleftarrows$	$H_2C(OH)_2 + CH_4$		
10.	$CH_3OCH_3 + H_2$	$\rightleftarrows$	$CH_4 + HOCH_2$	-22.76	10.99
11.	$HOCH_2OCH_3 + H_2$	$\rightleftarrows$	$CH_3OCH_3 + H_2O$		
12.		$\rightleftarrows$	$HOCH_2OH + CH_4$		
13.	$CH_3OCH_2OCH_3 + H_2$	$\rightleftarrows$	$HOCH_2OCH_3 + CH_4$		
14.		$\rightleftarrows$	$CH_3OH + CH_3OCH_3$		

C. Dehydrogenation

	Reactants		Products	ΔH^{o}_{500}	$\log K_{p500K}$
1.	$2H\overset{O}{\overset{\|}{C}}OCH_3$	$\rightleftarrows$	$CH_3O\overset{O}{\overset{\|}{C}}\overset{O}{\overset{\|}{C}}OCH_3 + H_2$		
2.		$\rightleftarrows$	$H\overset{O}{\overset{\|}{C}}OCH_2CH_2O\overset{O}{\overset{\|}{C}}H + H_2$		
3.		$\rightleftarrows$	$CH_3O\overset{O}{\overset{\|}{C}}CH_2O\overset{O}{\overset{\|}{C}}H + H_2$		
4.		$\rightleftarrows$	$CH_2O + H_2$	21.2	- 3.07

D. Hydration

1. $H_2CO + H_2O \rightleftarrows H_2C(OH)_2$

E. Hydrolysis

1. $HOCH_2OCH_3$ + H O $\rightleftarrows H_2C(OH)_2 + CH_3OH$
2. $CH_3O\overset{O}{\overset{\|}{C}}\overset{O}{\overset{\|}{C}}OCH_3 + H_2O \rightleftarrows CH_3O\overset{O}{\overset{\|}{C}}OH + HCOCH_3$
3. $CH_3\overset{O}{\overset{\|}{C}}OCH_3 + H_2O \rightleftarrows CH_3COOH + HOCH_3$
4. $CH_3O\overset{O}{\overset{\|}{C}}OCH_3 + H_2O \rightleftarrows CH_3O\overset{O}{\overset{\|}{C}}OH + HOCH_3$

F. Dehydration

	Reactants		Products	ΔH^o_{500}	$\log K_{p500K}$
1.	$2HOCH_3$	⇄	$CH_3OCH_3 + H_2O$	- 5.12	1.30
2.	$2HCOOH$	⇄	$OCHOCHO + H_2O$		
3.	$2H_2\overset{\overset{OH}{\mid}}{C}OCH_3$	⇄	$CH_3OCH_2OCH_2OCH_3 + H_2O$		
4.	$HOCH_3 + H\overset{\overset{O}{\|}}{C}OH$	⇄	$H\overset{\overset{O}{\|}}{C}OCH_3 + H_2O$	- 2.33	1.30
5.	$HOCH_2OCH_3 + HCOOH$	⇄	$H\overset{\overset{O}{\|}}{C}OCH_2OCH_3 + H_2O$		
6.	$2CH_2(OH)_2$	⇄	$HOCH_2OCH_2OH + H_2O$		
7.	$2H\overset{\overset{O}{\|}}{C}OCH_2CH_2OH$	⇄	$(H\overset{\overset{O}{\|}}{C}CCH_2CH_2)_2O + H_2O$		
8.	$2H\overset{\overset{O}{\|}}{C}CH_2OH$	⇄	$H_2\overset{\overset{O}{\|}}{C}CH_2OCH_2CH + H_2O$		
9.	$2H\overset{\overset{O}{\|}}{C}OCH_2OH$	⇄	$H\overset{\overset{O}{\|}}{C}OCH_2OCH_2OCH + H_2O$		
10.	$2HOCH_2OCH_3$	⇄	$CH_3OCH_2OCH_2OCH_3 + H_2O$		
11.	$2H\overset{\overset{O}{\|}}{C}OCH_2CH_2CCH\ OH$	⇄	$(H\overset{\overset{O}{\|}}{C}OCH_2CH_2OCH_2)\ O + H_2O$		
12.	$2HOCH_2OCH_2OCH_3$	⇄	$(CH_3OCH_2OCH_2)_2O + H_2O$		

G. Decarbonylation

	Reactants		Products	ΔH^o_{500}	$\log K_{p500K}$
1.	$H\overset{\overset{O}{\|}}{C}OCH_3$	⇄	$HOCH_3 + CO$	9.09	3.15
2.	$CH_3O\overset{\overset{O}{\|}}{C}\overset{\overset{O}{\|}}{C}OCH_3$	⇄	$CH_3C\overset{\overset{O}{\|}}{C}OCH_3 + CO$		
3.	$HCO\overset{\overset{O}{\|}}{C}H_2CH_2OCH$	⇄	$H\overset{\overset{O}{\|}}{C}OCH_2CH_2OH + CO$		
4.	$H\overset{\overset{O}{\|}}{C}OH$	⇄	$CO + H_2O$	- 2.75	6.65

Reaction			ΔH^o_{500}	$\log K_{p500K}$
5. $HCOCH_2CH_3$	$\rightleftarrows$	$HOCH_2CH_3 + CO$		
6. $HCOCH_2CH_2OH$	$\rightleftarrows$	$HOCH_2CH_2OH + CO$		
7. $HCOCH_2OH$	$\rightleftarrows$	$HOCH_2OH + CO$		
8. $HCOCH$	$\rightleftarrows$	$H_2CO + CO$		
9. $HCOCH$	$\rightleftarrows$	$HCOOH + CO$		
10. $HCOCH_2OCH_3$	$\rightleftarrows$	$HOCH_2OCH_3 + CO$		

H. Decarboxylation

Reaction			ΔH^o_{500}	$\log K_{p500K}$
1. $HCOCH_3$	$\rightleftarrows$	$CH_4 + CO_2$	-28.30	17.39
2. $CH_3OCCOCH_3$	$\rightleftarrows$	$CH_3COCH_3 + CO_2$		
3. $HCOCH_2CH_2OCH$	$\rightleftarrows$	$HCOCH_2CH_3 + CO_2$		
4. $HCOCH_3 + HCOOH$	$\rightleftarrows$	$HOCH_2OCH_3 + CO_2$		
5. CH_3OCOOH	$\rightleftarrows$	$CH_3OH + CO$		

I. Condensation

Reaction			ΔH^o_{500}	$\log K_{p500K}$
1. $2CH_2O$	$\rightleftarrows$	$HCOCH_3$	-56.59	4.21
2.	$\rightleftarrows$	$HOCH_2CHO$		
3. $H_2CO + HCOOH$	$\rightleftarrows$	$HCOCH_2OH$		
4. $H_2CO + HOCH_3$	$\rightleftarrows$	$HOCH_2CH_2OCH_3$		

			ΔH^o_{500}	$\log K_{p500K}$
J. Disproportionation				
1. 2CO	⇄	$C + CO_2$	-41.49	8.75
2. $H_2C(OH)OCH_3$	⇄	$H_2CO + CH_3OH$		
3.	⇄	$HCOOH + CH_4$		

From these tabulations, it is possible to identify desired and undesired reactions and to produce an overall reaction network.

DESIRED REACTIONS

I.
a) $HC(=O)OCH_3 + H_2 \rightleftarrows H_2C(OH)OCH_3$ — hydrogenation
b) $H_2COCH_3 + H_2 \quad 2CH_3OH$ — hydrogenolysis

II.
a) $HC(=O)OCH_3 + H_2 \rightleftarrows H_2CO + CH_3O$ — hydrogenolysis
b) $H_2CO + H_2 \rightleftarrows CH_3OH$ — hydrogenation

UNDESIRED REACTIONS

I. Reactions giving primary products

a) $HC(=O)OCH_3 + H_2O \rightleftarrows HC(=O)OH + CH_4$ — hydrogenolysis
b) $HC(=O)OCH_3 + H_2O \rightleftarrows HCOOH + CH_3OH$ — hydrolysis

Comment: Formic acid can be reduced to a diol, which could be dehydrated to give formaldehyde. This, in turn, could be hydrogenated to the wanted product, methanol. Unfortunately, this reaction scheme is disadvantageous thermodynamically.

c) $2HC(=O)OCH_3 \rightleftarrows HC(=O)OCH_2C(=O)OCH_3 + H_2$
d) $\rightleftarrows HC(=O)OCH_2CH_2OC(=O)H + H_2$
e) $\rightleftarrows CH_3OC(=O)C(=O)OCH_3 + H_2$

(c–e) dehydrogenation

Comment: All these dehydrogenation reactions are favoured at low pressure.

f. $HC(=O)OCH_3 \rightleftarrows CH_3OH + CO$ decarbonylation

g. $HC(=O)OCH_3 \rightleftarrows CO_2 + CH_4$ decarboxylation

Comment: Decarboxylation is catalysed by acids and bases, and is thermodynamically favourable.

h. $HC(=O)OCH_3 \rightleftarrows 2C + 2H_2O$ disproportionation

II. Unwanted reactions of desired intermediate products

a. $H_2C(OH)OCH_3 + H_2 \rightleftarrows CH_3OCH_3 + H_2O$ hydrogenolysis

Comment: Hydrolyses of dimethylether will give methanol, but this is an endothermic reaction. A competitive route involves hydrogenation of the ether to methanol and methane; methane formation is unwanted.

b. $H_2C(OH)\text{-}OCH_3 + H_2 \rightleftarrows CH_2(OH)_2 + CH_4$ hydrogenolysis

c. $H_2C(OH)OCH_3 \rightleftarrows CH_3OCH_2CH_3 + H_2O$ dehydration

d. $H_2CO \rightleftarrows CO + H_2$ dehydrogenation

III. Unwanted reaction of the desired product

a. $CH_3OH + H_2 \rightleftarrows CH_4 + H_2O$ hydrogenolysis

The overall reaction network can then be summarised as below.

The designers also chose to prepare a fairly complete description of the idea in thermodynamic terms. The target transformation

$$HCOOCH_3 + 2H_2 = 2CH_3OH$$

can be carried out with methyl formate and methanol in the gas or liquid states. Methyl formate boils at around 32°C, and reactor cooling would be needed for a liquid phase reaction unless a very active catalyst is available. As a result, the designers chose to consider a gas-solid reaction, remembering that methanol can condense at high pressures if the temperature is less than the critical

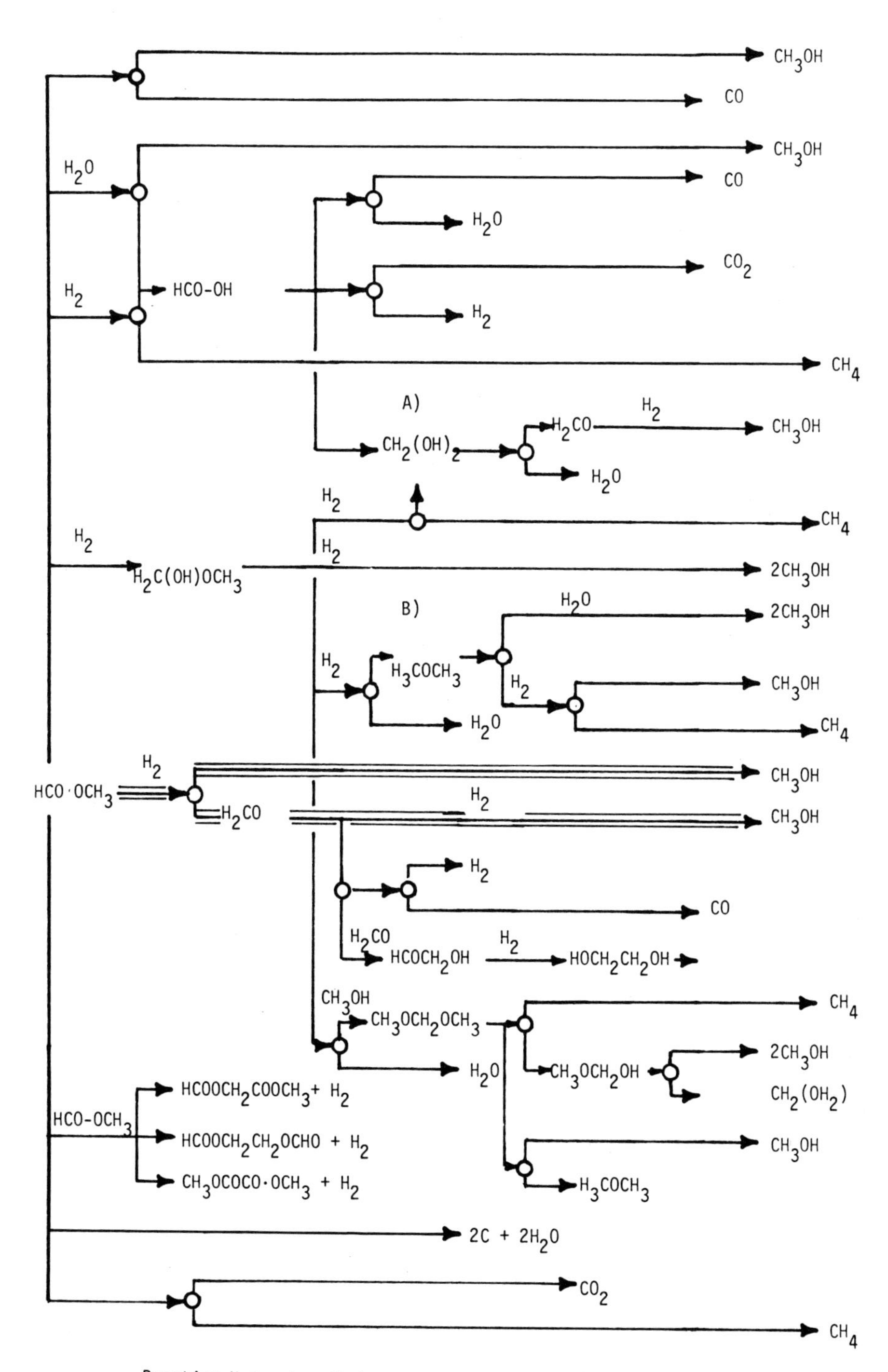

Reaction Network: Methanol from methyl formate

temperature. Examination of the literature showed that, at 500 K (227°C), methanol would not condense at less than 50 atm pressure, and this was selected as a desired operating temperature.

Values of the equilibrium constants listed above were then used to construct plots of conversion of methyl formate to methanol at various ratios of hydrogen: ester (Fig. 12.1) and variations of equilibrium constants for different reactions were plotted as a function of temperature (Fig. 12.2). From Figure 12.1 it is seen that the shade area is probably the most efficient area of operation, with a hydrogen:methyl formate ratio of 10 and a pressure of 10 atm offering easily accessible operating conditions. Increasing temperature has a relatively small effect on the yield of methanol, and many undesired reactions are favoured still more at higher temperatures (Fig. 12.2).

III. DESIGN OF PRIMARY COMPONENTS

The first guide lines used by the designers were <u>activity patterns</u> for related reactions. Based on the type of plots shown in Chapter 7, the following general indications were identified:

Hydrogenolysis:	Rh ≥ Ni ≥ Co ≥ Fe > Pd > Pt
Hydrogenation:	Ni > Co > Fe > Re ≥ Cu
Dehydrogenation:	Rh > Pt > Pd > Ni > Co ≥ Fe
Hydration:	Pt > Rh > Pd >> Ni ≥ W ≥ Fe

In addition, the designers were aware of the activity patterns for metal oxides reported in Chapter 7. Consideration of these patterns plus the reaction network led the designers to suggest

<u>Desired Phases</u>: Pt, Cu, Fe, W, Cr, Re, MnO, ZnO, SiO_2, Al_2O_3
<u>Undesired Phases</u>: Ni, Co, Rh, CoO, Cu_2O, MgO, Fe_3O_4, Cr_2O_3, TiO_2.

It is interesting to note that Cr_2O_3, which may well be present in "copper chromite", was excluded on the basis of the activity patterns.

The designers filed these suggestions, and moved on to consider the molecular mechanisms on the surface. They began by postulating possible <u>chemisorbed complexes</u>.

Methyl formate can be suggested to adsorb in four different ways. Representing an active site as an asterisk, these may be written:

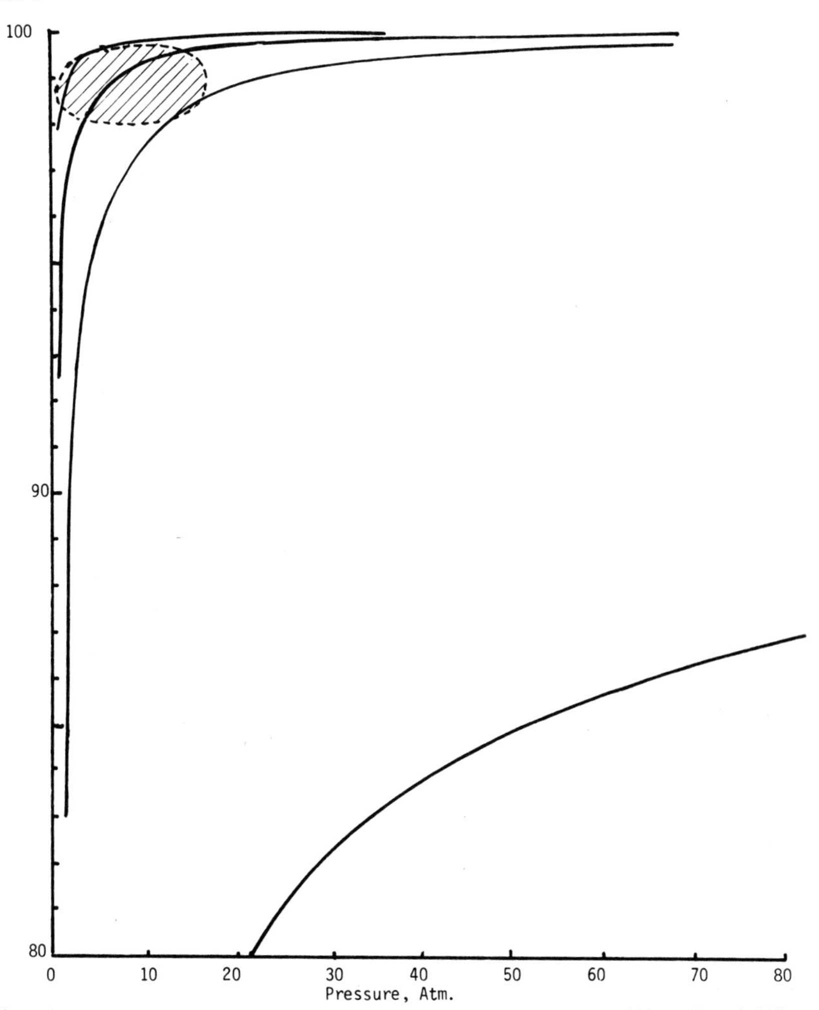

Figure 12.1 Conversion of methyl formate vs total pressure. Temperature: 500 K. K_p = 8.705. Mole ratio H_2:methyl formate 2:1, 5:1, 10:1, 30:1.

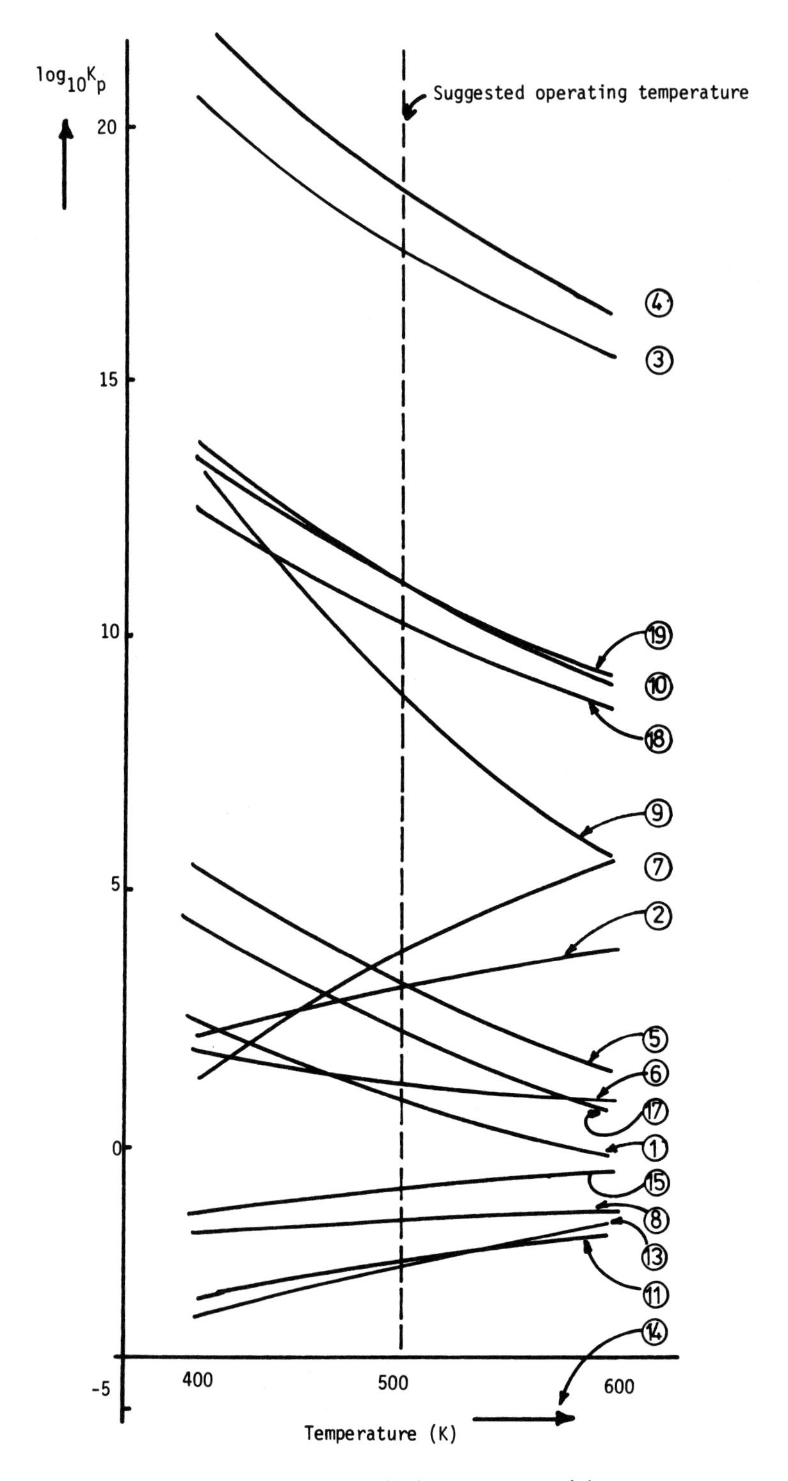

Fig. 12.2. Variations of equilibrium constants with temperature.

A. Dissociative Adsorption

a) H–C(=O)–* + CH_3–O–* b) H–C(=O)–* CH_3–*

Desired: can take place on metal, metal oxides and metal sulphides

Undesired: probably favoured by transition metals

B. Pi-bonded associative adsorption

O═C(OCH_3)(H), π-bonded via O═C to *

C. Sigma-bonded via the carbonyl oxygen

H, OCH_3
C
*

D. Sigma-bonded via the carbonyl carbon and oxygen

O - C(OCH_3)(H)
* *

Hydrogen may absorb in three different ways (refs. 1,2).

A. Dissociative

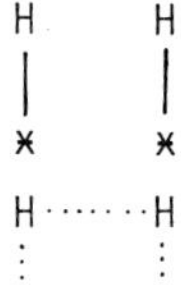

B. Associative

H·······H
* *

C. Associative "ionic"

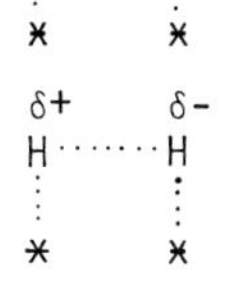

The possible adsorption of methanol was not considered, and this could be considered to be an error (see later).

Possible reaction mechanisms on the surface were then considered in terms of these postulated adsorbed species:

Mechanism A

Involving a reaction between dissociatively adsorbed ester and various forms of hydrogen, this may be written

a) With atomic hydrogen

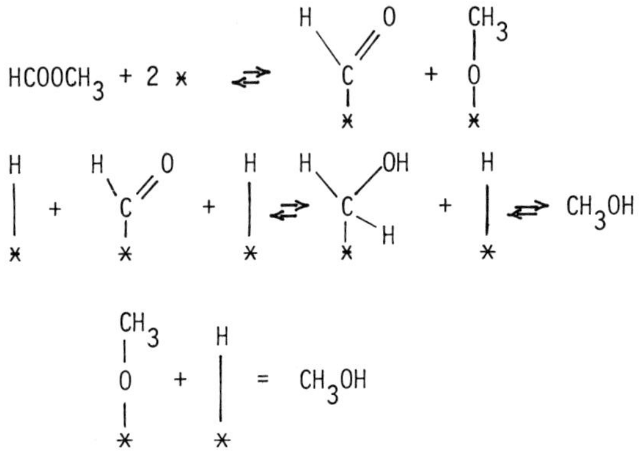

$$HCOOCH_3 + 2* \rightleftarrows H(O{=})C{-}* + CH_3{-}O{-}*$$

$$H{-}* + H(O{=})C{-}* + H{-}* \rightleftarrows H(OH)(H)C{-}* + H{-}* \rightleftarrows CH_3OH$$

$$CH_3{-}O{-}* + H{-}* = CH_3OH$$

b) With weakly adsorbed hydrogen

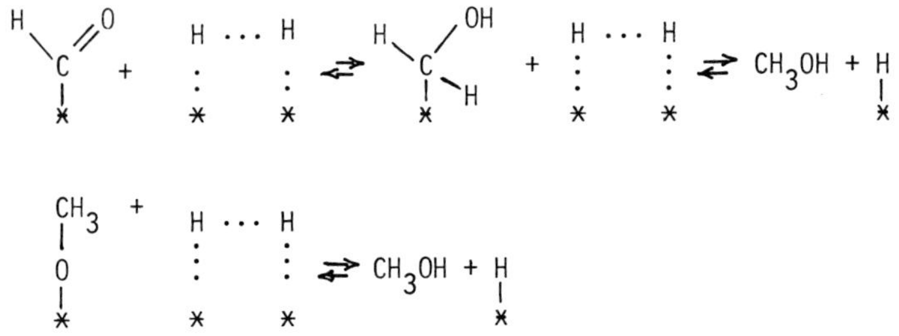

$$H(O{=})C{-}* + \underset{\vdots\ \ \ \ \ \vdots}{H\cdots H} \rightleftarrows H(OH)(H)C{-}* + \underset{\vdots\ \ \ \ \ \vdots}{H\cdots H} \rightleftarrows CH_3OH + H{-}*$$

$$CH_3{-}O{-}* + \underset{\vdots\ \ \ \ \ \vdots}{H\cdots H} \rightleftarrows CH_3OH + H{-}*$$

c) With "ionic" hydrogen

$$H(\overset{\delta-}{O}{=})\overset{\delta+}{C}{-}* + \overset{\delta+}{H}\cdots\overset{\delta-}{H}\ (\text{each}\ \vdots\ *) \rightleftarrows H(OH)(H)C{-}*$$

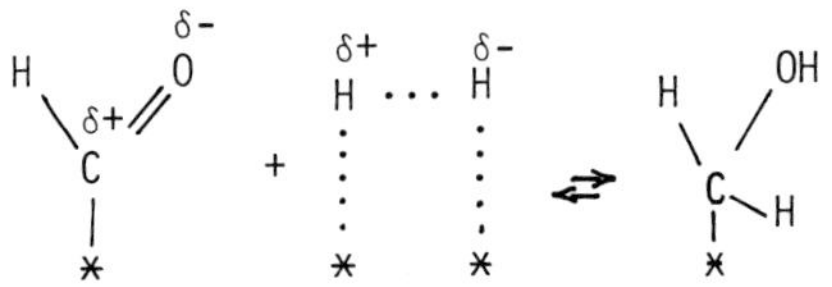

etc.

The major difficulty envisaged by the designers was to ensure dissociative adsorption producing the CH_3O fragment and not the CH_3 fragment. Writing the methyl formate molecule as

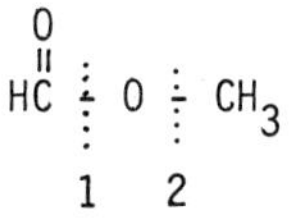

$$\overset{O}{\overset{\|}{HC}} \vdots_1\ O\ \vdots_2\ CH_3$$

they calculated that the bond energy at position 1 was 97 kcal $mole^{-1}$ while the bond energy at position 2 was 81 kcal $mole^{-1}$. As a result, dissociative adsorption to produce CH_3 could well be very significant.

Mechanism B

Involving the reaction between pi-bonded methyl formate and hydrogen the reaction was written as, for atomic hydrogen:

a) 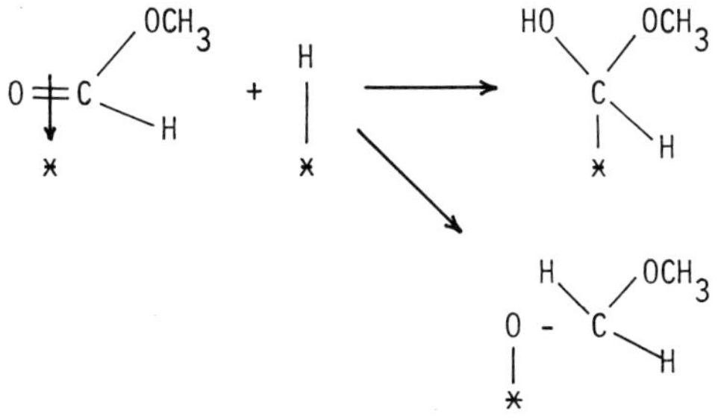 $O \mp C(OCH_3)(H)$ (π-bonded to *) + H–* → HO–C(OCH_3)(H)–* (i)

→ *–O–C(H)(OCH_3)(H) (ii)

–O–C(H)(H)–OCH_3 → 2 H– → H–O–* + CH_3OCH_3

–O–C(H)(H)–OCH_3 + H– → 2 *–O–CH_3 + 2 H–* → $2CH_3OH$

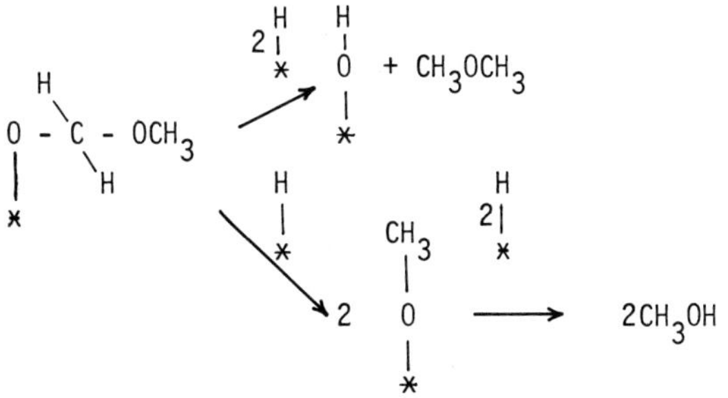

For weakly adsorbed hydrogen, the reaction was suggested to be

b) 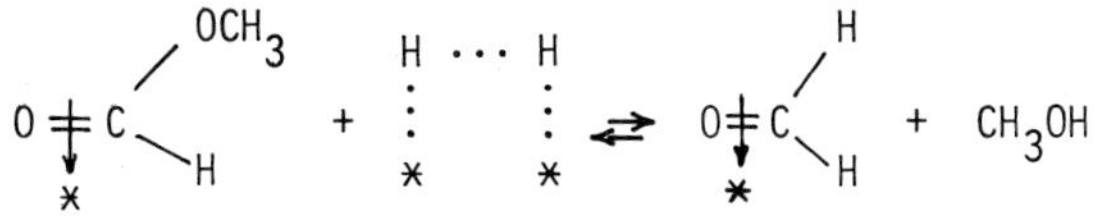 $O \mp C(OCH_3)(H)$ (π-bonded to *) + H···H (weakly adsorbed on * *) ⇄ $O \mp C(H)(H)$ (π-bonded to *) + CH_3OH

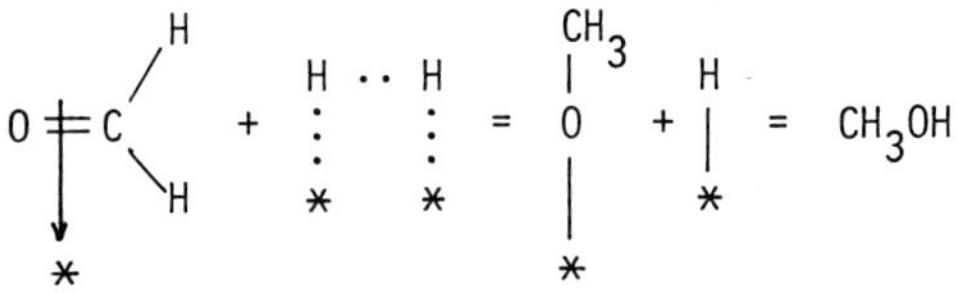

$$O{=}CH_2(\downarrow *) + H \cdots H(\vdots *\ \vdots *) = CH_3\text{-}O\text{-}* + H\text{-}* = CH_3OH$$

Considering "ionic" hydrogen, the reaction was written as

c) $$\overset{\delta-}{O}{=}\overset{\delta+}{C}(OCH_3)H(\downarrow *) + \overset{\delta+}{H} \cdots \overset{\delta-}{H}(\vdots *\ \vdots *) \rightleftarrows HO\text{-}C(OCH_3)(H)\text{-}* + H\text{-}*$$

$$\rightleftarrows H(OCH_3)C(H)\text{-}O\text{-}* + H\text{-}*$$

$$H(OCH_3)C(H)\text{-}O\text{-}* + H\text{-}* \rightleftarrows 2\,CH_3\text{-}O\text{-}* + 2\,H\text{-}* \rightleftarrows 2CH_3OH$$

$$+ \overset{\delta+}{H} \cdots \overset{\delta-}{H}(\vdots *\ \vdots *) \rightleftarrows CH_3OH + CH_3\text{-}O\text{-}* + H\text{-}* \rightleftarrows CH_3OH$$

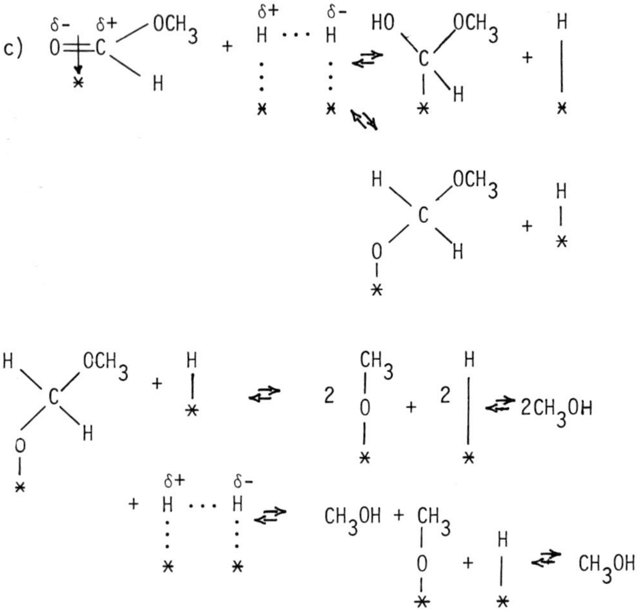

Mechanism C

Involving the sigma-bonded associative adsorption of methyl formate through oxygen, the mechanism was described as

a) with atomic hydrogen

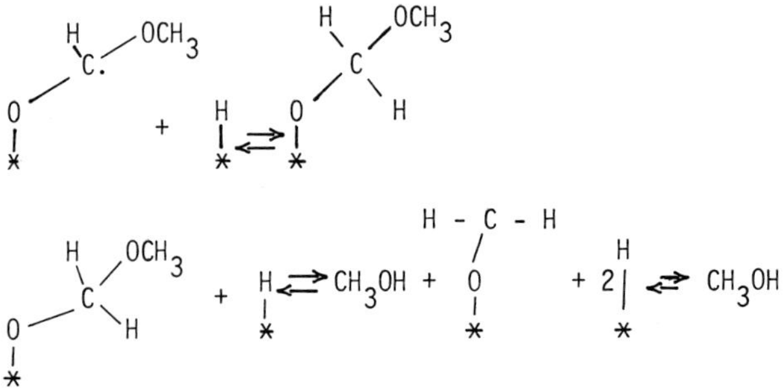

$$H(OCH_3)\dot{C}\text{-}O\text{-}* + H\text{-}* \rightleftarrows H(OCH_3)C(H)\text{-}O\text{-}*$$

$$H(OCH_3)C(H)\text{-}O\text{-}* + H\text{-}* \rightleftarrows CH_3OH + H\text{-}C(H)\text{-}O\text{-}* + 2\,H\text{-}* \rightleftarrows CH_3OH$$

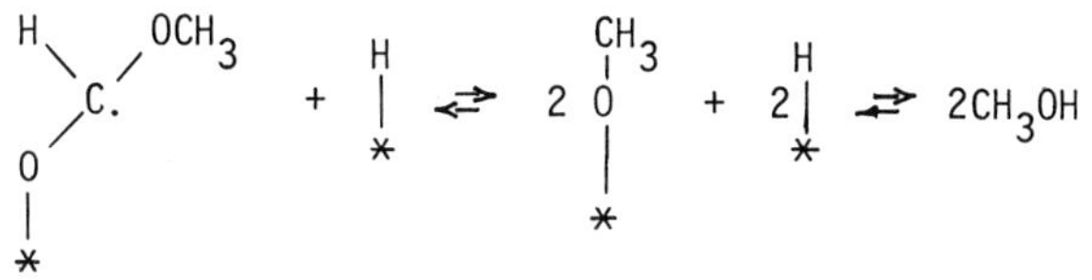

$$H(OCH_3)\dot{C}{-}O{-}* \; + \; H{-}* \; \rightleftarrows \; 2\,CH_3{-}O{-}* \; + \; 2\,H{-}* \; \rightleftarrows \; 2CH_3OH$$

b) with weakly adsorbed hydrogen

$$H(OCH_3)\dot{C}{-}O{-}* \; + \; \underset{\vdots}{H} \cdots \underset{\vdots}{H} \; (* \;\; *) \; \rightleftarrows \; CH_3OH \; + \; H_2\ddot{C}{-}O{-}*$$

$$H_2\ddot{C}{-}O{-}* \; + \; \underset{\vdots}{H} \cdots \underset{\vdots}{H} \; (* \;\; *) \; \rightleftarrows \; CH_3OH$$

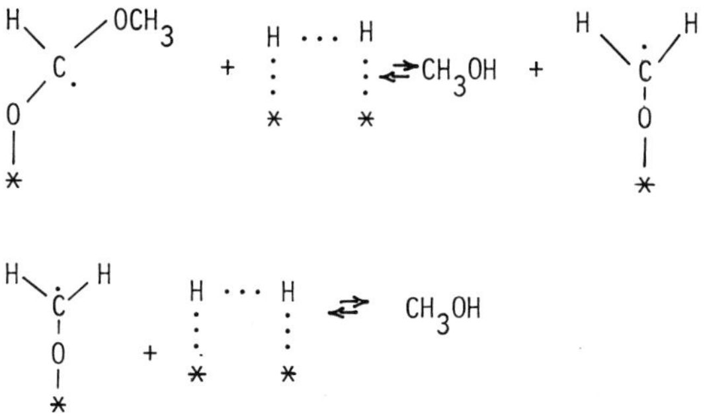

Reaction with "ionic" hydrogen was difficult to evaluate, in that the electron distribution within the sigma-adsorbed methyl formate would depend on the centre at which it was adsorbed. The nature of the metal and the presence or absence of a metal oxide or sulphide would influence the electron distribution and, as a result, the reaction products.

Having written down the possible mechanisms, the designers decided that mechanism A should be avoided if possible, since dissociative adsorption leading to adsorbed methyl fragments seemed to be a probable, but undesired, reaction path.

In considering mechanism B, they recognised that pi-bonded associative adsorption would be favoured on elements and compounds of the transition metals. During the reaction, the surface bonding of methyl formate derived species changed from pi to sigma. They also recognised that mechanisms B(a) and B(c) could lead to the production of CH_3OCH_3 (which was thermodynamically favourable), and they chose to regard this as undesirable. It is, in fact, arguable that hydration of the dimethyl compound could lead to methanol, and that the inclusion of a component in the catalyst that would favour hydration would make these routes more desirable. However, such a component could also favour the hydrolysis of methyl formate (see overall reaction network) leading to decreased selectivity for methanol.

Similar arguments hold with respect to mechanism C(a). As a result, the designers argued that mechanisms B(b) or C(b) were the most desirable routes: both of these involve the reaction of weakly adsorbed hydrogen. For reaction with pi-bonded methyl formate, a primary component having 0,1,2,3,8,9 and 10 d electrons is favoured. On these grounds, possible components included

Metals Sc, Ti, V, Ni, Cu, Zn, Y, Zr, Rh, Pd, Ag, Cd, Pt, Au

Metal ions (oxides or sulphides)
Sc^{3+}, Y^{3+}, Ti^{4+}, Zr^{4+}, Hf^{4+}, Th^{4+}, V^{3-5+}, Nb^{3-5+}, Cr^{3+}, Mo^{3-6+}, W^{6+}, U^{4-6+}, Mn^{3-4+}, Re^{4+}, Ni^{2+}, Pd^{2+}, Pt^{2+}, Cu^{2+}, Ag^{+}, Au^{3+}.

Of the metals, it would be impracticable to consider the use of Sc, Ti, V, Y, Zr, Cd and Au.

Sigma-bonded adsorption of methyl formate could involve species with 4,5,6,7 or 8 d electrons and 1 or 2 s electrons.

For metals, these include

Cr, Mn, Fe, Co, Ni, Cu, Zn, Mo, Ru, Rh, Ag, Cd, Re, Os, Ir, Pt and Au.

For metal ions, a list includes
Mn^{2+}, Re^{2+}, Fe^{2+}, Fe^{3+}, Ru^{2-4+}, Os^{4+}, Co^{2-3+}, Ru^{1-3+}, Ir^{3-4+}, Ni^{2+}, Pd^{2+}, Pt^{2-4+}.

Again, for metals, the use of Cr, Mn, Mo, Cd and Au is impracticable.

Bearing in mind that both desired reaction sequences are designed to avoid atomically adsorbed hydrogen, Ni, Rh, Pd, Pt, Fe, Co, Ru, Rh, Re and Ir should be avoided in the metallic form, although oxides or sulphides may be more suitable (e.g. metal sulphide). This leaves Cu or Ag (on which hydrogen adsorption is very weak if it exists at all) as suitable metals for consideration.

In considering metal oxides, the designers recognised that some oxides would be reduced under the conditions of the reaction, and that strong hydrogenolysis should be avoided. Most of the transition metal oxides would be reduced to the corresponding metal, as would copper, silver and gold oxides.

As a result, on the basis of chemisorbed complexes desired in the proposed reaction schemes, the following primary components were suggested:

Mechanism B(b): reaction between pi-adsorbed methyl formate and weakly adsorbed hydrogen.

Metals: Cu, Zn, Ag.

Metal oxides/sulphides : U^{3+}, U^{5+}, Cr^{3+}, Mo^{6+}, W^{6+}, U^{6+}, Mn^{4+}.

Mechanism C(b): reaction of sigma-bonded methyl formate with weakly bonded hydrogen.

Metals: Cu, Zn, Ag.

Metal oxides/sulphides : Mn^{2+}, Co^{2+}, Co^{3+} (may be reducible).

Filing this information, the designers then chose to consider the geometric approach. Although they worked as a team, the particular memeber assigned to this problem chose to apply the reasoning to mechanisms A, B and C.

Mechanism A

Desired and undesired dissociative adsorption of methyl formate may be represented as

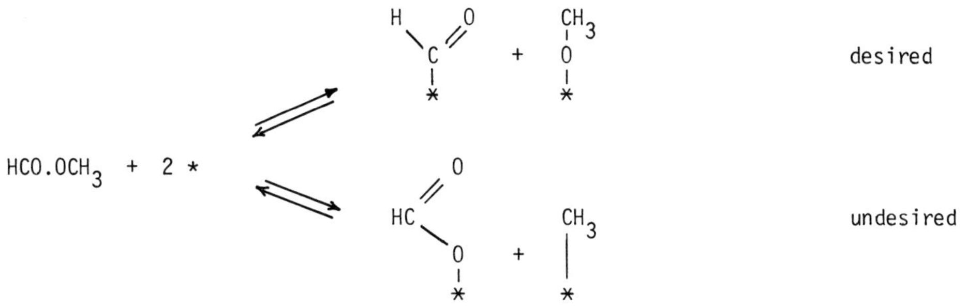

$$HCO.OCH_3 + 2* \rightleftharpoons \underset{*}{\overset{|}{HC(=O)}} + \underset{*}{\overset{|}{OCH_3}} \quad \text{desired}$$

$$HCO.OCH_3 + 2* \rightleftharpoons \underset{*}{\overset{|}{HC(=O)O}} + \underset{*}{\overset{|}{CH_3}} \quad \text{undesired}$$

Bond distances in the methyl formate were found to be

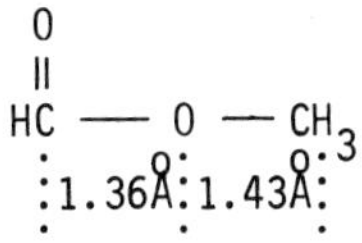

$$\overset{O}{\overset{\|}{HC}} — O — CH_3$$

$$:1.36Å:1.43Å:$$

Assuming that the bond should be stressed on adsorption by 20% (an arbitrary figure), the desired distances were then calculated

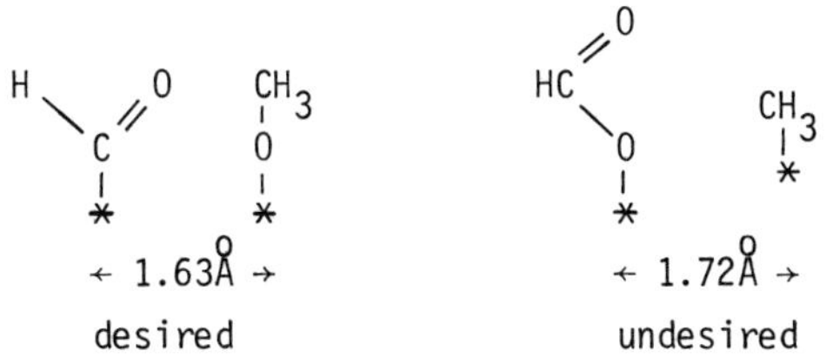

H–C(=O)	CH₃–O	HC(=O)–O	CH₃
*	*	*	*
← 1.63Å →		← 1.72Å →	
desired		undesired	

Inspection of lattice distances (Chapter 7) then suggested the following compounds as being possible absorbers, <u>on geometric grounds only</u>:

αB_2O_3 (1.45 Å), BeO (1.65Å), C diamond (1.54Å), C graphite (1.42 Å), αSiO_2 (1.61Å), βSiO_2 (1.54Å), V_2O_8 (1.54-2.81Å).

With the exception of vanadium oxide and, possibly, of carbon, none of these compounds would appear to be of interest as catalysts.

Mechanism B

Accepting that this mechanism involves pi adsorption converting to sigma adsorption

$$\underset{\underset{*}{|}}{O}-C\begin{smallmatrix}CH_3\\ H\end{smallmatrix} + \underset{\underset{*}{|}}{H} \rightleftarrows \underset{\underset{*}{|}}{O}-CH_2-OCH_3$$

followed by further reaction

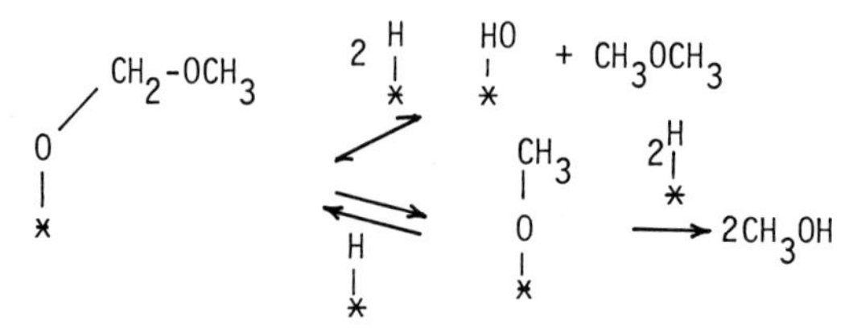

The designers assumed that the second reaction was more important and was open to possible control on geometric grounds. Based on bond distances in the desired and undesired products they calculated

$$\underset{\underset{*}{|}}{H}\ \ \underset{\underset{*}{|}}{O}(-CH_2OCH_3)\ \ \underset{\underset{*}{|}}{H} \longrightarrow CH_3OCH_3$$

0.90Å 2.03Å

$$\underset{\underset{*}{|}}{O}(-CH_2-OCH_3)\ \ \underset{\underset{*}{|}}{H} \longrightarrow 2\ \underset{\underset{*}{|}}{\overset{CH_3}{\overset{|}{O}}}$$

2.03Å

From tables of lattice distances (Chapter 7), most of the transition metals were found to be suitable while, for metal oxides, the list could be narrowed to

Cr_2O_3 (2.01Å), Mn_2O_3 (2.02Å), MoO_3 (2.00Å), NbO (2.10Å), NbO_2 (2.02 Å), Nb_2O_5 (2.03Å), TiO (2.12Å), TiO_2 (1.95Å), Ti_2O_3 (2.03Å), UO_3 (2.08Å), VO (2.05Å), V_2O_3 (2.05Å), VO_2 (1.76-2.05Å) and WO_2 (2.00Å).

<u>Mechanism C</u>

Writing the initial adsorption as

$$H-\overset{\overset{O-CH_3}{|}}{C}=O + * \rightleftarrows H-\overset{\overset{OCH_3}{|}}{\underset{\underset{\underset{*}{|}}{O}}{\underset{|}{C}}}\cdot$$

they recognised that formation of a radical-like species would lead to rapid acceptance of hydrogen. As a result, they suggested that an ionic intermediate might be preferred:

$$\underset{\underset{O}{\|}}{H\diagdown C \diagup OCH_3} + M^{n+} \rightleftarrows \begin{matrix} + \\ H - C - OCH_3 \\ | \\ O \\ | \\ M^{(n-1)+} \end{matrix} \rightleftarrows \begin{matrix} \cdot \\ H - C - OCH_3 \\ | \\ O \\ | \\ M^{n+} \end{matrix}$$

As with mechanism B, the desired lattice distance is 2.03Å, and they selected oxides on the grounds of lattice distance and redox potential

Oxide	Lattice distance (Å)	Redox potential (eV)
Cr_2O_3	2.01	- 0.41
NiO	2.09	- 0.23
Ti_2O_3	2.03	- 2.0
V_2O_3	2.05	- 0.255

NiO would be reduced under reaction conditions, and they argued that the redox potential of Ti_2O_3 was so large that irreversible adsorption could occur. Surprisingly, they did not consider other oxides which had been already identified (Mechanism B) as being of desired geometry.

Examination of the melting points of the oxides proposed led to rejection of NbO, NbO_2 and VO as being unstable and MoO_3, Nb_2O_5, Ti_2O_3 and VO_2 as being reducible.

IV. SUMMARY

The designers summarised their conclusions on the design of the primary components in the form of the table given below.

They recognised that transition metals (and particularly Ni) may be too strong hydrogenation catalysts and that Cr_2O_3 has been found to be an effective catalyst for the decomposition of formic acid.

They also, at this late stage in the design, remembered that weak adsorption of hydrogen was desirable. As a result, and without devoting significant attention to this aspect, they recommended Cu, Au, Ag, Zn and Cd as possible catalysts. Recognising that adsorption of hydrogen on copper was poor below 300°C, they suggested that an alloy of Cu with another metal could be advantageous.

TABLE 12.1

Summary: Design of primary components for methyl formate to methanol

Based on	Activity Patterns		Surface Chemistry		Geometric Factors		Proposal	
	Metals	Metal Oxides	Metals	Metal Oxides	Metals	Metal Oxides	Metals	Metal Oxides
	Ni	MnO	Ni	V_2O_3		V_2O_3	Ni	V_2O_3
	Pt	CoO		V_2O_5				
	Cu	Cu_2O	Cu	Cr_2O_3		Cr_2O_3	Cu	Cr_2O_3
	Co	ZnO	Co	MoO_3			Co	
	Fe	SiO_2	Fe	WO_3		SiO_2	Fe	SiO_2
	W	Al_2O_3	Pd	UO_3				
	Rh		Ag					
	Cr							
	Re							

V. DESIGN OF THE SUPPORT

The designers chose to consider supports that are readily available, and their primary choice was based on the physical properties of the materials. Recognising that over-hydrogenation to methane was a possibility, pores of diameter 100-500 Å were considered essential. They then listed supports that could meet this requirement:

γ-Al_2O_3 : Mean pore radius ca. 200 Å
Surface area ~ 200-300 m^2g^{-1}
Chemically inert, although reaction with oxides such as chromia could occur.

α-Al_2O_3 : Mean pore radius ca. 500 Å
Surface area ~ 10 m^2g^{-1}
α-Al_2O_3 is a dehydration catalyst and could favour ether formation.

SiO_2 : Mean pore radius 10-20 Å
Surface area ~ 300-1000 m^2g^{-1}
Highly porous, the support is not mechanically strong.

MgO : Dissolves in water

MnO : Rejected as it is a dehydration catalyst.

ZnO : Rejected as a dehydration catalyst.

Cr_2O_3 : Recognition as a hydrogenation catalyst.

As a result, either αAl_2O_3 or Cr_2O_3 was suggested as a support, with SiO_2 being an alternative if the mechanical stress could be kept low.

In the knowledge that Cu had been proposed as a possible catalyst, the designers recognised that a spacer material could be necessary to stabilise the catalyst. Al_2O_3, Cr_2O_3, ZnO and MgO were considered as possible spacers.

VI. COMMENTS

In many ways this was a good design, but the designers did make some errors. Their logical use of the design procedure was excellent although, partially through lack of time and lack of experience, they did not extend the procedure as far as would be desired.

The five main errors in the design are all reasonably obvious. Despite the recognition of the importance of weakly bonded hydrogen, little attention was paid to the effect of this requirement on the design. Secondly, they tended to ignore the possibility of sulphide catalysts, although recognising that they could be used. This fault is quite common, since data is not so readily available on sulphides. However, some data is available concerning hydrogenation reactions (ref. 3) and, in particular, it is known that nickel sulphide is a less active and more selective catalyst than nickel.

The third fault is also quite common in a design, and it is a major error. Having written down plausible reactions on the surface, the designers decided which were the desired reaction sequences and concentrated their attention on these. They did not pay any attention to the possibility of stopping undesired reaction sequences. Thus, for example, they recognise the need to avoid the production of adsorbed methyl fragments or of carbon but they selected Ni as a potential catalyst. It is well known that Ni will favour methanation and carbon formation (although not at the temperatures suggested for reaction) but, in selecting Ni, this fact was totally ignored.

This is a very common fault in a design, and it is easily detected during experimental testing. Nonetheless, precaution is often less time consuming than cure!

The fourth fault in the design was related to the above, in that the designers did not consider possible adsorption or reactions of the product, methanol. Recognition that dehydration catalysts could favour the conversion of methanol to dimethyl ether was achieved, but mainly in the connection with reactions leading to methanol rather than with reactions of the product.

Finally, the designers did not check that heats of adsorption data could not be used as pointers to the design. Although such data is not common, hydrogenation/dehydrogenation reactions are one of the few cases where some data exists.

Otherwise, there were some small errors that arose because of the lack of experience of the designers and through lack of time with the design. Thus, for example, insufficient attention was paid to the known fact that transition metals favour chemisorption of atomic hydrogen, and the reasoning behind the rejection of some metal oxide catalysts was not always made as clear as it should have been.

It is interesting to note that the design did suggest a catalyst that is known to be active i.e. "copper chromite" or, in terms of the design, Cu on Cr_2O_3. No experimental testing was carried out, and the design of secondary components was not undertaken as a result of the absence of data from the testing.

REFERENCES

1 D.A. Dowden, La Chimica e l'Industria, 55, 639 (1973).
2 D.A. Dowden, Chem. Eng. Progr. Symp., 63, No. 73, 90 (1967).
3 O. Weisser and S. Landa, Sulphide Catalysts: their properties and applications Pergamon, Oxford, and Vieweg, Braunschweig, 1973.

CHAPTER 13

THE DESIGN OF A CATALYST FOR THE SELECTIVE HYDROGENATION OF ACETYLENE IN THE PRESENCE OF ETHYLENE

I. THE IDEA

The production and polymerisation of olefins must be regarded as one of the most important reaction sequences in chemical industry. It is difficult to imagine a world which did not depend significantly on polyethylene and polypropylene, despite the fact that this was the case only some thirty years ago. As a result, it could be argued that the reactions involved have received detailed scientific attention to the point where most of the problems have been solved.

This is far from the case. It is true to say that efficient reaction sequences have been developed, but it is not true to suggest that these could not be improved. Thus, for example, the production of ethylene by steam cracking (ref. 1) is a "blunderbuss" operation which requires considerable energy input and produces a wide range of products. More selective processes operating at lower temperatures would be highly desirable.

Given the source of the ethylene, it is not surprising that organic impurities may be present. Of these, acetylene is particularly troublesome in that it is difficult to separate from ethylene to the levels desired (ca. $\leq$ 20 ppm in polymerisation feedstocks). As a result, a catalytic hydrogenation process has been developed in which acetylene is hydrogenated to ethylene and the production of ethane is minimised. Although this sequence has received considerable attention over the years (refs. 2,3) the designers chose to look anew at possible catalysts for this selective hydrogenation.

II. THE DESCRIPTION OF THE IDEA

The designers chose to describe the reaction in the more complete terms favoured by Dowden (ref. 4). They first wrote down most of the possible reactions.

The target transformation:

$$CH \equiv CH + H_2 \rightleftarrows CH_2 = CH_2$$

<u>The characteristic chemistry</u>

a) primitive processes (PP)

$$CH \equiv CH \rightleftarrows 2C + H_2$$

b) Self-interactions (SI)

$2CH{\equiv}CH \rightleftarrows CH{\equiv}C{-}CH{=}CH_2$

$\rightleftarrows CH_2{=}CH_2 + 2C$

$\rightleftarrows CH_4 + 3C$

$\rightleftarrows CH{\equiv}C{-}C\ CH + H_2$ (i)

c) Cross-interactions (CI)

$CH{\equiv}CH + H_2 \rightleftarrows CH_2{=}CH_2$

d) Derived primitive processes (DPP)

$CH_2{=}CH_2 \rightleftarrows 2C + 2H_2$

$\rightarrow CH_4 + C$

$CH_4 \rightarrow C + 2H_2$ Not thermodynamically feasible below $800^{o}K$

$CH{\equiv}C{-}CH{=}CH_2 \rightleftarrows CH_4 + 3C$

$\rightleftarrows CH{\equiv}C{-}CH_3 + C$

$\rightleftarrows CH_2{=}C{=}CH_2 + C$ unlikely

$\rightleftarrows CH_2{=}C{=}C{=}CH_2$ unlikely

$CH{\equiv}C{-}C{\equiv}CH \rightleftarrows CH{\equiv}CH + 2C$ (i)

$CH{\equiv}C{-}C{\equiv}CH \rightleftarrows 4C + H_2$

e) Derived self-interactions (DSI)

$2CH_4 \rightleftarrows CH_3{-}CH_3 + H_2$

$\rightleftarrows CH_2{=}CH_2 + 2H_2$

$\rightleftarrows CH{\equiv}CH + 3H_2$

$2CH_2{=}CH_2 \rightleftarrows CH_2{=}C{=}C{=}CH_2 + 2H_2$ unlikely

$\rightleftarrows CH{\equiv}C{-}C{\equiv}CH + 2H_2$ (i)

$\rightleftarrows CH_2{=}CH{-}CH{=}CH_2 + H_2$ (i)

$\rightleftarrows CH_2{=}CH{-}CH_2{-}CH_3$ stable below $775^{o}K$

$\rightleftarrows CH_2{=}C(CH_3){-}CH_3$ stable below $840^{o}K$

(i) unlikely on thermodynamic grounds (see later)

$$\rightleftarrows \begin{array}{l} CH_2-CH_2 \\ | \quad\;\; | \\ CH_2-CH_2 \end{array}$$ stable below 450°K

$$\rightleftarrows \begin{array}{l} CH=CH \\ | \quad\;\; | \\ CH_2-CH_2 \end{array} + H_2$$ (ii)

$$\rightleftarrows CH_3-\underset{CH_2}{CH-CH_2}$$ (ii)

$$\rightleftarrows CH_3-C{\equiv}C-CH_3 + H_2$$ (ii)

$$2CH{\equiv}C-CH{=}CH_2 \rightleftarrows C_6H_5-CH{=}CH_2$$

f) Sequential reactions (SR)

$$CH_2{=}CH_2 + H_2 \rightleftarrows CH_3-CH_3$$

$$CH{\equiv}C-CH{=}CH_2 + H_2 \rightleftarrows CH_2{=}CH-CH{=}CH_2$$ (i)

$$CH{\equiv}C-C{\equiv}CH + H_2 \rightleftarrows CH_2{=}CH-C{\equiv}CH$$ (i)

$$CH_2{=}CH-CH{=}CH_2 + H_2 \rightleftarrows CH_2{=}CH-CH_2-CH_3$$ feasible below 975°K

g) Sequential cross-interactions (SCI)

$$CH{\equiv}CH + CH_2{=}CH_2 \rightleftarrows CH_2{=}CH-CH{=}CH_2$$

h) Interjacent primitive processes (IPP)

$$CH_3-CH_3 \rightleftarrows 2C + 3H_2$$ feasible below 465°K

$$CH_2{=}CH-CH{=}CH_2 \rightleftarrows CH_2{=}C{=}C{=}CH_2 + H_2$$ unlikely

$$CH_2{=}CH-CH_2-CH_3 \rightleftarrows CH_2{=}C{=}CH-CH_3 + H_2$$ unlikely

$$\rightleftarrows CH_2{=}C{=}C{=}CH_2 + 2H_2$$ unlikely

$$CH_2{=}\underset{CH_3}{\underset{|}{C}}-CH_3 \rightleftarrows \underset{CH_2}{CH-CH}-CH_3 + H_2$$ (ii)

$$C_6H_5-CH{=}CH_2 \rightleftarrows C_6H_5-C{\equiv}CH + H_2$$

(ii) Not stable on thermodynamic grounds: see later.

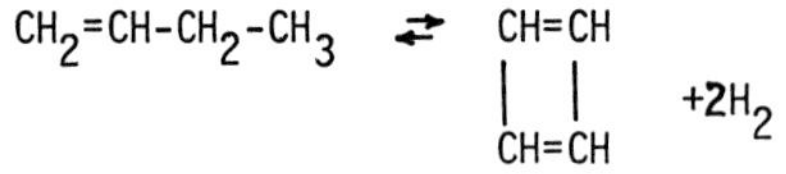

$$CH_2{=}CH{-}CH_2{-}CH_3 \rightleftarrows \begin{matrix} CH{=}CH \\ | \quad\; | \\ CH{=}CH \end{matrix} + 2H_2$$

$$CH_2{=}CH{-}CH_2{-}CH_3 \rightleftarrows CH{\equiv}C{-}CH_2{-}CH_3 + H_2$$

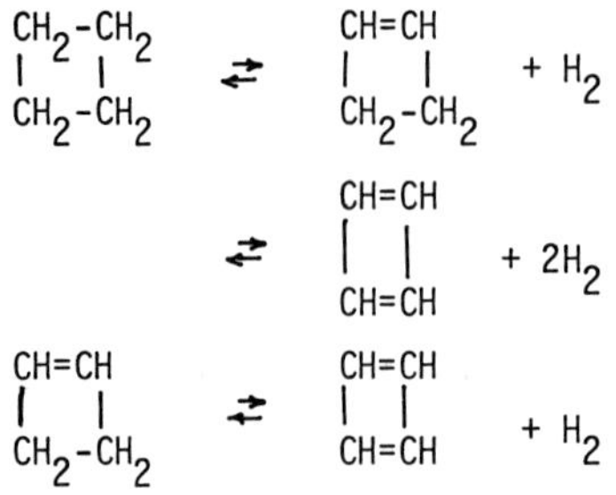

$$\begin{matrix} CH_2{-}CH_2 \\ | \qquad\; | \\ CH_2{-}CH_2 \end{matrix} \rightleftarrows \begin{matrix} CH{=}CH \\ | \quad\; | \\ CH_2{-}CH_2 \end{matrix} + H_2$$

$$\rightleftarrows \begin{matrix} CH{=}CH \\ | \quad\; | \\ CH{=}CH \end{matrix} + 2H_2$$

$$\begin{matrix} CH{=}CH \\ | \quad\; | \\ CH_2{-}CH_2 \end{matrix} \rightleftarrows \begin{matrix} CH{=}CH \\ | \quad\; | \\ CH{=}CH \end{matrix} + H_2$$

$$CH_3{-}C{\equiv}C{-}CH_3 \rightleftarrows 4C + 3H_2$$

i) Interjacent self-interactions (ISI)

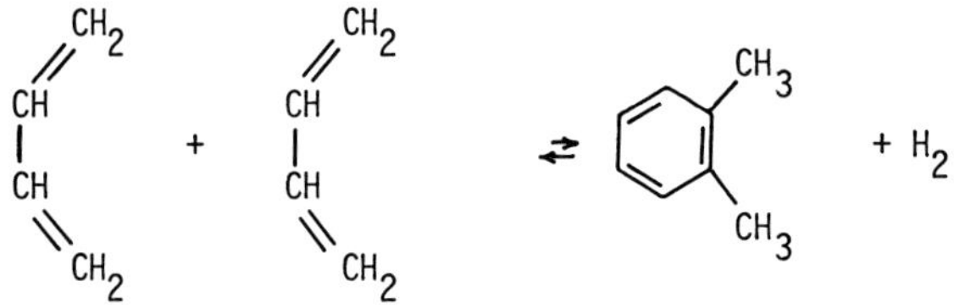

$$\begin{matrix} CH_2 \\ /\!/ \\ CH \\ | \\ CH \\ \backslash\!\backslash \\ CH_2 \end{matrix} + \begin{matrix} CH_2 \\ /\!/ \\ CH \\ | \\ CH \\ \backslash\!\backslash \\ CH_2 \end{matrix} \rightleftarrows \begin{matrix} CH_3 \\ CH_3 \end{matrix} + H_2$$

$$2CH_2{=}CH{-}CH{=}CH_2 \rightleftarrows \text{Polymers}$$

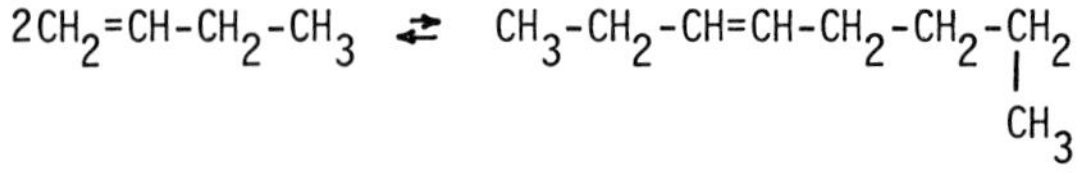

$$2CH_2{=}CH{-}CH_2{-}CH_3 \rightleftarrows CH_3{-}CH_2{-}CH{=}CH{-}CH_2{-}CH_2{-}CH_2(CH_3)$$

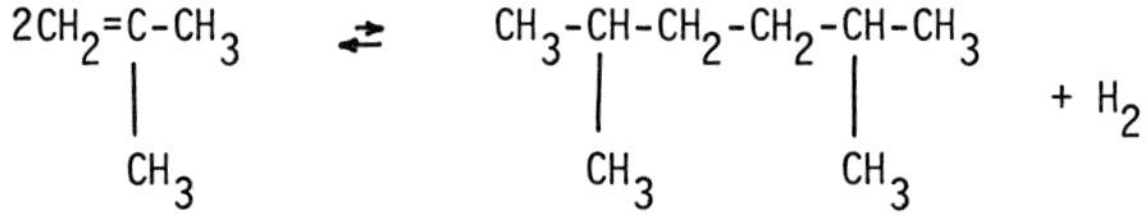

$$2CH_2{=}C(CH_3){-}CH_3 \rightleftarrows CH_3{-}CH(CH_3){-}CH_2{-}CH_2{-}CH(CH_3){-}CH_3 + H_2$$

$$2CH_2{=}CH{-}C\ CH \rightleftarrows CH\ C{-}CH_2{-}CH{=}CH{-}CH_2(C{=}CH)$$

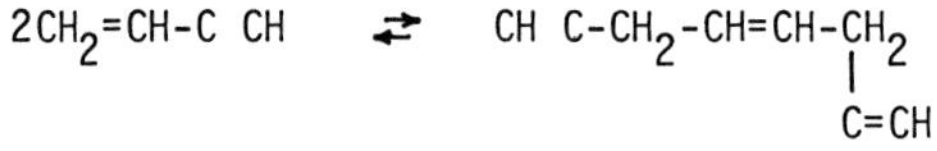

j) Interjacent cross-interactions (ICI)

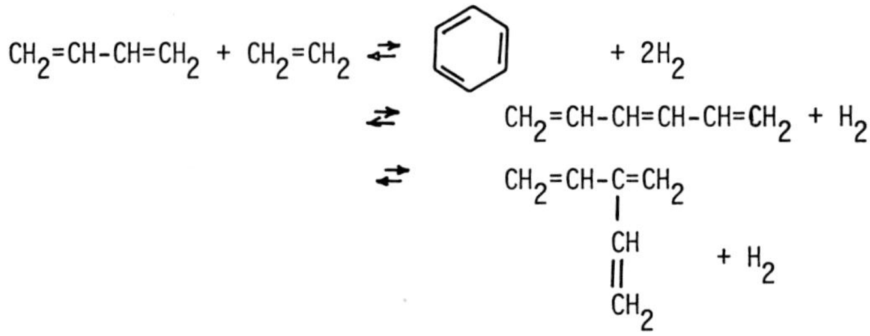

$$CH_2{=}CH{-}CH{=}CH_2 + CH_2{=}CH_2 \rightleftarrows + 2H_2$$

$$\rightleftarrows CH_2{=}CH{-}CH{=}CH{-}CH{=}CH_2 + H_2$$

$$\rightleftarrows \begin{matrix} CH_2{=}CH{-}C{=}CH_2 \\ | \\ CH \\ \| \\ CH_2 \end{matrix} + H_2$$

$C_6H_4(CH_3)_2$ + $CH_2{=}CH{-}CH{=}CH_2$ $\rightleftarrows$ $C_{10}H_6(CH_3)_2$ + H_2 and so on.

$CH_2{=}CH{-}CH_2{-}CH_3 + CH_2{=}CH{-}CH{=}CH_2 \rightleftarrows CH_3{-}CH_2{-}CH_2{-}CH{=}CH{-}CH_2{-}CH(CH_2)$

$CH_2{=}CH_2 + CH_2{=}CH{-}CH_2{-}CH_3 \rightleftarrows CH_3{-}CH_2{-}CH_2{-}CH{=}CH{-}CH_3$

$\rightleftarrows CH_2{=}C(CH_2{-}CH_3){-}CH_2{-}CH_3$

$CH_2{=}CH_2 + CH_2{=}CH{-}C{\equiv}CH \rightleftarrows CH{\equiv}C{-}CH_2{-}CH{=}CH{-}CH_3$

$\rightleftarrows CH{\equiv}C{-}C({=}CH{-}CH_3){-}CH_3$ etc.

$CH_2{=}C(CH_3){-}CH_3 + CH_2{=}CH{-}C{\equiv}CH \rightleftarrows CH{\equiv}C{-}CH_2{-}CH{=}CH{-}CH(CH_3){-}CH_3$

$\rightleftarrows CH{\equiv}C{-}C({=}CH{-}CH(CH_3){-}CH_3){-}CH_3$

The designers then allocated these reactions according to general types, and obtained thermodynamic data for most of them (refs. 5,6).

I. Cracking

Reaction			ΔG_j^o kJ/mol at temperature (K) 300	600	1000
1) $CH{\equiv}CH$	$\rightleftarrows$	$2C+H_2$	-209.07	-191.78	-169.97
2) $2CH{\equiv}CH$	$\rightleftarrows$	$CH_2{=}CH_2 + 2C$	-349.91	-296.06	-221.81
3) $2CH{\equiv}CH$	$\rightleftarrows$	$CH_4 + 3C$	-468.82	-406.63	-320.85
4) $CH_2{=}CH_2$	$\rightleftarrows$	$2C + 2H_2$	- 68.23	- 87.50	-118.14
5)	$\rightleftarrows$	$CH_4 + C$	-118.91	-110.57	- 99.05
6) CH_4	$\rightleftarrows$	$C + 2H_2$	50.68	23.07	- 19.09

	Reaction	ΔG_j^o kJ/mol at temperature (K) 300	600	1000
7)	$CH{\equiv}C{-}CH{=}CH_2 \rightleftarrows CH_4 + 3C$	-494.40	-436.88	-354.65
8)	$\rightleftarrows CH{\equiv}C{-}CH_3 + C$	-249.28	-207.78	-148.05
9)	$CH{\equiv}C{-}C{\equiv}CH \rightleftarrows CH{\equiv}CH + 2C$	-234.65	-222.03	-203.77
10)	$\rightleftarrows 4C + H_2$	-443.72	-413.81	-373.74
11)	$CH_3{-}CH_3 \rightleftarrows 2C + 3H_2$	32.59	- 24.91	-109.26
12)	$CH_3{-}C{\equiv}C{-}CH_3 \rightleftarrows 4C + 3H_2$	-185.65	-229.79	-295.93

II. Cyclisation

	Reaction	300	600	1000
13)	$CH_2{=}C(CH_3){-}CH_3 \rightleftarrows$ CH=CH-CH₃ ring closed through CH₂ (methylcyclopropene) $+ H_2$	55.85	66.14	76.5
14)	$CH_2{=}CH{-}CH{=}CH_2 \rightleftarrows$ cyclo(CH=CH / CH=CH) $+ H_2$	161.9	79.8	- 3.5

III. Aromatisation

	Reaction	300	600	1000
15)	$2CH{\equiv}C{-}CH{=}CH_2 \rightleftarrows C_6H_5{-}CH{=}CH_2$	-397.74	-331.45	-241.60
16)	$2CH_2{=}CH{-}CH{=}CH_2 \rightleftarrows$ o-$C_6H_4(CH_3)_2 + H_2$	- 28.20	38.92	134.35

IV. Hydrogenation

	Reaction	300	600	1000
17)	$CH{\equiv}CH + H_2 \rightleftarrows CH_2{=}CH_2$	-140.84	-104.28	- 51.83
18)	$CH_2{=}CH_2 + H_2 \rightleftarrows CH_3{-}CH_3$	-100.82	- 62.59	- 8.89
19)	$CH{\equiv}C{-}CH{=}CH_2 + H_2 \rightleftarrows CH_2{=}CHCH{=}CH_2$	77.77	121.81	185.13
20)	$CH{\equiv}C{-}C{\equiv}CH + H_2 \rightleftarrows CH_2{=}CH{-}C{\equiv}CH$	- 32.94	- 24.97	- 13.50
21)	$CH_2{=}CH{-}CH{=}CH_2 + H_2 \rightleftarrows CH_2{=}CH{-}CH_2{-}CH_3$	- 79.18	- 45.55	2.90

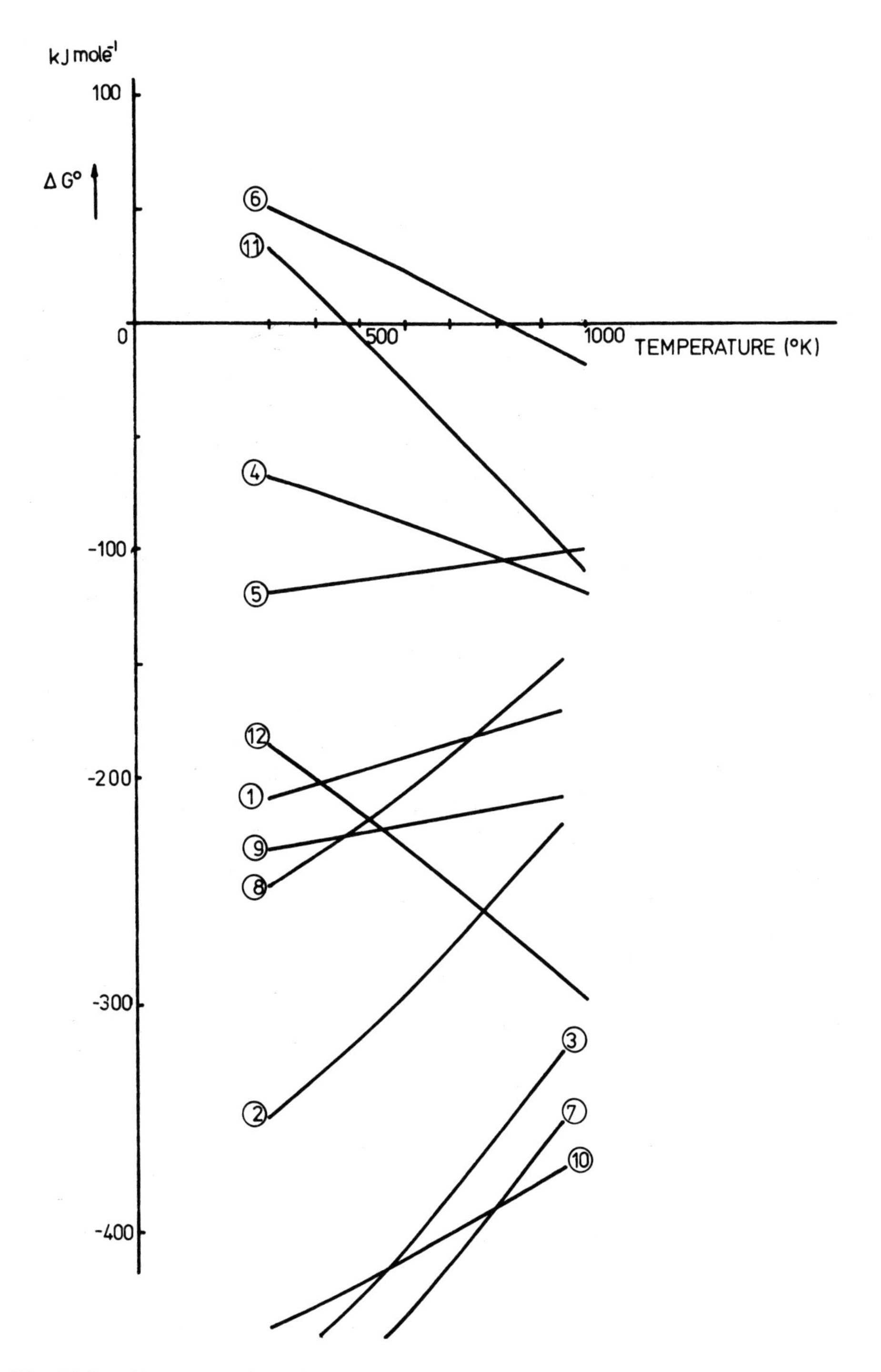

Fig.13.1. Free energies of cracking reactions as a function of temperature.

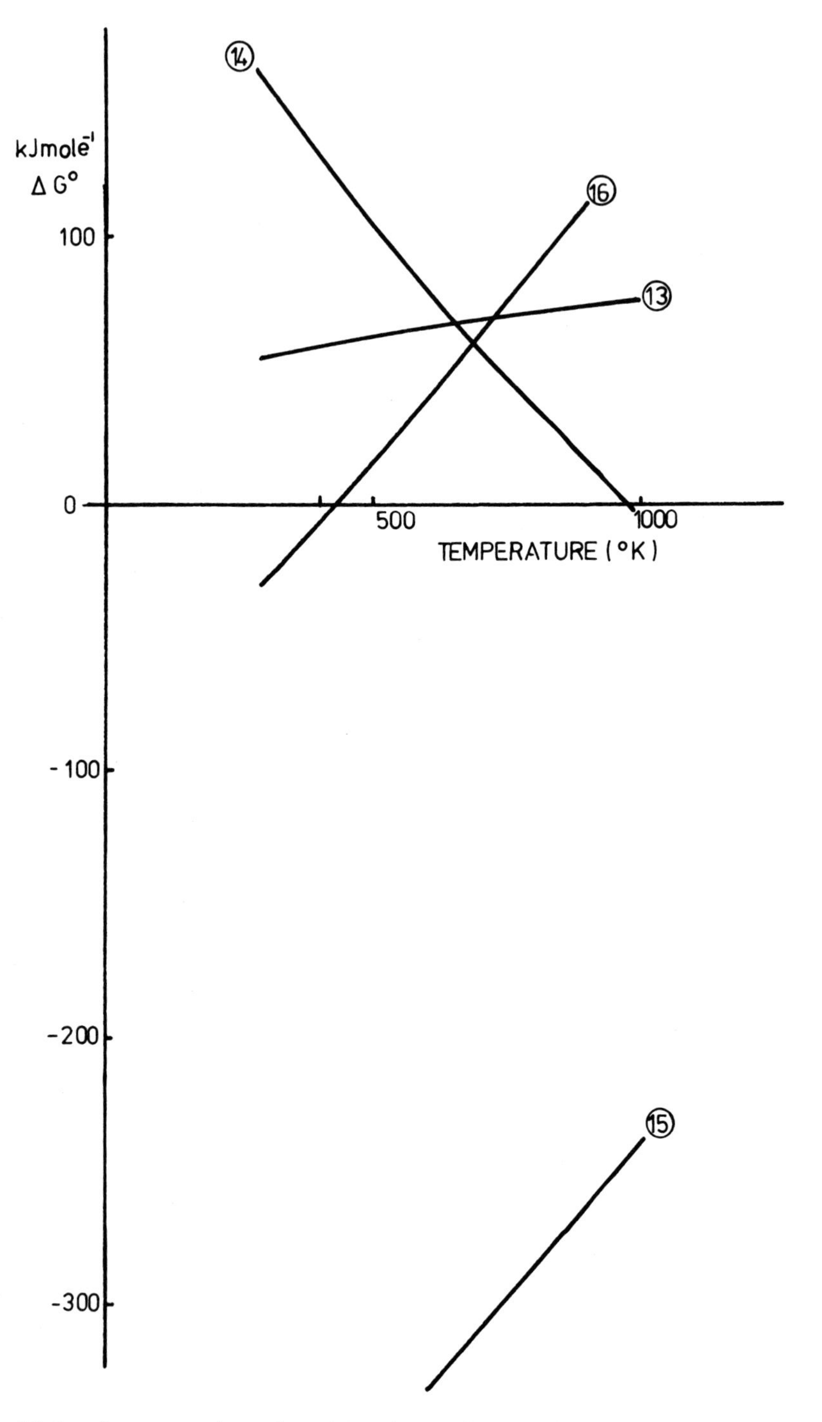

Fig. 13.2. Free energies of cyclisation and aromatisation as a function of temperature.

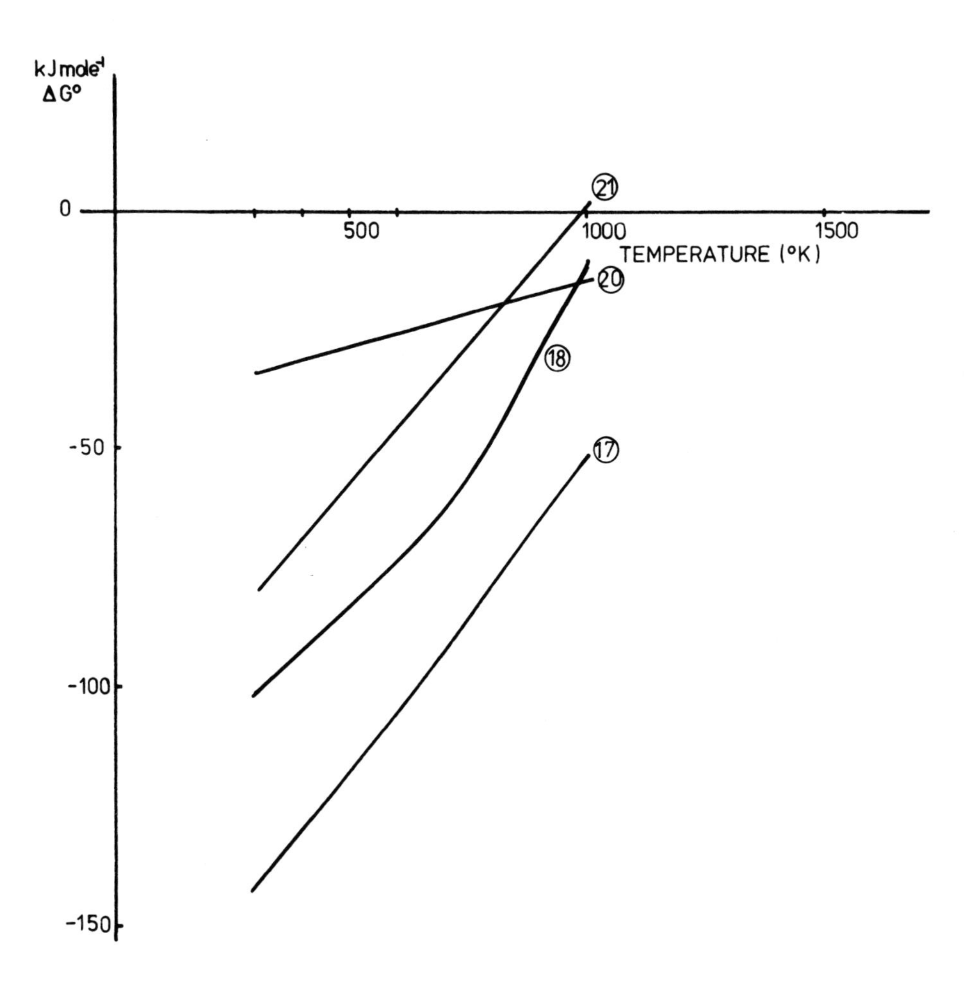

Fig. 13.3. Free energies of hydrogenation as a function of temperature.

V. Dehydrogenation

	ΔG° kJ/mol at temperature (K) 300	600	1000
22) $2CH{\equiv}CH \rightleftarrows CH{\equiv}C{-}C{\equiv}CH + H_2$	25.8	30.25	33.80
23) $2CH_2{=}CH_2 \rightleftarrows CH{\equiv}C{-}C{\equiv}CH + 2H_2$	307.26	238.81	137.46
24) $\rightleftarrows CH{=}CH{-}CH{=}CH_2 + H_2$	14.44	20.76	24.69
25) $\rightleftarrows$ cyclo-$(CH{=}CH{-}CH_2{-}CH_2) + H_2$	38.53	50.69	64.00
26) $\rightleftarrows CH_3{-}C{\equiv}C{-}CH_3 + H_2$	49.19	54.79	59.65
27) $CH_3{-}CH_3 \rightleftarrows 2C + 3H_2$	32.59	- 24.91	-109.26
28) $CH_2{=}C(CH_3){-}CH_3 \rightleftarrows CH{-}CH{-}CH_3$ (ring closed by CH_2) $+ H_2$ calculated	55.85	66.14	76.59
29) $C_6H_5{-}CH{=}CH_2 \rightleftarrows C_6H_5{-}C{\equiv}CH + H_2$	147.8	113.28	64.52
30) $CH_2{=}CH{-}CH{=}CH_2 \rightleftarrows$ cyclo-$(CH{=}CH{-}CH{=}CH) + H_2$	82.70	34.24	- 0.6
31) $CH_2{=}CH{-}CH_2{-}CH_3 \rightleftarrows CH{\equiv}C{-}CH_2{-}CH_3 + H_2$	130.54	93.38	41.01
32) cyclo-$(CH_2{-}CH_2{-}CH_2{-}CH_2) \rightleftarrows$ cyclo-$(CH{=}CH{-}CH_2{-}CH_2) + H_2$	64.43	23.47	- 33.76
33) cyclo-$(CH_2{-}CH_2{-}CH_2{-}CH_2) \rightleftarrows$ cyclo-$(CH{=}CH{-}CH{=}CH) + 2H_2$	123.04	27.78	- 73.67
34) cyclo-$(CH{=}CH{-}CH_2{-}CH_2) \rightleftarrows$ cyclo-$(CH{=}CH{-}CH{=}CH) + H_2$	58.60	4.30	- 39.91
is a cracking reaction only!			
35) $CH_3{-}C{\equiv}C{-}CH_3 \rightleftarrows 4C + 3H_2$	-185.65	-229.79	-295.93
36) $2CH_2{=}CH{-}CH{=}CH_2 \rightleftarrows C_6H_4(CH_3)_2 + H_2$	- 28.20	38.92	134.35
37) $2CH_2{=}C(CH_3){-}CH_3 \rightleftarrows CH_3{-}CH(CH_3){-}CH_2{-}CH_2{-}CH(CH_3){-}CH_3 + H_2$	-105.2	- 20.1	93.5

	Reaction	ΔG^{0} kJ/mol at temperature (K) 300	600	1000
38)	$CH_2{=}CH{-}CH{=}CH_2 + CH_2{=}CH_2 \rightleftarrows$ (benzene ring) $+ 2H_2$	- 89.2	-100.6	-118.6
39)	$\rightleftarrows CH_2{=}CH{-}CH{=}CH{-}CH{=}CH_2 + H_2$	41.0	40.0	37.2
40)	$\rightleftarrows CH_2{=}CH{-}C({-}CH{=}CH_2){=}CH_2 + H_2$	34.9	34.9	32.5
41)	(o-xylene ring, two CH_3) + $CH_2{=}CH{-}CH{=}CH_2 \rightleftarrows$ (naphthalene ring with two CH_3) $+ H_2$; and so on $\rightleftarrows$ C	- 16.0	217.2	506.1
42)	$2CH_4 \rightleftarrows CH_3{-}CH_3 + H_2$	68.77	71.05	71.08
43)	$\rightleftarrows CH_2{=}CH_2 + 2H_2$	169.6	133.6	78.0
44)	$\rightleftarrows CH{\equiv}CH + 3H_2$	519.5	429.7	301.8

VI. Dimerisation

	Reaction	300	600	1000
45)	$2CH{\equiv}CH \rightleftarrows CH{\equiv}C{-}CH{=}CH_2$	-112.22	- 74.20	- 22.68
46)	$\rightleftarrows CH{\equiv}C{-}C{\equiv}CH + H_2$	25.58	30.25	33.80
47)	$2CH_2{=}CH_2 \rightleftarrows CH{\equiv}C{-}C{\equiv}CH + 2H_2$	307.26	238.81	137.46
48)	$\rightleftarrows CH_2{=}CH{-}CH{=}CH_2 + H_2$	14.44	20.76	24.69
49)	$\rightleftarrows CH_2{=}CH{-}CH_2{-}CH_3$	- 64.74	- 24.79	27.59
50)	$\rightleftarrows CH_2{=}C(CH_3){-}CH_3$	- 77.94	- 34.65	21.96
51)	$\rightleftarrows$ cyclo-($CH_2{-}CH_2{-}CH_2{-}CH_2$)	- 25.90	27.22	97.76
52)	$\rightleftarrows$ cyclo-($CH{=}CH{-}CH_2{-}CH_2$) $+ H_2$	38.53	50.69	64.00
53)	$\rightleftarrows CH_3{-}CH{-}CH_2$ (ring closed via CH_2)	unstable		

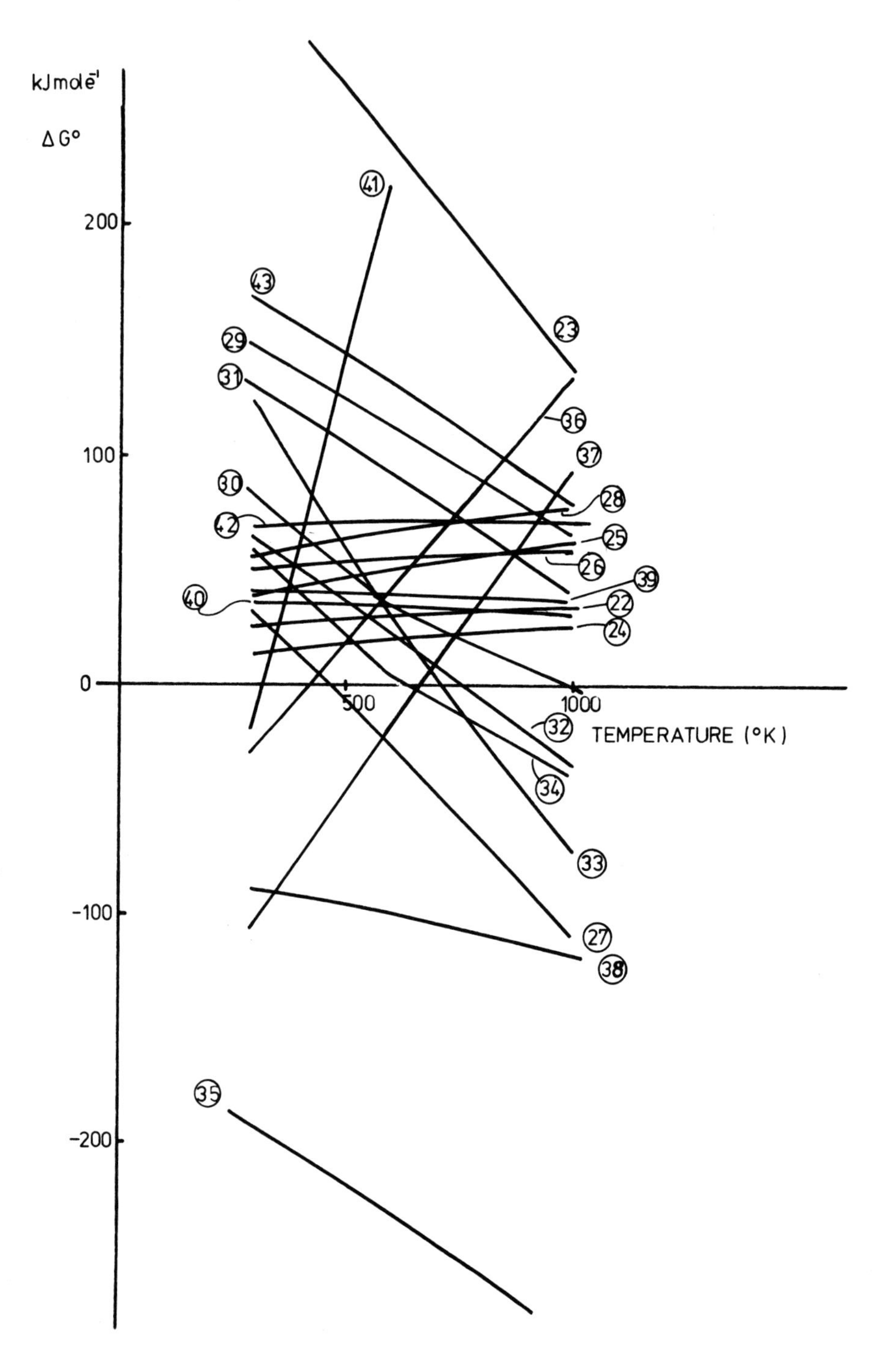

Fig. 13.4. Free energies of dehydrogenation as a function of temperature.

Reaction	ΔG° kJ/mol at temperature (K) 300	600	1000
54) $2CH_2=CH_2 \rightleftarrows CH_3\text{-}C\equiv C\text{-}CH_3 + H_2$	49.19	54.79	59.65
55) $2CH\equiv C\text{-}CH=CH_2 \rightleftarrows C_6H_5\text{-}CH=CH_2$	-397.74	-331.45	-241.60
56) $CH\equiv CH + CH_2=CH_2 \rightleftarrows CH_2=CH\text{-}CH=CH_2$	-126.40	- 83.52	- 27.14
57) $CH_2=CH\text{-}CH=CH_2 + CH_2=CH\text{-}CH=CH_2 \rightleftarrows C_6H_4(CH_3)_2$ (1,2-) $+ H_2$	- 28.20	38.92	134.35
58) $2CH_2=CH\text{-}CH=CH_2 \rightleftarrows CH_2=CHCH=CHCH_2CH_2CH=CH_2$	- 24.5	10.6	58.6
59) $2CH=CH\text{-}CH_2\text{-}CH_3 \rightleftarrows CH_3\text{-}CH_2\text{-}CH=CH\text{-}CH_2CH_2\text{-}CH_2\text{-}CH_3$	- 38.10	5.63	61.73
60) $CH_2=CH_2 + CH_2=CH\text{-}CH_2\text{-}CH_3 \rightleftarrows CH_3\text{-}CH_2\text{-}CH_2\text{-}CH=CH\text{-}CH_3$	- 62.94	- 21.18	33.89
61) $\rightleftarrows CH_2=C(CH_2CH_3)\text{-}CH_2\text{-}CH_3$	- 59.21	- 15.45	82.63
62) $CH_2=CH_2 + CH_2=CH\text{-}C\equiv CH \rightleftarrows CH\equiv C\text{-}CH_2\text{-}CH=CH\text{-}CH_3$	- 83.24	85.62	84.45
63) $\rightleftarrows CH\equiv C\text{-}C(=CH\text{-}CH_3)\text{-}CH_3$	68.86	72.95	74.53
64) $CH_2=CH\text{-}CH=CH_2 + CH_2=CH\text{-}C\equiv CH \rightleftarrows CH\equiv C\text{-}CH_2\text{-}CH=CH\text{-}CH_2\text{-}CH=CH_2$	- 48.53	- 7.60	71.44
65) $\rightleftarrows CH\equiv C\text{-}C(=CH\text{-}CH_2\text{-}CH=CH_2)\text{-}CH_3$	- 45.52	- 31.03	- 16.37

	Reaction	ΔG^o kJ/mol at temperature (K): 300	600	1000
66)	$CH_2=CH-CH_2-CH_3 + CH_2=CH-C\equiv CH \rightleftarrows CH\equiv C-CH_2-CH=CH-CH_2-CH_2-CH_3$	- 58.12	- 11.79	47.24
67)	$\rightleftarrows CH\equiv C-C(=CH-CH_2-CH_2-CH_3)-CH_3$	- 72.43	- 24.46	36.33
68)	$CH_2=C(CH_3)-CH_3 + CH_2=CH-C\equiv CH \rightleftarrows CH\equiv C-CH_2-CH=CH-CH(CH_3)-CH_3$	- 48.18	- 2.51	55.7
69)	$\rightleftarrows CH\equiv C-C(=CH-CH(CH_3)-CH_3)-CH_3$	- 61.3	- 15.2	44.8
70)	$2CH_4 \rightleftarrows CH_3-CH_3 + H_2$	68.77	71.05	71.08
71)	$\rightleftarrows CH_2=CH_2 + 2H_2$	169.6	133.6	78.0
72)	$\rightleftarrows CH\equiv CH + 3H_2$	519.5	429.7	301.8
73)	$2CH_2=C(CH_3)-CH_3 \rightleftarrows CH_3-CH(CH_3)-CH_2-CH_2-CH(CH_3)-CH_3 + H_2$	-105.2	- 20.1	93.5
74)	$2CH_2=CH-C\equiv CH \rightleftarrows CH\equiv C-CH_2-CH=CH-CH_2-C\equiv CH$	- 72.6	- 25.9	34.0
75)	$CH_2=CH-CH=CH_2 + CH_2=CH_2 \rightleftarrows$ (benzene) $+ 2H_2$	- 89.2	-100.6	-118.6
76)	$\rightleftarrows CH_2=CH-CH=CH-CH=CH_2 + H_2$	41.0	40.0	37.2
77)	$\rightleftarrows CH_2=CH-C(-CH=CH_2)=CH_2 + H_2$	34.9	34.9	32.5

		300	600	700
78)	(o-xylene: benzene ring with CH_3, CH_3) + $CH_2=CH-CH=CH_2$ $\rightleftarrows$ (naphthalene ring with CH_3, CH_3) $+ H_2$	- 16.0	217.2	506.1

and so on $\rightleftarrows$ C

	ΔG^o kJ/mol at temperature (K) 300	600	700
79) $CH_2=CH-CH_2-CH_3 + CH_2=CH-CH=CH_2 \rightleftarrows$ $CH_3-CH_2-CH_2-CH=CH-CH_2-CH=CH_2$	-104.1	6.4	59.6

To conserve space, values of ΔG^o have not been plotted for these reactions.

From this thermodynamic analysis, the designers were able to suggest the following possibilities:

Unwanted reactions (400^oK-500^oK)

Cracking:
Reactions 1,2,3,4,5,7,8,10,11

Cyclisation:
None

Aromatisation:
Reactions 15 (large negative ΔG^o) and 16 (small negative ΔG^o)

Hydrogenation:
Reactions 18,20,21

Dehydrogenation:
Reactions 23,27,35,37,38

Dimerisation:
45,49,50,51,55,56,57,58,59,60,61,64,65,66,67,68,69,73,75,79

Desired reactions (400^oK-500^oK)

Cracking:	None	
Cyclisation:	None	
Aromatisation:	None	
Hydrogenation:	17 (target transformation)	
Dehydrogenation:	42,43,44 70,71,72	Not thermodynamically feasible over this temperature range
Dimerisation:	70,71,72	Not thermodynamically feasible over this temperature range

Although this thermodynamic analysis was very thorough, it is interesting to note that the designers gave no reason for selection of the 400-500 K operating temperature. In fact, bearing in mind that the target transformation is very exothermic (ΔH^{o}_{300} = -174.44 KJ/mole^{-1}), there are good reasons to suggest that the reactor temperature be maintained at ca. 300 K in order that the catalyst bed temperature should not rise above ca. 500 K. This limit is dictated by the increasing importance of polymerisation and carbon forming reactions at higher temperatures. The designers did recognise the need for control of temperature, but did not pay particular attention to this point at the later stages of the design (see below). In fact, as we shall see, from this point onwards the design degenerated.

III. THE DESIGN OF THE PRIMARY COMPONENTS

The designers began by recognising that selectivity was as important as activity in this design. Over-hydrogenation to produce ethane would be extremely undesirable, and they suggested that metallic catalysts could be made less active and more selective by, for example, sulphidation. Although lower activities may mean higher operating temperatures, metal sulphides are known to be less affected by the deposition of carbonaceous residues.

As a result, the activity patterns of metals, metal oxides and metal sulphides were all taken into account (Chapter 7). For metals, the activity was known (generally) to increase on moving from left to right of the periodic table and to increase when moving vertically in each Group 8 triad. In addition, it was known that Pd and Ni can selectively hydrogenate triple bonds in the presence of double bonds (ref. 4).

The activity patterns for hydrogenation over oxides and sulphides are reported in Chapter 7 and were used to select possible catalysts. As usual, this was done by analogy, since activities of oxides had been established for the hydrogen-denterium exchange and of sulphides for the hydrogenolysis of carbon disulphide.

On the basis of these patterns, the following selection was made:

<u>Metals:</u> Pt > Pd > Ni, Rh > Co > Fe > Cu > Ru

<u>Metal oxides, sulphides:</u> Cr_2O_3, Co_3O_4, NiO, Cr_2S_3, Co_4S_3, NiS

The designers did not realise that NiO would certainly be reduced to Ni and that Co_3O_4 could well be reduced under the conditions of reaction.

They also recognised the need to apply activity pattern arguments to the undesired reactions of polymerisation and cracking. Polymerisation can be effected by strong bases, but is more often catalysed by acids. Cracking is also favoured by acids and can also occur on metals.

Inspection of the literature showed that polymerisation of butadiene did not occur over nickel or Raney nickel, but did occur on a nickel on silica-alumina catalyst (ref.7). Similarly, a palladium on ZnO-Cr_2O_3-diatomite catalyst was active for polymerisation mainly in proportion to the acidity of the support, and polymerisation/carbon formation could be minimised by the use of high hydrogen:acetylene ratios (ref. 8). As a result, it was concluded that polymerisation and carbon formation could be minimised by avoiding acid catalysts and by increasing the hydrogen:acetylene ratio.

An activity pattern for the cracking of 1,1,3-trimethyl-cyclopentane on transition metals was also found in the literature (ref. 9). The catalytic activity for cracking was found to decrease in the order

Ni > Mo > Rh > Fe >> Pt, Pd.

Considering all these factors together, the designers decided, on the basis of activity patterns, that the following materials were worthy of further attention:

Pt,Pd,Ni,Rh,Fe,Co_3O_4,Co_4S_3,NiS.

The oxides and sulphides of chromium were eliminated on the grounds that they could be too acidic.

This information was filed, and the designers moved on to consider whether heats of adsorption could provide a second guide line. Surprisingly, the team discovered some less obvious data and failed to use data which has been reported in many texts.

They summarised heats of adsorption data as shown in Table 13.1,

TABLE 13.1

Thermochemical data for adsorption on metals

Quantity (kcal/mol)	Fe	Co	Ni	Cu	Pd	Pt	Ru	Rh
ΔH_{ads} (H_2)	32-36	24	29-32	28	27	28	26	26
Bond Strength (M-H)	67.6-69.6	63.6	66.1-69.6	65.6	65.1	65.6	64.6	64.6
ΔH_{ads}(C_2H_2)			67	19				
ΔH_{ads}(C_2H_4)	68		58					

while completely ignoring information on heats of adsorption of ethylene and hydrogen which has been reported graphically by Bond (refs. 2,10). As a result, they were forced to turn to general correlations of adsorption (ref. 2) as summarised in Table 13.2.

TABLE 13.2
Probability of adsorption on metals (ref. 2)

Metals	C_2H_2	C_2H_4	H_2
Mo,W,F,Re,Ni,Co,Rh,Pd,Pt,Ir	A	A	A
Cu,Au	A	A	NA
Ag,Zn,Cd,Ge,Sn,Pb,As,Sb,Bi, Se,Te	NA	NA	NA
A = adsorbed	NA = not adsorbed		

Of the metals that adsorb acetylene, ethylene and hydrogen, Mo and W are hard to reduce. Knowing that Ni and Pd are active catalysts for the hydrogenation, the team then decided that a heat of adsorption for hydrogen of ca. 27-29 kcal mole^{-1} was desirable (Table 13.1) and, as a result, they selected Ni,Rh,Pd,Pt, Ir and Re as catalysts worthy of further study. Cu was ignored owing to the large difference in the heat of adsorption of acetylene as compared with Ni, and as a result of the fact that copper-acetylene complexes can be explosive.

Given that not all of the available data had been recognised, this choice was reasonable. Perhaps more important, however, was the recognition that the heat of adsorption of acetylene was greater than that for ethylene (Table 13.1). As a result, they argued that acetylene could displace ethylene from the surface, thereby increasing selectivity but introducing the possibility of a self-poisoned reaction. Following up this point, they reported that the kinetics of acetylene hydrogenation on metals had been found to be roughly first order in hydrogen and zero or negative order in acetylene (ref. 2). As a result, they suggested that the slow step in the reaction involved a low concentration of adsorbed hydrogen reacting with strongly adsorbed acetylene.

This was an important step in the design and one which was largely ignored. Two mistakes were made. Given this finding, it should have been obvious that efforts to decrease the adsorption of acetylene and increase the adsorption of hydrogen should result in a better catalyst. This aspect was ignored. What should have been done would have been to explore the possibility of poisoning some of the more active sites on the catalyst by, for example, pre-adsorption of a different gas or pre-treatment by a mild poison.

The second mistake was that the designers moved from this position to consider, in detail, results which had been published concerning the deuteration of acetylene and ethylene. Although extremely interesting in their own right, these results threw no light on the design as it stood at that time.

In many ways, this is a classic mistake for designers working with a well known system. In colloquial terms, it can be very difficult to distinguish the wood from the trees. The feeling of delight on discovery that information is available on a given system is rapidly replaced by a desire to use the information at all costs, no matter what its relevance to the aspect of design under consideration.

Attention was then focused on writing down the mechanism on the surface, but things began to go really wrong at this stage. No attempt was made to recognise all of the possible species on the surface, and some elementary thinking was applied only in the context of the possible adsorbed intermediate

```
H       H
 \     /
  C=C
 /     \
X       X
```

Some geometric arguments were applied to this species but these need not be reproduced.

Some thought was also devoted to the selection of the support. From the nature of the undesired reactions, it is obvious that acidity should be avoided. In addition, it was recognised that selective hydrogenation to ethylene rather than to ethane was required and, as a result, that large pores were desired. The designers suggested that α-Al_2O_3 which had been neutralised by titration with a base could be used or, alternatively, that CeO_2 could be suitable. Insufficient attention was paid to the possibility of undesired effects due to heat liberation by the reaction, although the designers pointed out that the supports were thermally stable.

IV. COMMENTS

This was a design which started out well but finished badly. The stoichiometric statement was used to recognise desired and undesired reaction types, and activity patterns were used to select possible catalysts. The possibility of using heats of adsorption data was recognised and used, despite the fact that all available data was not found. From that point, however, the design collapsed.

Perhaps the most valuable lesson from this design is the ease with which it is possible to be distracted from the framework set up. It is very easy to follow an interesting lead in a given area, forgetting the basic purpose of the exercise. Of course, most information is of use, but recognition of the stage in the design when it should be used is always important.

REFERENCES

1 S.B. Zdonik, E.J. Green and L.P. Harlee, Manufacturing Ethylene, Petroleum Publishing Corp., Tulsa, Okl., 1970.
2 G.C. Bond, Catalysis by Metals, Academic Press, New York, 1962.
3 J.E. Germain, The Catalytic Conversion of Hydrocarbons, Academic Press, New York, 1969.
4 D.A. Dowden, La Chimica e l'Industria, 55, 639 (1973).
5 D.R. Stull, E.F. Westrum Jr. and G.C. Sinke, "The Chemical Thermodynamics of Organic Compounds, John Wiley, New York, 1969.

6 G.J. Janz, Thermodynamic Properties of Organic Compounds, Academic Press, New York, 1967.
7 T. Matsumoto and A. Onishi, Kogyo Kagaku Zasshi, 71, 1709 (1968).
8 M. Chauda and S.S. Ghosh, J. Indian Inst. Sci., 51, 180 (1969).
9 J.J. Muller and F.G. Gault, Bull Soc. Chim., 7r, 416, (1970).
10 G.C. Bond, Heterogeneous Catalysis Principles and Applications, Clarendon Press, Oxford, 1974.

CHAPTER 14

THE MANUFACTURE OF TERPENES

One interesting design exercise to consider is based on a catalyst to produce terpenes. It is interesting because it illustrates what can be done by a single designer. The man concerned completed the design in three months, working three afternoons a week. When he started the design his knowledge of catalysis was minimal, and the design involved both education and development. The design shows evidence of inexperience, but it is a good example of what can be achieved in a comparatively short time.

I. THE IDEA

Despite the fact that there are many cheaper solvents on the market, turpentine oil is often preferred for particular applications. Consisting of a mixture of terpenes (hydrocarbons containing ten carbon atoms with boiling points in the range 150-200^{0}C), turpentine is produced by steam distillation of soft wood resins (refs. 1,2,3). As a result, only comparatively small amounts of the material are available. This is used either as a solvent or as a source for the production of particular terpenes for cosmetic or flavour purposes.

The three main types of terpene are acyclic, monocyclic or bicyclic molecules, but isomerisation between monocyclic and bicyclic terpenes is known to take place quite readily (refs. 3,4). α-Pinene (structure I) is the main component of most turpentine oils, and camphane (structure II) is also important. Limonene (structure III) is not a major component, but it could be argued that isomerisation to a more important terpene would be easy (refs. 3,4).

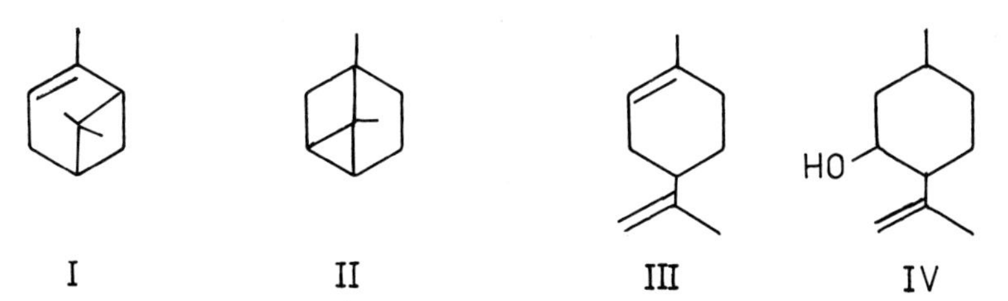

In view of the complexity of the pinene and camphane molecules, attention was focused on limonene and, as part of the development of the idea, possible reaction routes were described and costed.

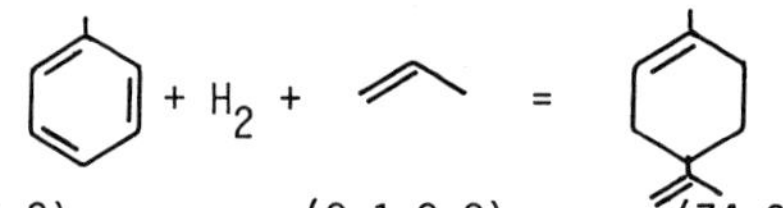

+ H_2 + = ΔG^o = + 3.9 kcal mole^{-1}

(16.8) (8.1-8.8) (74.96)

2 = ΔG^o = -27.7 kcal mole^{-1}

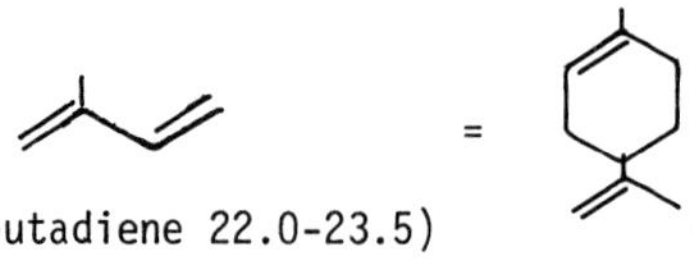

(butadiene 22.0-23.5) (74.96)

+ CH_4 = ΔG^o = +31.8 kcal mole^{-1}

(37.0-38.3) (74.96)

Thermodynamic calculations were based on the group contribution method (refs. 5,6), and may be open to some error for more complicated molecules. Prices of reactants and products (refs. 7,8) were converted to a U.S. dollar/k mole basis, and the range listed reflects the purity of the material (values given above in brackets). It is obvious that a process to produce limonene would be economically attractive and, were it possible to isomerise limonene to α-pinene (perfumery grade 674.6 $/k mole) or to convert it to menthol (structure IV: 2134.4 $/k mole), then the designer could order a Rolls Royce.

II. THE DESCRIPTION OF THE IDEA

The designer began by considering possible reaction mechanisms without reference to the surface and starting with reasonably easily obtained raw materials:-

(i) + H_2 → + →

$-H_2O$

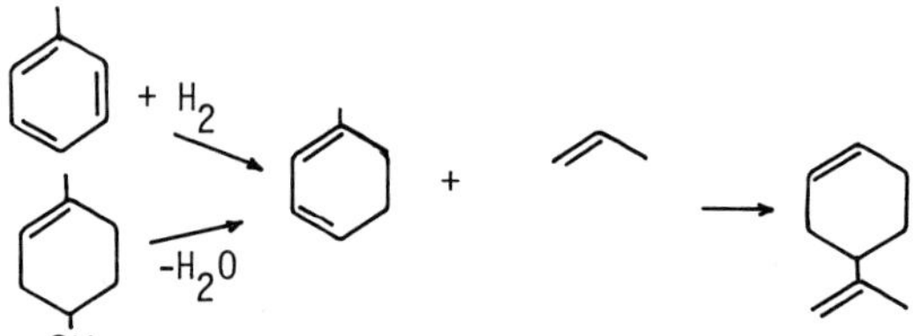

(ii) OH

(iii)  + → + H_2

(iv) 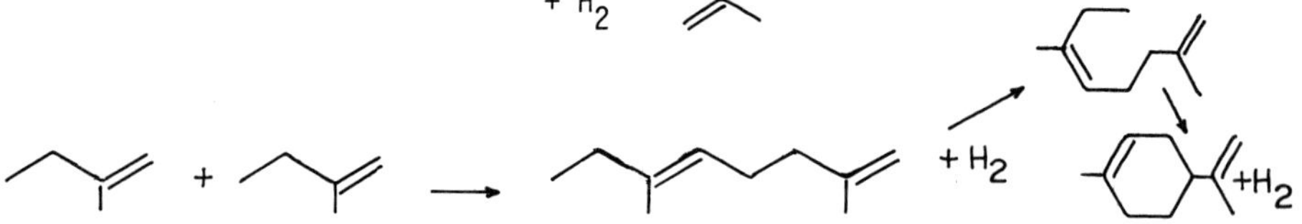 + → + H_2 → → +H_2

(v)

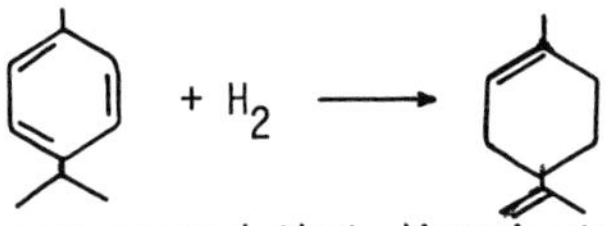

$$+ H_2 \longrightarrow$$

It was argued that dimerisation/cyclisation reactions would be less desirable because of the geometric requirements: this may well be true for reaction iii, but may be less accurate for reaction iv. Reaction v would introduce obvious selectivity problems and, as a result, reaction i or ii was chosen as the target reaction route. Considering the possibilities, the table below was prepared.

TABLE 14.1

Reaction	Pressure	Temp.	ΔG^o_{450}	Feed Price
a) 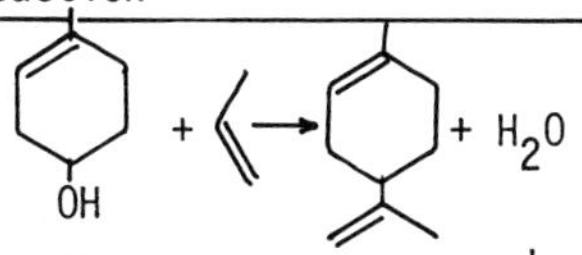 OH + → + H_2O	atm.	rel. high	-13.07	rel. high
b) 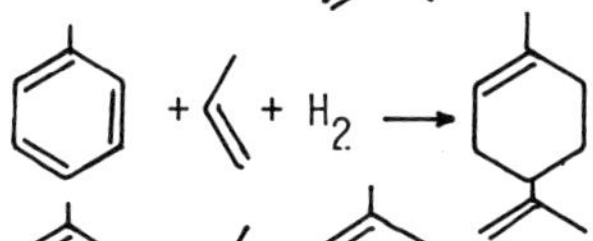 + + H_2 →	high	rel. low	+ 3.9	low
c) 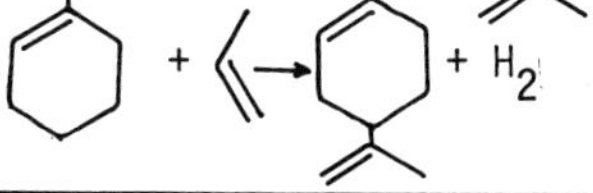 + → + H_2	atm.	rel. high	+ 8.9	rel. high

The possibility of carrying out reaction (a) in two stages was not considered, and - as written - the preferred path would probably involve alkylation in the ring rather than replacement of the hydroxy group. Because of the high price of the feedstocks, reaction (c) was rejected, leaving reaction (b) as the target transformation.

Through inexperience, the designer did not recognise the difficulty in producing methylcyclohexadiene from toluene. With the exception of a few esoteric liquid phase catalysts, this is a very difficult conversion (ref. 4). Nonetheless, it is interesting to follow his approach to the design.

The target transformations were then written down, together with more obvious undesired reactions:

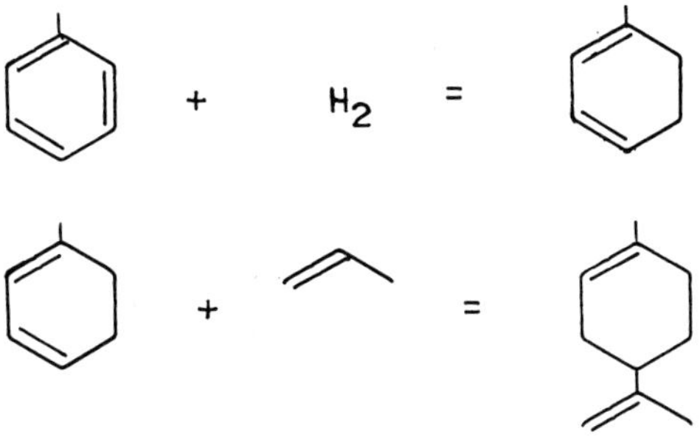

$$+ \quad H_2 \quad =$$

$$+ \quad =$$

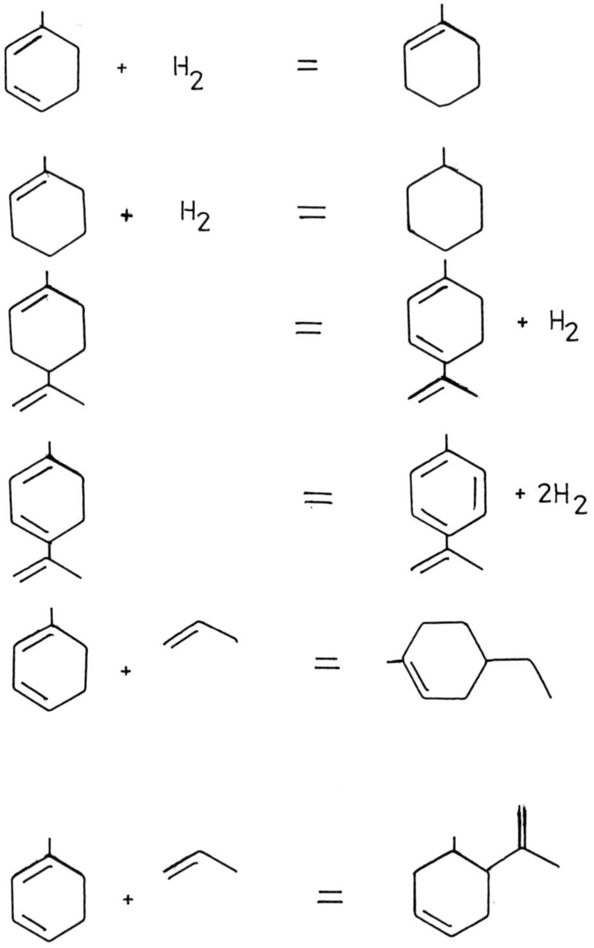

+ H_2 =

+ H_2 =

= + H_2

= + $2H_2$

+ =

+ =

III. DESIGN OF THE PRIMARY COMPONENTS

The designer considered each of the desired transformations in turn. He adopted the somewhat unusual procedure of recognising a general class of catalyst for the reaction, and considering the surface reactions on that class.

Hydrogenation

Most group viii transition metals are active for the hydrogenation of aromatics, as are metal oxides with a d^3 or d^8 configuration (refs. 9,10,11). The oxides are less active than the metals (refs. 10,11).

Turning first to the metals, guidance was sought from heats of adsorption data. Heats of adsorption of ethylene on various metals have been reported (refs. 9,12), and there is a general tendency for the heat of adsorption to decrease in any given period as one moves from left to right in the periodic system. Heats of adsorption of hydrogen are also reported (refs. 9,12) but nothing could be found on toluene. By analogy with ethylene, and in the knowledge that only partial hydrogenation was required, it was suggested that a metal from groups V-VII or a poisoned group VIII metal could be suitable.

The designer chose to consider hydrogenation in terms of a pi-adsorbed species, in that he found that this was also desirable for the second coupling reaction. It was recognised that several transition metals can pi-adsorb toluene and can absorb hydrogen, and the choice of catalyst was deferred until after consideration of the coupling reaction.

This was a major error, which arose because of the designer's inexperience. He did not appreciate the difficulty of partially hydrogenating toluene and, as a result, failed to pay sufficient attention to this aspect of the design.

Coupling

On the assumption that a methyl cyclohexadiene could be produced, the next stage was to consider combination of this molecule with propylene. It was recognised that an olefin could adsorb either via a pi-bond or, if a hydrogen could be split off, via a pi-allyl structure.

H_3C-CH=CH_2 → H_2C ‡ CH-CH_3 (x) pi

H_3C-CH=CH_2 —(-H)→ H_2C-CH-CH_2 (*) pi-allyl

(methyl cyclohexadiene) → pi

(methyl cyclohexadiene) —(-H)→ pi-allyl

The most obvious reaction path involves a two site mechanism involving these adsorbed species. Different forms are favoured on different catalysts, and a reaction mechanism involving pi-bonded adsorption was suggested on metals while pi-allyl species were suggested to be more important on oxides.

Alkylation on metallic catalysts

For metallic catalysts, geometric considerations were taken as the main requirement - given that several transition metals can pi adsorb the two olefins. These geometric requirements were calculated for pi adsorption and pi-allyl adsorption at the same time, on the basis of standard bond lengths reported in the literature (ref. 13) and of desired and undesired products.

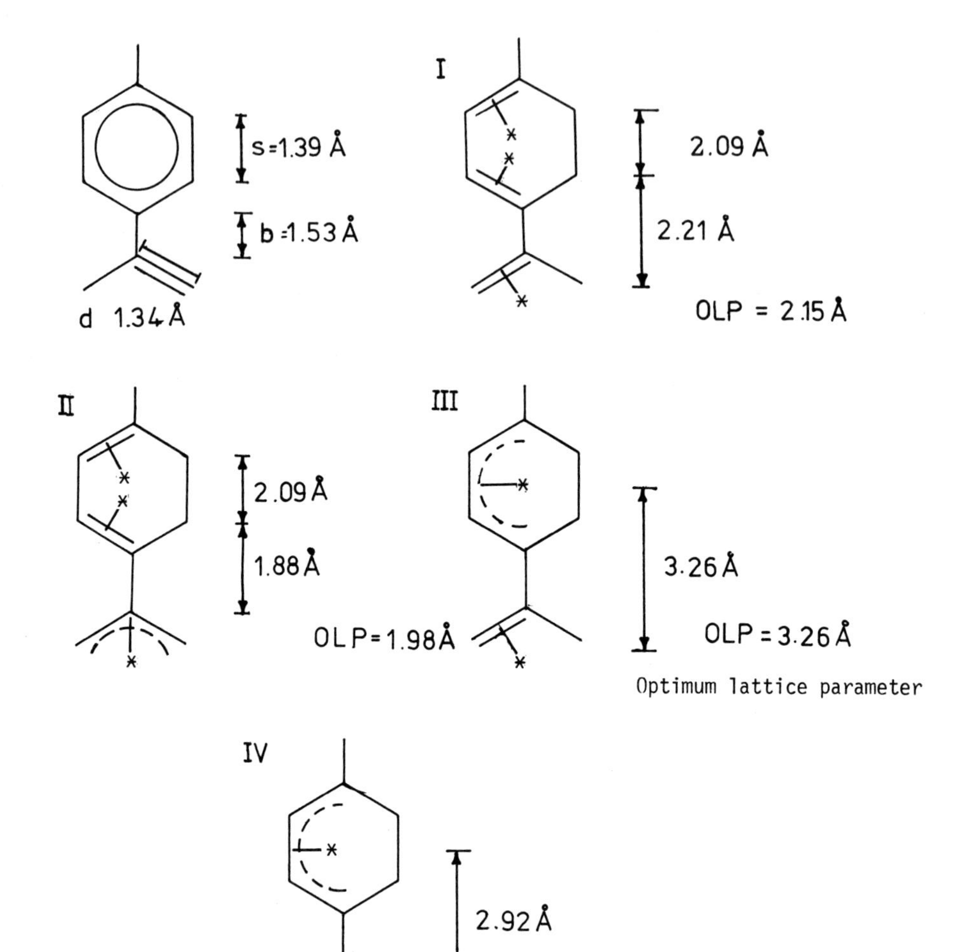

The possibility of producing undesired σ- and m-substituted products was also considered. For symmetry reasons, only the σ-complexes are drawn:

1.46 Å

3.84 Å

4.33 Å

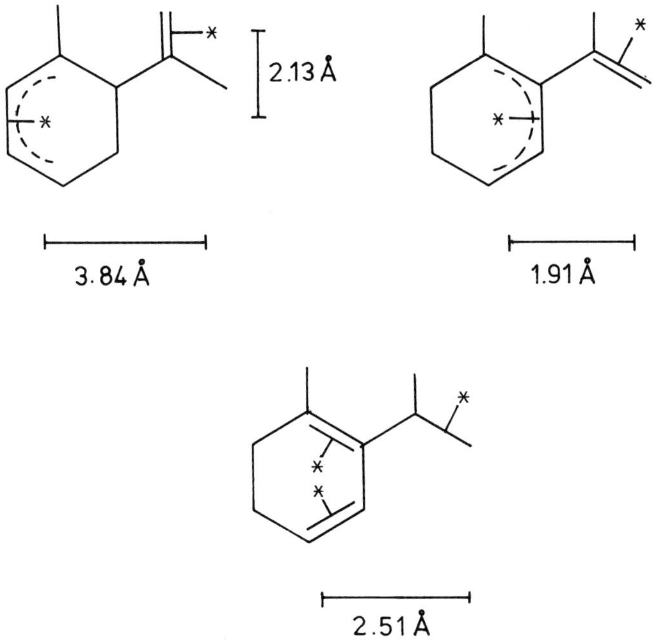

2.13 Å

3.84 Å

1.91 Å

2.51 Å

Inherent in these calculations, although unstated, is the suggestion that the reaction mechanism should be written, for example, as

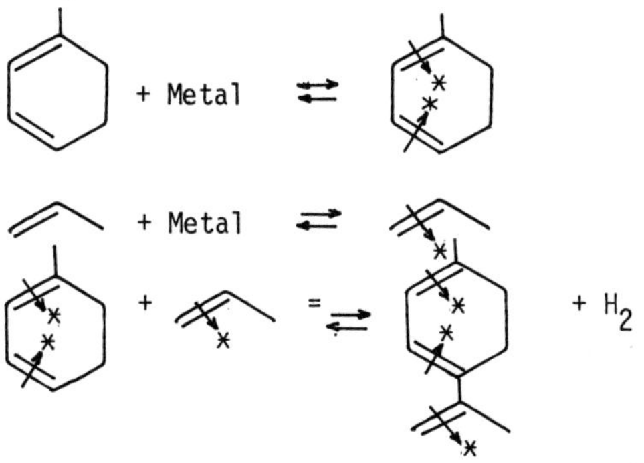

+ Metal ⇄

+ Metal ⇄

+ ⇄ + H_2

Exactly how the adsorbed product left the surface or became partially hydrogenated was left to the imagination.

Nonetheless, using the predicted values of lattice parameters for desired and undesired reactions, comparisons were drawn with values for different metals (Chapter 7) to give Table 14.2.

TABLE 14.2

Lattice parameters of transition metals correlated to desired (——) and undesired values (----) for limonene production

2.92 3.26 3.84 3.33

2.6 2.9 3.2 3.5 3.8 4.1 4.4 4.7

lattice parameter (Å)

Element	l.parameter
Ru	2.70
Os	2.73
Re	2.76
Fe	2.86
Cr	2.89
Ti	2.95
Cd	2.97
V	3.02
Mo	3.14
W	3.16
Hf	3.20
Zr	3.22
Nb	3.29
Ta	3.30
Ni	3.52
Rh	3.80
Pd	3.88
Pt	3.92

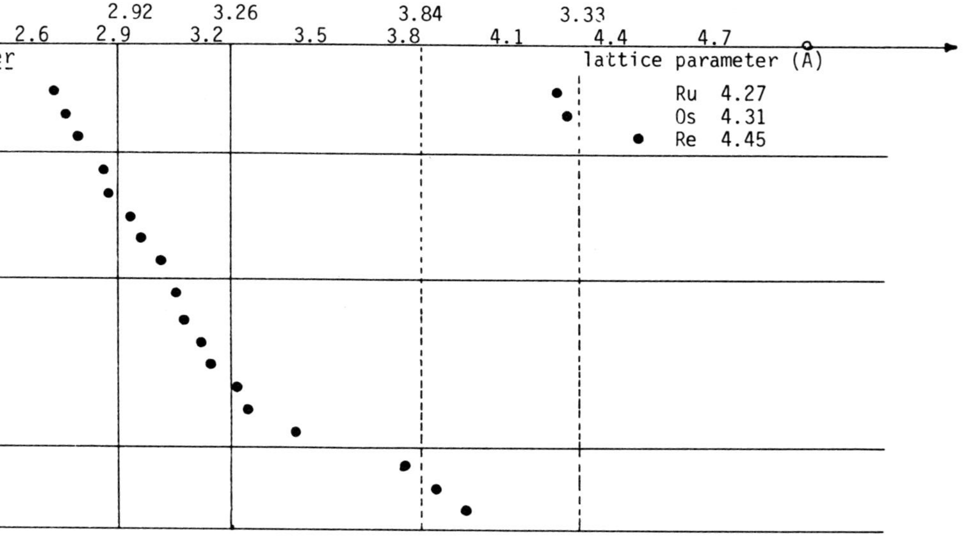

Element	l.parameter
Ru	4.27
Os	4.31
Re	4.45

From the Table it is apparent that Rh, Pd, Pt, Ru, Re and Os are unsuitable, while Zr, Nb, Mo, Hf, Ta, W and (possibly) Ni possess an appropriate geometry to favour the desired reaction.

Alkylation via an oxidation route

Turning attention to the coupling reaction on oxides, the designer distinguished between an oxidative process and an acid catalysed process. In the first case, a surface reaction mechanism was proposed as given below:

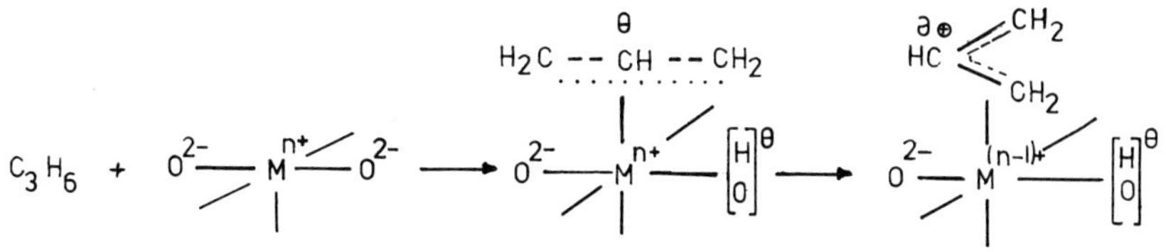

$$C_3H_6 + O^{2-}—M^{n+}—O^{2-} \longrightarrow O^{2-}—M^{n+}(H_2C\text{--}\overset{\ominus}{C}H\text{--}CH_2)—[HO]^{\ominus} \longrightarrow O^{2-}—M^{(n-1)+}(HC^{\partial\oplus}(CH_2)(CH_2))—[HO]^{\ominus}$$

Propylene is dissociatively chemisorbed as symmetrical pi-allyl species (refs. 11,14,15). From the orbital calculations of Jorgensen and Salem (ref. 15), the pi-allyl intermediate could well be bent.

The cyclohexadiene can be suggested to pi-adsorb in the same manner to give:

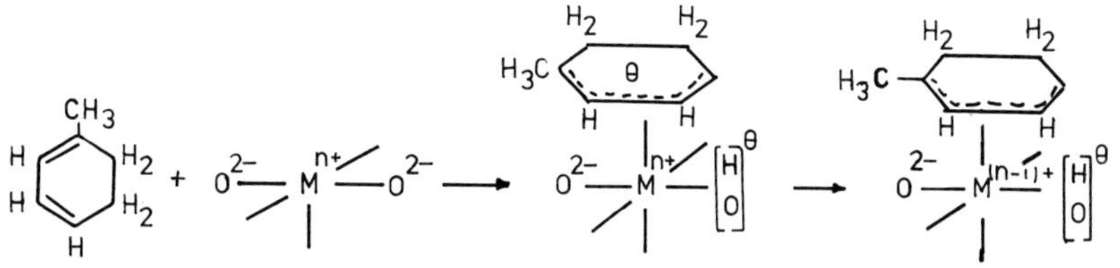

$$\text{CH}_3\text{-cyclohexadiene} + O^{2-}—M^{n+}—O^{2-} \longrightarrow O^{2-}—M^{n+}(H_3C\text{-}C_6H_4H_2^{\ominus})—[HO]^{\ominus} \longrightarrow O^{2-}—M^{(n-1)+}(H_3C\text{-}C_6H_4H_2)—[HO]^{\ominus}$$

If the diene and the propene molecules are pi-adsorbed on adjacent sites on a catalyst with the appropriate lattice spacing, the coupling may now take place. This process can be thought to proceed as follows:

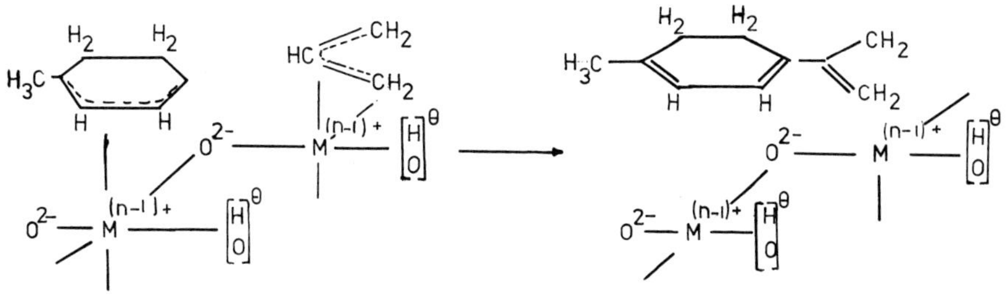

$$O^{2-}—M^{(n-1)+}(H_3C\text{-}C_6H_4H_2)—[HO]^{\ominus} \;+\; O^{2-}—M^{(n-1)+}(HC(CH_2)(CH_2))—[HO]^{\ominus} \longrightarrow H_3C\text{-}C_6H_4H_2\text{-}C(CH_2)(CH_2) + O^{2-}—M^{(n-1)+}—[HO]^{\ominus} + O^{2-}—M^{(n-1)+}—[HO]^{\ominus}$$

This figure shows the overall geometry and charge transfer aspects of the reaction, which are the most important factors to be examined for the catalyst design. The coupling may well involve several intermediate steps. Further, the above product will have to be hydrogenated to give limonene, and the desorption should not take place until this step is completed. This will be discussed later.

The surface can be restored by reoxidizing metal ions with gaseous oxygen, and by the combination of adsorbed hydroxyl ions to give water as a net product. The overall reaction for this process will be:

$$2M^{(n-1)+} + 2(OH)^- + O_2(g) \longrightarrow 2M^{n+} + 2O^{2-} + H_2O(g)$$

In a formal language, the oxidative coupling mechanism leading to limone ($C_{10}H_{18}$) may then be written as:

$$C_7H_{10} + M^{n+} + O^{2-} \longrightarrow [C_7H_9\text{-}M]^{n+} + (OH)^- + e^-$$

$$\longrightarrow [C_7H_9\text{-}M^{(n-1)+}] + (OH) \qquad (1)$$

$$C_3H_6 + M^{n+} + O^{2-} \longrightarrow [C_3H_5\text{-}M]^{n+} + (OH)^- + e^-$$

$$\longrightarrow [\overset{\delta+}{C_3H_5}\text{-}M^{(n-1)+}] + (OH) \qquad (2)$$

$$[C_7H_9\text{-}M^{(n-1)+}] + [\overset{\delta+}{C_3H_5}\text{-}M^{(n-1)+}] \longrightarrow C_{10}H_{14} + 2M^{(n-1)+} \qquad (3)$$

$$\tfrac{1}{2}O_2(g) + 2M^{(n-1)+} \longrightarrow 2M^{n+} + O^{2-} \qquad (4)$$

$$2(OH)^- \longrightarrow H_2O + O^{2-} \qquad (5)$$

$$C_{10}H_{14} + H_2 \longrightarrow C_{10}H_{16} \qquad (6)$$

The oxidation performed in this manner leads to a cyclic compound stabilized by conjugated double bonds. This compound must be partly hydrogenated to give limonene, which may be a difficult task. Over a hydrogenation catalyst, the following reactions may take place:

(I) —isom.→ (II) $\qquad \Delta G^o_{450} = -32.6 \frac{kcal}{mole}$

(I) —$+ H_2$→ (III) $\qquad \Delta G^o_{450} = +13.3 \frac{kcal}{mole}$

(I) —$+ H_2$→ (IV) $\qquad \Delta G^o_{450} = -0.8 \frac{kcal}{mole}$

Thermodynamics (refs. 5,6) show that p-cymene (II) is by far the most probable product, but limonene (IV) is more likely to be formed than is the conjugated cyclodiene (III), possibly due to the inductive effect of the methyl substituent.

In addition, there is some chance for the coupling to take place at the end of the propylene molecule. Selection of a catalyst with the appropriate geometry should minimize the tendency for this reaction to take place.

IV. CATALYST CONSIDERATIONS

It appears from the assumed mechanism that the surface cation must be able to pi-absorb both in the M(n) and in the M(n-1) valence states. Dowden (ref. 16) gives a table in which the ability of common metals in different valence states to give pi- and sigma-type of adsorption is listed, from which the following set of potential catalysts may be desired:

TABLE 14.3
Possible catalysts

Metal Atom	Suitable Valency States	Potential Oxides
V	+4 +3 +2	V_2O_3
Cr	+5 +4 +3 (+2)	CrO_2, (Cr_2O_3)
Mo	+5 +4	
W	+5 +4	
Ni	(+3) +2	Ni_2O_3
Cu	+2 +1	

Turning to geometric arguments, and using the desired and undesired distances calculated above, the designer recognised that different sites could be involved. Using a two-dimensional representation of an oxide lattice, AB adsorption and AC adsorption were considered possible.

A two-dimensional representation of the oxide lattice is:

```
M     0     M     0     M
      A           B
0     M*    0     M*    0
            C
M     0     M*    0     M
0     M     0     M     0
```

Assuming that chemisorption of the reactants on both types of site will lead to the product, the geometry calculations presented earlier can be compared with the oxide lattice presented above. The results can be summarised as in Table 14.4 of desired and undesired distances between cations and anions:

TABLE 14.4

Geometry and adsorption

Coupling position of propylene to diene	Mode of adso.		Calculated Me-Me dist. (A)	Cation-anion (Me-O) distance (A)	
	Diene	Propylene		"AB"-site assumed	"AC"-site assumed
para	pi-allyll	pi-allyll	2.92 x	1.46	2.06
	pi-allyll	pi-adso	3.26 x	1.63	2.41
	pi-adso	pi-adso	2.51	1.31	1.78
ortho	pi-allyll	-	3.84	1.92	2.62
	pi-allyll	pi-adso	4.33	2.17	3.06

x - desired values

Cation-anion distances listed in Chapter 7 were then correlated to the above calculated lattice parameters for the oxides selected from electronic considerations (Table 14.5).

Table 14.5 shows that the most suitable oxide catalysts, with respect to geometry, include

VO	Cr_2O_3		NiO		ZnO
V_2O_3	MoO_2			Ag_2O	CdO
(VO_2)	WO_2	ReO_2			HgO

Knowing that the cyclodiene was produced by the oxidative coupling, the designer then turned his attention to the final hydrogenation step to produce limonene. Failing, again, to recognise the selectivity problem involved in saturating part of an organic ring in the presence of an unsaturated side chain, he returned to the arguments used to design a catalyst for the partial hydrogenation of toluene. As a result, he suggested a bifunctional catalyst to be used both for hydrogenation and for oxidative coupling. A moment's thought would have saved him from this mistake, in that the oxygen - necessary for the coupling reaction - would convert the metal catalysts to metal oxides, and totally alter their activity.

The possibility of using metal oxide hydrogenation catalysts was considered and, based on the activity patterns reported in Chapter 7, V_2O_3, Cr_2O_3, MoO_2 and WO_2 were chosen as possible active species. Again, however, the designer failed to realise that these oxides could not be maintained in their oxidation state in the presence of oxygen. It would be possible to use a two stage process, in which hydrogen was bled into the gas stream between reactors, but this would add to the complexity and cost of the operation.

Alkylation via a carbonium ion route

The designer also recognised that stoichiometric oxides could catalyse alkylation of the diene by a route involving a carbonium ion. Such an ion may be formed

TABLE 14.5

Cation-anion distances, in A, of oxides correlated to desired (———) and undesired values (------) for limonene production

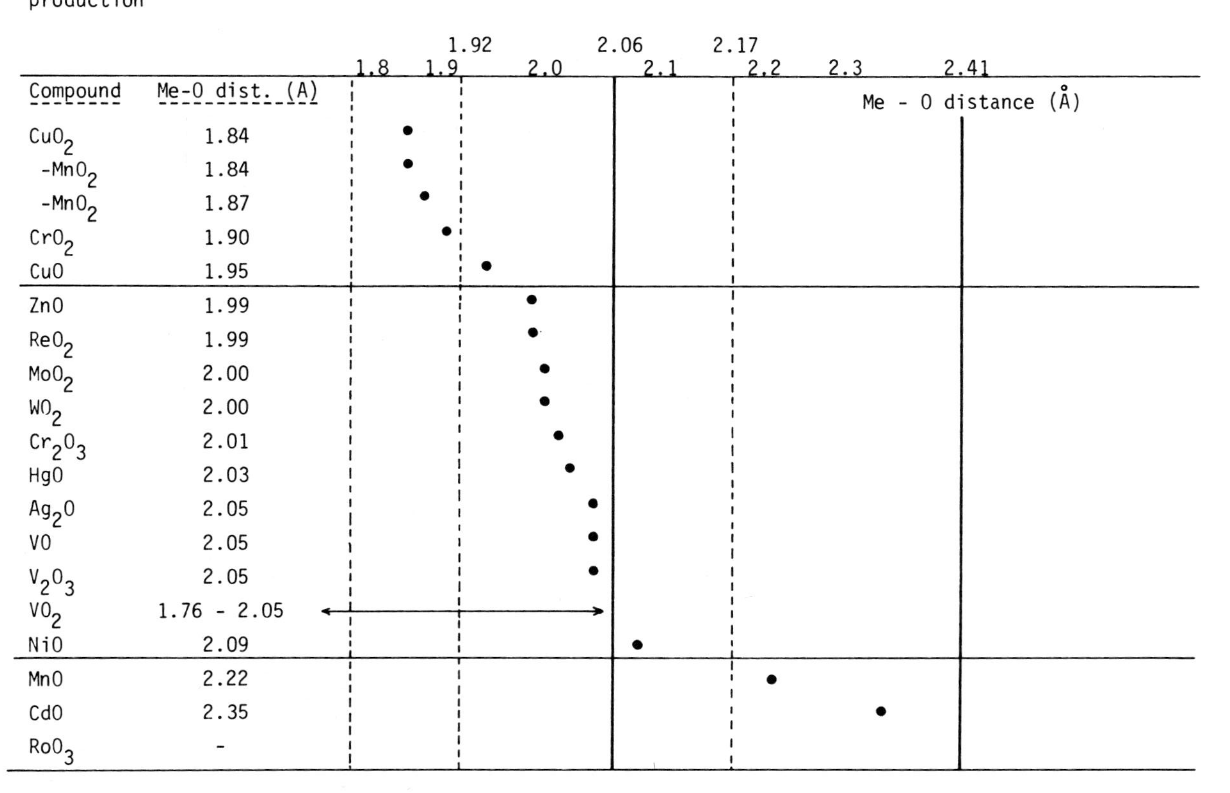

Compound	Me-O dist. (A)
CuO_2	1.84
$-MnO_2$	1.84
$-MnO_2$	1.87
CrO_2	1.90
CuO	1.95
ZnO	1.99
ReO_2	1.99
MoO_2	2.00
WO_2	2.00
Cr_2O_3	2.01
HgO	2.03
Ag_2O	2.05
VO	2.05
V_2O_3	2.05
VO_2	1.76 - 2.05
NiO	2.09
MnO	2.22
CdO	2.35
RoO_3	-

from propylene

$$CH_3 - CH = CH_2 + H^+ \rightleftarrows H_3C\text{-}\overset{+}{C}H\text{-}CH_3$$

or, in various position, from methyl cyclohexadiene

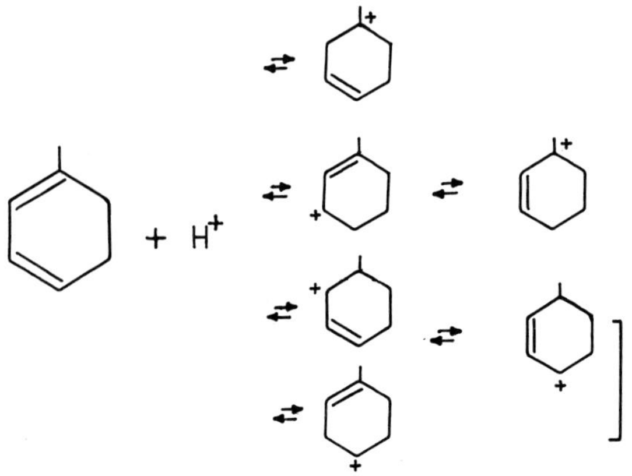

+ H^+

The next steps in the sequence can be illustrated by some possible reactions:

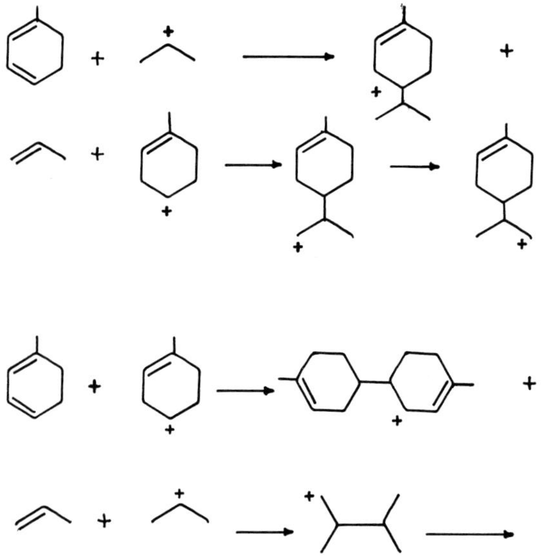

If the catalyst geometry is carefully selected, it may be possible to suppress the undesired side reactions c) and d).

The desired (and undesired) products may then be obtained by a proton shift:

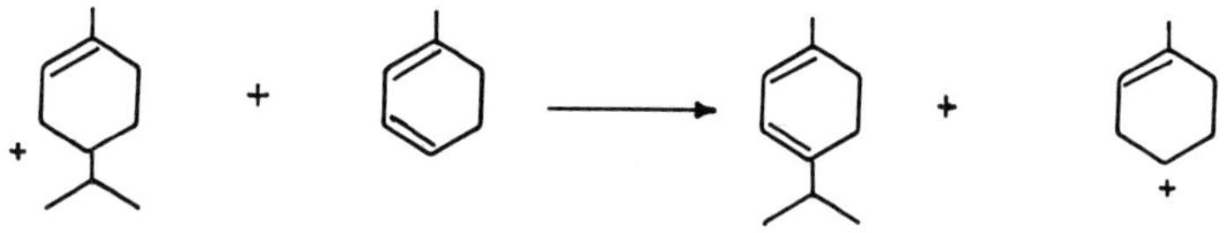

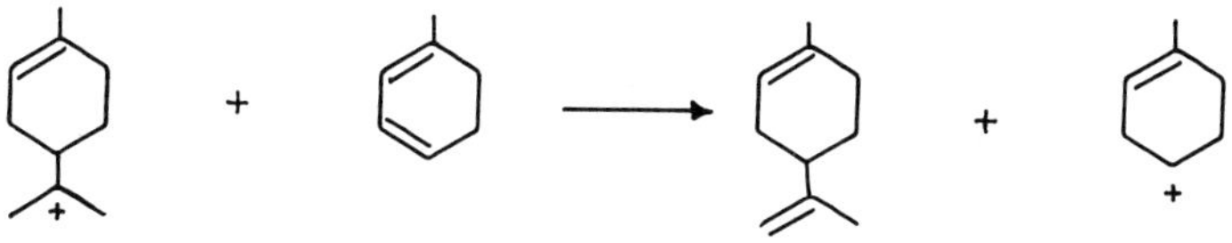

In practice, an alternative to this formulation may be more realistic, as the above reactions involve free migration of the complex at the surface. In this case, the ion may be thought to give a proton back to the catalyst, which immediately will pass it over to freshly adsorbed species according to step 1.

In order to maximize the selectivity towards limonene, it is clear that the formation of propyl carbonium ions should be as low as possible. This must probably be effected by taking propylene as the limited reactant. In this way, dialkylation of the cyclic species is also kept low.

The biggest problem with this approach lies in the number of reactions that a carbonium ion can be involved in. Considering, for example, the combination of the two species, we may write:

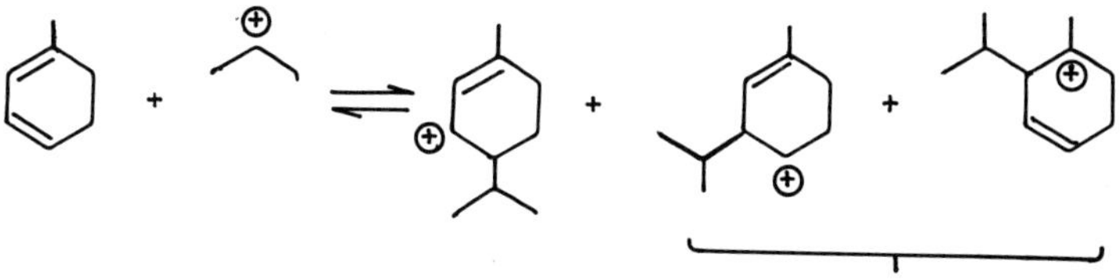

may be avoided by the correct catalyst geometry

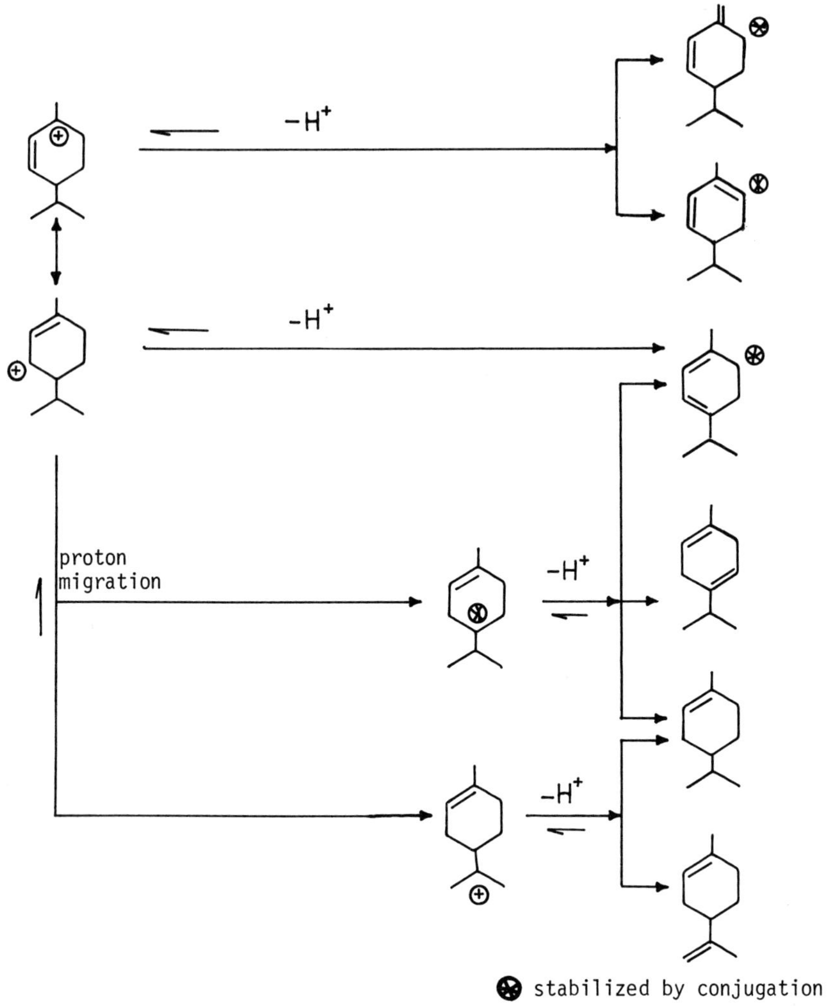

-H⁺

-H⁺

proton
migration

-H⁺

-H⁺

stabilized by conjugation

It seems obvious, even from this limited consideration of possible reactions, that the selectivity of an acid catalysed reaction towards limonene will be small. As a result, the approach was abandoned.

V. DESIGN OF THE REST OF THE CATALYST

One useful feature of this design lay in the fact that attempts were made to recognise potential problems. Thus, for example, it was recognised that excess acidity in the catalyst would promote polymerisation, and that over-oxidation to carbon dioxide could be a problem. Metallic catalysts could also be poisoned by sulphur or nitrogen. No experimental testing was carried out, and no attempt was made to design secondary components of the catalyst.

In considering the physical state of the catalyst, the designer recognised that limonene was a bulky molecule. Writing diffusion coefficients as (ref. 18):

Knudsen diffusion $D_K = \frac{2}{3} r_e u$

Bulk diffusion $D_B = \frac{1}{3} \beta u$

where u is the Maxwellian velocity, r_e is the mean pore radius and β is the mean free path, he recognised that, for Knudsen diffusion to be insignificant,

$$2r_e \gg \beta$$

Under ideal conditions, the mean free path, β, may be written as (ref. 18):

$$\beta = \frac{RT}{2\pi d^2 PN}$$

where P is the total pressure, N the Avogadro number, d the diameter of the molecule, R the gas constant and T the temperature. Taking a value of the limonene molecules diameter as 6 Å gives β = 38 Å. As a result, it was suggested that the pore diameter should be greater than ca. 100 Å

VI. COMMENTS

The design reflects both the industry and the inexperience of the designer. Considering the reaction

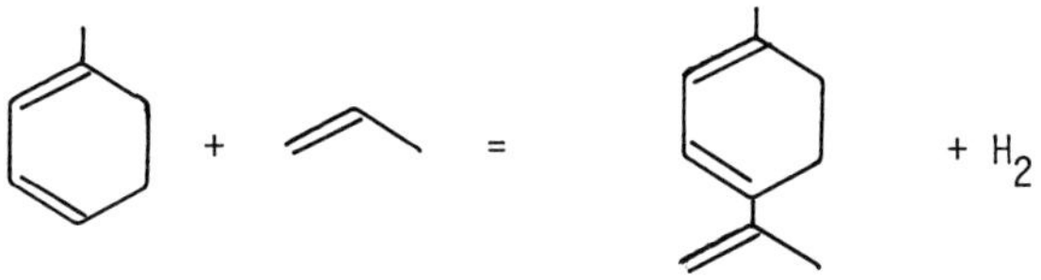

+ = + H_2

it is clear that a good design was carried out and that this suggested an oxidative route.

There is also no doubt that, in dealing with the hydrogenation reactions required to effect the reaction:

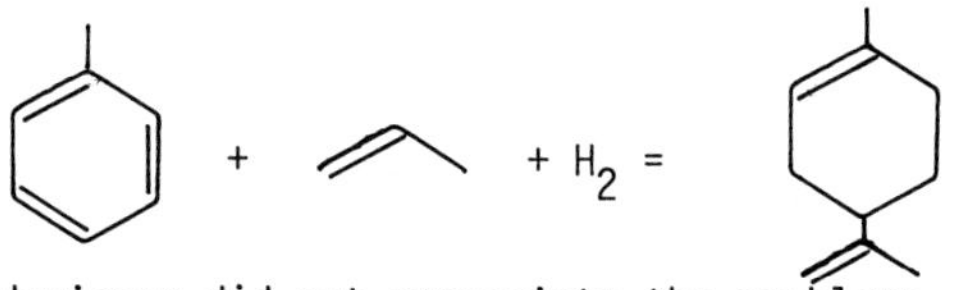

+ + H_2 =

the designer did not appreciate the problems, particularly those arising from the desired selectivity.

Nonetheless, the designer carried out an excellent design, bearing in mind the circumstances under which the exercise was carried out. The design illustrates what can be achieved in a relatively short time by an inexperienced designer.

REFERENCES

1 E. Guenther, "The essential oils" Vol. 2, Van Nostrand, New York, 1949.
2 A. Wiessberger, E.S. Proskanev, "Organic Solvents", Second Edition, Interscience, New York, 1955.
3 W. Templeton, "An Introduction to the Chemistry of Terpenoids and Steroids", Butterworths, London, 1969.
4 J.E. Germain, "Catalytic Conversion of Hydrocarbons", Academic Press, London, 1968.
5 D.W. Van Krevelen and H.A.G. Chermin, Chem. Eng. Sci. 1, 66 (1951).
6 G.J. Janz, "Estimation of Thermodynamic Properties of Organic Compounds", Second Edition, Academic Press, New York, 1967.
7 Chem. Marketing Reporter, August 30 (1976).
8 European Chem. News, 29, No. 753, Sept. 17 (1976).
9 J.R. Anderson, "Structure of Metallic Catalysts", Academic Press, New York, 1975.
10 D.A. Dowden, Catal. Revs. Sci. Eng. 5, 1 (1972).
11 O.V. Krylov, "Catalysis by Non-Metals", Academic Press, London, 1969.
12 G.C. Bond, "Catalysis by Metals", Academic Press, London, 1962.
13 R.C. West, (Ed.), "Handbook of Chemistry and Physics", 56th Edition, CRC Press, Cleveland, 1975.
14 J.M. Peacock, M.J. Sharp, A.J. Barker, P.G. Ashmore and J.A. Hockey, J. Catal. 15, 373,379,387,398 (1969).
15 W.L. Jorgensen and L. Salem, "The Organic Chemist's Book of Orbitals", Academic Press, New York, 1973.
16 D.A. Dowden, Chem. Eng. Progr. Symp. Sci. 63, No. 73, 90 (1967).
17 B.S. Greensfelder, H.H. Voge and G.M. Good, Ind. Eng. Chem. 41, 2573 (1949).
18 J.M. Thomas and W.J. Thomas, "Introduction to the Principles of Heterogeneous Catalysis", Academic Press, London, 1970.

CHAPTER 15

DESIGN OF A METHANATION CATALYST

One of the features of the nineteen seventies has been the realisation that supplies of fuels such as oil or natural gas are limited. As a result there has been increasing attention focused on the conversion of fuels from a less to a more desirable form. In this connection, perhaps the best known process involves the partial oxidation of coal to carbon monoxide and hydrogen (refs. 1,2) followed by the recombination of these gases to form methane (refs. 1,2,3). Despite the fact that the initial gasification is probably more critical, the methanation reaction has been studied in detail (refs. 3,4,5,6).

This interest was reflected by two catalyst designs, completed in 1974 and 1978. The two designs are of interest, both as individual exercises and in comparison, since the sophistication of the design and the amount of background information had increased considerably in the intervening years. As before, they are presented as they were carried out, and comments are made at the end of both designs.

In view of the large amount of recorded information on methanation, appropriate studies are quoted with reference to relevant parts of the designs rather than in an introduction.

A. METHANATION (1974)

I. DESCRIPTION OF THE IDEA

The methanation reaction has been well known for many years. The overall process may be represented by the reactions

$$CO + 3H_2 \rightleftarrows CH_4 + H_2O \quad (1)$$

$$2CO + 2H_2 \rightleftarrows CH_4 + CO_2 \quad (2)$$

$$CO_2 + 4H_2 \rightleftarrows CH_4 + 2H_2O \quad (3)$$

$$CO + H_2O \rightleftarrows CO_2 + H_2 \quad (4)$$

Other reactions which may occur may be undesired or desired. Reactions producing carbon

$$2CO \rightleftarrows CO_2 + C \quad (5)$$

$$CO + H_2 \rightleftarrows H_2O + C \quad (6)$$

$$CH_4 \rightleftarrows 2H_2 + C \quad (7)$$

are not desired, since carbon can deposit on the catalyst to cause loss of activity.

Fischer-Tropsch reactions, on the other hand, may not be totally undesired since the products may be valuable in their own right or may add to the calorific value of methane. Typical reactions include

$$(Zn+1)H_2 + nCO \rightleftharpoons C_nH_{2n+2} + nH_2O \tag{8}$$

$$(n+1)H_2 + ZnCO \rightleftharpoons C_nH_{2n+2} + nCO_2 \tag{9}$$

$$ZnH_2 + nCO \rightleftharpoons C_nH_{2n} + nH_2O \tag{10}$$

$$nH_2 + ZnCO \rightleftharpoons C_nH_{2n} + nCO_2 \tag{11}$$

$$ZnH_2 + nCO \rightleftharpoons C_nH_{(Zn+1)}OH + (n-1)H_2O \tag{12}$$

A thorough review of the thermodynamics of the reactions was undertaken, both with respect to methanation and to the Fischer-Tropsch reactions. Methanation reactions were found to be exothermic with high heats of reaction (ref. 3): as a result, the removal of heat from the reactor could be an important aspect of design. Methane yields were found to be highest at low temperatures and high hydrogen: carbon monoxide ratios and the production of carbon (the major undesired product) is minimised under the same conditions.

The thermodynamics of the Fischer-Tropsch reaction was also examined (refs. 3, 7), with the general conclusions

i. Reactions that form carbon dioxide have larger equilibrium constants than those that form water and are, as a result, thermodynamically favourable at higher temperatures.
ii. The free energy changes per carbon atom are more negative for reactions producing methane than for reactions producing higher hydrocarbons. Below 500°C, reactions producing graphite are more negative than reactions yielding higher hydrocarbons but less negative than reactions producing methane.
iii. The practical upper limit of the formation of higher hydrocarbons is about 500°C.

The designers concluded from this study of the thermodynamics, that

i. High conversion to methane could be attained at pressures of the order of 25 atm.
ii. Carbon deposition would be favoured if the hydrogen:carbon monoxide ratio was less than 2.5, the pressure was low and the temperature was high.
iii. Methanation would not be favoured at above ca. 620°C.

iv. The production of heavy hydrocarbons would not be favoured above ca. 400°C.

II. HEATS OF ADSORPTION AND CATALYST DESIGN

The designers next moved directly to the consideration of adsorption in the system.

i) Hydrogen

Based on the argument that activity could be related to strength of adsorption, the designers first identified elements that could adsorb hydrogen, and then sought values of binding energies. These were presented in a tabular form (ref. 8) as

V	Cr	Fe	Co	Ni	Cu	
	74	68	64	64	56	kcal mole^{-1}
Nb	Mo	Ru	Rh	Pd	Ag	
	72	65	65	65	53	kcal mole^{-1}
Ta	W	Os	Ir	Pt	Au	
75	75		65	65	~50	kcal mole^{-1}

ii) Carbon monoxide

Heats of adsorption of carbon monoxide on evaporated films were found in the literature (ref. 8) and reproduced without comment, except to note that intermediate values of heats of adsorption were obtained on iron, cobalt, nickel, rhodium, palladium and platinum. It was also noted that McKee (ref. 9) had studied the co-adsorption and interaction of carbon monoxide and hydrogen on unsupported noble metals maintained at 100-200°C. Carbon monoxide was found to be preferentially adsorbed on platinum, rhodium and iridium, and the interaction with hydrogen was small. Adsorbed carbon monoxide could be completely removed from ruthenium by reduction with hydrogen at 150°C, methane being the only product observed.

In considering the nature of the bonding that could be involved in methanation, an analogy with metal carbonyls was used in the first instance. Bonding of carbon monoxide to a transition metal atom proceeds by the donation of the lone pair of electrons on the carbon atom to vacant d orbitals of the metal, with stabilisation of the bond being obtained by back donation of electrons from filled metal d orbitals to the vacant π^* orbitals on the carbon monoxide.

The designers also recognised that two (bridged and linear) forms of adsorbed carbon monoxide may be present (refs. 1,5,8). It was suggested that the bridge form might be less important on ruthenium, because the metal-metal distance in the lattice was greater than, for example, nickel.

III. PROPOSED MECHANISMS

Several possible mechanisms were considered by the designers, based on either bridged or linear adsorbed carbon monoxide. They recognised that the desired reaction

$$CO + 3H_2 \rightleftharpoons CH_4 + H_2O \tag{1}$$

involved production of water, and assumed (wrongly, as we shall see below) that silica-alumina, used as a dehydration catalyst (ref. 10), would favour the reaction.

Mechanism (a).

A bridged CO molecule was suggested to be hydrogenated and protonised:

$$\text{H{-}M} + (\text{M})_2\text{C{=}O} + \text{H{-}M} \longrightarrow (\text{M})_2\text{C(H)(OH)} + \text{H}^+\text{A}^- \rightleftharpoons (\text{M})_2\text{C(H)(OH}_2^{\,+}) + \text{A}^-$$

$$(\text{M})_2\text{C(H)(OH}_2^{\,+}) \rightleftharpoons \text{H}_2\text{O} + (\text{M})_2\text{C(H)}^{\oplus} + \text{H{-}M} \longrightarrow (\text{M})_2\text{CH}_2 \tag{13}$$

$$(\text{M})_2\text{CH}_2 + \text{H}_2 \longrightarrow \text{H{-}M} + \text{M{-}CH}_3 \longrightarrow \text{M{-}M} + \text{CH}_4$$

The designers realised the fundamental fault in this mechanism (apart from the general feasibility, which is open to question) that the protonated intermediate remaining after extraction of water had mysteriously disappeared. As a result, they proposed a related, but more probable, mechanism

$$\text{H{-}M} + (\text{M})_2\text{C{=}O} + \text{H{-}M} \longrightarrow (\text{M})_2\text{C(H)(OH)} + \text{H{-}M} \longrightarrow (\text{M})_2\text{C}^{\bullet}\text{H} + \text{M} + \text{H}_2\text{O} \tag{14}$$

$$(\text{M})_2\text{C}^{\bullet}\text{H} + \text{H{-}M} \longrightarrow (\text{M})_2\text{CH}_2 + \text{H}_2 \longrightarrow \text{CH}_4 + \text{M{-}M}$$

Mechanism (b).

An alternative mechanism was proposed, based on arguments put forward by McKee (ref. 9) for methanation on unsupported ruthenium

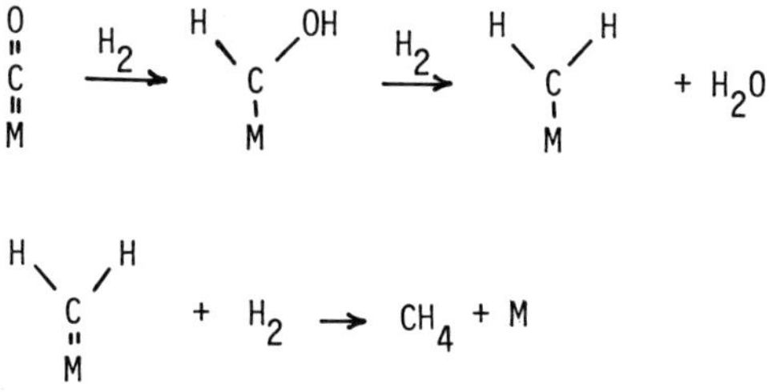

$$M{=}C{=}O \xrightarrow{H_2} M{-}C(H)(OH) \xrightarrow{H_2} M{-}CH_2 + H_2O \qquad (15)$$

$$M{=}CH_2 + H_2 \rightarrow CH_4 + M$$

As they correctly pointed out, hydrogen insertion almost certainly proceeds via chemisorbed hydrogen rather than via hydrogen gas. As a result, they transcribed the reactions in terms of possible surface arrangements.

Mechanism (a) : surface

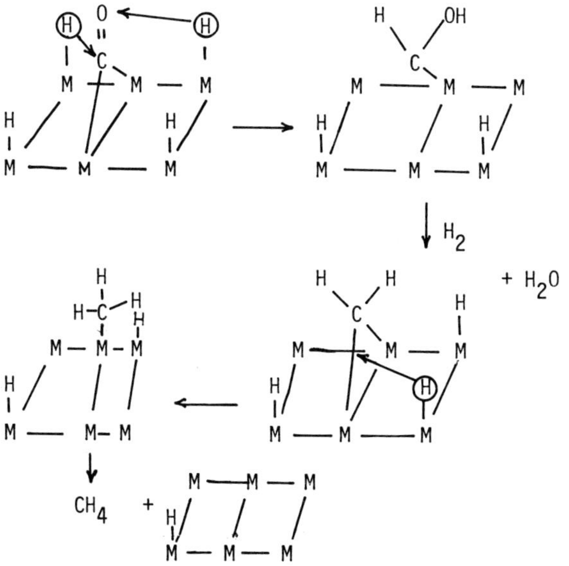

M M M M M M M M M M M M — H, O, C, OH, H_2, $+ H_2O$, CH_4 (16)

Mechanism (b) : surface

H O H H M M M C H M M M — OH, $+ H_2$

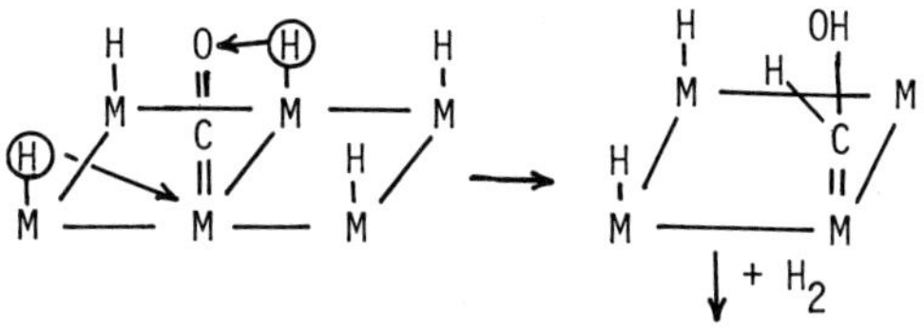

```
                    H
                    |                 1+
                    M ---------- M
  H                /  H     H  /
  |               /    \   /  /
  M -- M  + CH4 <- H    C   /      + H2O        (17)
 /    /            |    ||/
M -- M             M ---- M
```

Accepting, for the moment, these proposed mechanisms, the designers turned their attention to the main undesired reaction -- the formation of carbon. They suggested that the Boudouard reaction

$$2CO \rightleftarrows CO_2 + C \qquad (5)$$

was probably the most serious problem, and searched the literature for information. It has been reported that carbon formation from carbon monoxide occurs on iron, nickel and cobalt single crystals (ref. 11), but does not occur on Cu, Ag, Cr, Mo, Pd and Rh. They missed the significance (see later) of the fact that carbon was formed on Co cubic single crystals at $550^{o}C$ and hexagonal single crystals at $410^{o}C$: apart from anything else, this would indicate that carbon formation was influenced by geometric factors. Comparing the three metals, carbon was formed on iron almost twice as fast as on Ni and about four times as fast as on cobalt (ref. 12). Iron-nickel and iron-cobalt alloys allowed more carbon formation than pure iron, but cobalt-nickel alloys produced carbon as would be expected from the component metals. Iron was also known to be an efficient Fischer-Tropsch catalyst (ref. 7), and could favour the production of hydrocarbons and alcohols of higher molecular weight than methane. As a result, iron was rejected as a possible methanation catalyst.

The designers then returned to the mechanisms proposed on the surface, and made a fundamental mistake in allowing themselves to be influenced by what they knew about catalysts which had been reported in the literature (refs. 4,5,6). In particular, they accepted that the catalyst had to be a metal (see later) and they allowed themselves to be influenced by the fact that ruthenium and nickel are known to be active catalysts (refs. 4,5,6).

As a consequence of the first assumption, they related the mechanisms proposed to geometric properties of various metals, ignoring the possibility of metal oxide catalysts. Based on the fact that C-H and O-H bonds must be formed, the preferred geometric distances (based on bond lengths) were compared with lattice parameters of metals. This showed that tantalum, nickel or rhodium were possible catalysts.

As a result, they finalised their primary design at this point, using the information given below.

IV. SUMMARY

i. Adsorption strength

Hydrogen : iron, cobalt, nickel, ruthenium, rhodium, palladium, iridium and platinum have 'intermediate' binding energies for hydrogen.

Carbon Monoxide : iron, cobalt, nickel, rhodium, palladium and platinum have 'intermediate' values of heats of adsorption.

ii. Geometric factors

Arguments based on bond lengths and the mechanisms proposed suggest tantalum, nickel or rhodium as possible catalysts.

iii. Carbon formation

The desire to avoid excessive carbon formation suggests that iron should not be used. No comments were made concerning the activity of nickel for carbon formation (refs. 11,12).

iv. Supports

The production of water in the reaction led to the suggestion that silica-alumina would be a good support, since this could favour dehydration.

We shall see, in the discussion of the two designs, that very many errors were made in this part of the design, even though the primary constituents emerging from the design do include two catalysts which are known to be very active (nickel and ruthenium). However, in this particular exercise, the students extended the design to chemical engineering aspects of the problem, and here they produced a much better prediction.

V. CHEMICAL ENGINEERING

Aspects of design: heat and mass transfer in catalyst pellets

Using the rate expression reported by Rostrup-Nielsen (ref. 13), calculations were carried out to show that the rate was very dependent on temperature. In addition, the thermodynamic study had shown that the reactions were very exothermic, and that carbon formation would be favoured at higher temperatures (see above). It is also known that sintering of nickel catalysts may be expected above about $600^{o}C$ (ref. 14).

The designers started by calculating the thickness of the pellet in terms of the maximum allowed temperature ($600^{o}C$). Using a continuous slab of catalyst of thermal conductivity, λ, the heat balance equation was written

$$\frac{d^2T}{dZ^2} = -\frac{\Delta H}{\lambda} \cdot R \tag{18}$$

where R is the rate of reaction and ΔH is the heat of reaction. Comparing the rate at the inlet temperature with the rate at $600^{o}C$, this equation can be solved to show that the thickness of catalyst should lie between 10^{-3} and 10^{-5} cm i.e. that a supported catalyst must be used.

The arguments were extended to consider a catalyst bed (ref. 15), where the major problems were found to be boundary layer diffusion effects and heat transfer through the catalyst bed. As a result, it was suggested that the catalyst could either be coated on the wall of a heat exchanger (as has been done in practice (ref. 16)) or that a fluidised bed reactor (with improved heat transfer properties) should be considered.

B. METHANATION (1978)

A considerable amount of work had been carried out on the methanation reaction in the period between the two designs. As before, this is probably best discussed at appropriate places in the text, but two lines of study proved particularly relevant.

The first of these involved an intensive study of the kinetics of methanation over various metals by Vannice (refs. 6,17,18). The activity of different metals based on turnover numbers was compared, both with respect to the formation of methane and to the formation of higher hydrocarbons. The average molecular weight of hydrocarbon products formed was found to decrease over various catalysts in the order

Ru > Fe > Co > Rh > Ni > Ir > Pt > Pd

The specific activity for methanation was found to be related to the heats of adsorption of carbon monoxide and hydrogen (refs. 6,17,18), activity decreasing with increasing heat of adsorption of carbon monoxide on different metals and increasing heat of adsorption of hydrogen.

A second, very important, line of study has been focused on the mechanism of reaction. Early work had suggested that methanation did not involve the formation and subsequent hydrogenation of carbon on the surface, on the grounds that carbon hydrogenation rates were considerably slower than methanation rates (ref. 5). The use of isotopic tracers showed, in fact, that this was incorrect (refs. 19-22), and that the reaction proceeds mainly by the sequence

$$CO + 2M = MO + MC \qquad (19)$$

$$MC + 2H_2 = M + CH_4 \qquad (20)$$

$$MO + H_2 = M + H_2O \qquad (21)$$

The discrepancy arises mainly because of the previously undetected presence of an active carbon layer on the surface. Adsorption of carbon monoxide produces such a deposit which, in time, reorganises to the less active carbidic or graphitic form of carbon (ref. 23). Following the detection of the multi-step reaction path, considerable attention has been focused on the role of surface ensembles in methanation and related reactions (refs. 19-22). Bond and Turnham (ref. 24) suggest that an ensemble of four Ru sites is needed to catalyse methanation.

I. DESIGN OF PRIMARY COMPONENTS.

The design was based on the short description of the desired reaction as reproduced below:

$$CO + M(2M) = \underset{(linear)}{COM} \text{ or } \underset{(bridged)}{COM_2} \qquad (22)$$

$$H_2 + 2M = 2H\text{-}M \qquad (23)$$

$$COM \text{ (or } COM_2) = CM + OM \qquad (24)$$

$$CM + 4H\text{-}M = CH_4 + 5M \qquad (25)$$

$$OM + 2H\text{-}M = H_2O_{ads} + 3M \qquad (26)$$

$$H_2O_{ads} = H_2O_{gas} \qquad (27)$$

Adsorption of carbon monoxide in the linear or bridged form could be possible as a precursor to formation of C-M. As a result the reaction was suggested to involve (a) adsorption of reactants (reactions 22,23), (b) dissociation of carbon monoxide (reaction 24), (c) the desired reactions (reactions 24,25 and 26) and (d) desorption of products (reaction 27). It was realised that the active site, M, could be a metal, a metal oxide or a metal sulphide, although the former was more likely.

(a) Heats of adsorption approach.

Since the proposed mechanism involves adsorbed carbon monoxide and adsorbed hydrogen, potentially active metals must be limited to W, Ta, Mo, Ti, Zr, Fe, Ca, Ba, Ni, Pt, Rh and Pd (ref. 8). Linear and bridged forms of adsorbed carbon monoxide have been detected on most of these metals (ref. 8), with up to six forms of adsorption being detected on nickel (ref. 5).

From reactions 25 and 26, it would be expected that a catalyst adsorbing more hydrogen than carbon monoxide would be desirable. However, as shown by Vannice (ref. 6), high activity is related to high heats of adsorption of hydrogen and to low heats of adsorption of carbon monoxide. This is presumably a consequence of the necessity of adsorption in the correct forms.

If this is the case, then some advantage could result from the use of a two-component system, in which one component favours the dissociative adsorption of carbon monoxide and the other acts to supply the large amounts of hydrogen required. From the work of Vannice (ref. 6) Ni or Ru is the first component, but several arguments can be applied to the choice of the second component.

It is obviously necessary for the component to adsorb large amounts of hydrogen and, as such, palladium is an obvious alternative. Other "hydrogen reservoirs" that have been developed include magnesium nickel, iron titanium, lanthanum nickel and misch-metal (ref. 25), in which the gas occupies interstitial spaces within the crystal lattices. Carbides such as TC and WC have also been shown to adsorb hydrogen preferentially from a CO/H_2 mixture (ref. 6), and borides/borohydrides are also of considerable interest in this respect (ref. 26).

However, it is necessary to have both an adequate source of hydrogen and a means of delivery to the desired site. As a result, it is necessary to consider factors such as spillover. Sermon and Bond (ref. 27) have shown that hydrogen spillover is high on Pd/Al_2O_3 and Boudart et al. (ref. 28) have shown a similar effect on Pt/C. In addition, the ease of reduction of metal oxides by hydrogen spilt over from various metals has been reported (ref. 27). It is interesting to note that, in many cases, other metals are more effective than Pd or Pt in promoting oxide reduction: thus, for example, a mechanical mixture of MoO_3 with several metals, gives the efficiency of additives in increasing the rate of oxide reduction as (ref. 27):

$$Cu,Ag > Co > Ir > Ti > Fe > (Re,Ta,Rh,Ru) > (Os,Cr,Ni) > (W,Pd,Pt)$$

and, for metal oxides used to promote the reduction of NiO,

$CuCr_2O_4$ > Pd/Al_2O_3 > $CoMoO_4$ > $NiMoO_4$

This information is useful, but must be treated with care since it reflects (a) the adsorption of hydrogen by the additive, (b) the transport of hydrogen to the active site, and (c) the reducibility of the oxide. This latter step could well be different from the reducibility of carbon monoxide or of fragments derived therefrom. Nonetheless, the studies provide a useful pointer which we will return to at a later stage.

To summarise, then, the heats of adsorption approach suggests the following:

(i) Specific activities for methanation will be highest on materials for which the heat of adsorption of carbon monoxide is low and that of hydrogen is high (ref. 6). Suitable metals include Ru and Ni (refs. 4,5,6).

(ii) Active catalysts may involve one component capable of dissociatively adsorbing large amounts of hydrogen (materials suggested include Pd, magnesium nickel, iron titanium, lanthanum nickel, misch-metal, TC, WC and borides/borohydrides).

(iii) Activity may be enhanced by spillover effects. Hydrogen spillover is high on Pd/Al_2O_3 and Pt/C. Based upon the reducibility of oxides, the following additives have been found to accelerate spillover/reduction (ref. 27).

TABLE 15.1
Efficiencies of additives intended to favour hydrogen spillover and oxide reduction

Oxide to be reduced	T (^oC)	Efficiency of additive in increasing rate of reduction
MoO_3	515-765	Cu,Ag > Co > Ir > Ti > Fe > (Re,Ta,Rh,Ru) > (Os,Cr,Ni) > (W,Pd,Pt)
NiO	211-232	Cu > Pd,Pt
NiO	283	$CuCr_2O_4$ > Pd/Al_2O_3 > $CoMoO_4$ > $NiMoO_4$
NiO	300	Mo_2C > NbC > ZrC
NiO	200	Ir,Os > Rh > Cu > Ru > Pd > Pt

These results will be returned to at a later stage in the design.

(b) Design based on the dissociative adsorption mechanism

The second approach to design was based firmly on the dissociative adsorption mechanism for methanation (refs. 19.22). Writing the reaction as

$$CO + 2M = MC + MO \tag{19}$$

$$MC + 4H\text{-}M = 5M + CH_4 \quad (25)$$

$$MO + 2H\text{-}M = 3M + H_2O \quad (26)$$

then the rate determining step may be the dissociative adsorption (reaction 19) or the hydrogenation of fragments (reactions 25 and 26) to generate free sites on which further adsorption can take place. In addition, several pertinent questions can be asked:

(i) If reaction (19) is rate determining, is the dissociation of CO rate determining or is it the formation of MC or MO?

(ii) Why should the two sites used in reaction (19) be the same? Would it not be possible to write

$$CO + M + M' = MC + M'O \quad (19')$$

(iii) Would there be an advantage with respect to the case of reactions (25) or (26) if the two sites were different?

As a result, differing possibilities amenable to investigation were considered individually.

(b)i. Formation and hydrogenation of M,C

There are obvious problems in this approach. It is easy to select a metal (such as Ta or W) that will readily form a carbide (and, presumably, accelerate reaction 19), but the resulting carbide will be hard to hydrogenate (slow reaction 25) and the overall reaction will be slow. What is needed is a centre which will favour dissociative adsorption but will allow easy hydrogenation. The difficulty is, however, that we know next to nothing about the form of carbon on the surface - except that it is not a carbide and is not amorphous or graphitic carbon. All we can say is that it is produced on nickel (ref. 19) and is, presumably, produced on ruthenium. As an aid to design, this is certainly not illuminative.

(b)ii. Formation and hydrogenation of M'O

In this case, we are on much stronger ground. In terms of reaction (19') it is possible to state that M must be Ni or Ru (see above) but that M' could accelerate reaction by favouring adsorption and subsequent hydrogenation of oxygen. As a result, there may be a correlation between methanation activity and ease of acceptance of O by M' and/or ease of hydrogenation of M'O. In assessing this, factors such as the heat of adsorption of oxygen on metals or metal oxides or the ease of reducibility of metal oxides are obvious possibilities, and correlations with less direct measures of the oxidisability/reducibility of metals or metal oxides (such as redox potentials) may also be revealing. In addition, however, we may go further. Since it is known that Ni (M = M' = Ni) and Ru(M = M' = Ru) are active catalysts, some measure of the desired properties of M' may be obtained.

As a result of arguments such as these, experimental testing of the proposed correlations was initiated (ref. 29). Using the minimum temperature at which methanation could be initiated as a guide to catalytic activity, combinations of nickel and other components were prepared and tested. The results were plotted against various functions expected to reflect the ease of production/ease of reducibility of oxides of the second component as shown in Figures 15.1-15.5 (see later). Note that there is no necessity for the second component to be a metal, since oxides can also adsorb oxygen (e.g. $Cr^{2+} \rightarrow Cr^{3+}$) and be subsequently reduced to the original oxide. In some cases, due to a lack of data, there is some uncertainty as to which oxide is present under ambient conditions, and the appropriate function is plotted as a bar between values for possible oxides.

The results show that there is, indeed, a correlation between activity and factors that would be expected to reflect the ease of O adsorption/ease of reduction, but that there is no clear indication of which reaction is rate controlling. In an attempt to resolve this, experiments were carried out at different temperatures. Noting that the minimum temperature at which the overall methanation reaction could be carried out was ca. 110°C, stepwise experiments were completed. In the first, carbon monoxide was passed over a nickel molybdate catalyst at 180°C, the flow was stopped and the temperature dropped to 80°C. Hydrogen was then passed over the catalyst but methane was not produced. Secondly, carbon monoxide was passed over the catalyst at 80°C and the bed was then rinsed with nitrogen. The temperature was raised to 180°C and hydrogen was passed to yield methane. As a result, it would appear that the hydrogenation reactions (25 and 26) are rate determining, even though the overall reaction has been shown not to involve an isotope effect (ref. 30).

(c) Experimental testing

Using the minimum temperature at which methanation could be observed as a guide to catalytic activity, a series of catalysts were prepared and tested. No attempt was made to optimise catalyst preparation, metal salts being precipitated on γ-alumina, calcined in air at 500°C and reduced overnight before use.

Vannice has shown that catalytic activity can be related to the heats of absorption of carbon monoxide and hydrogen (ref. 6), and Bond has shown that heats of absorption can be related to periodic group number (ref. 8). As a result, a check on the accuracy of the results can be made by plotting activities against periodic group number. As shown in Figure 15.1, catalytic activity is, indeed, related to periodic group number.

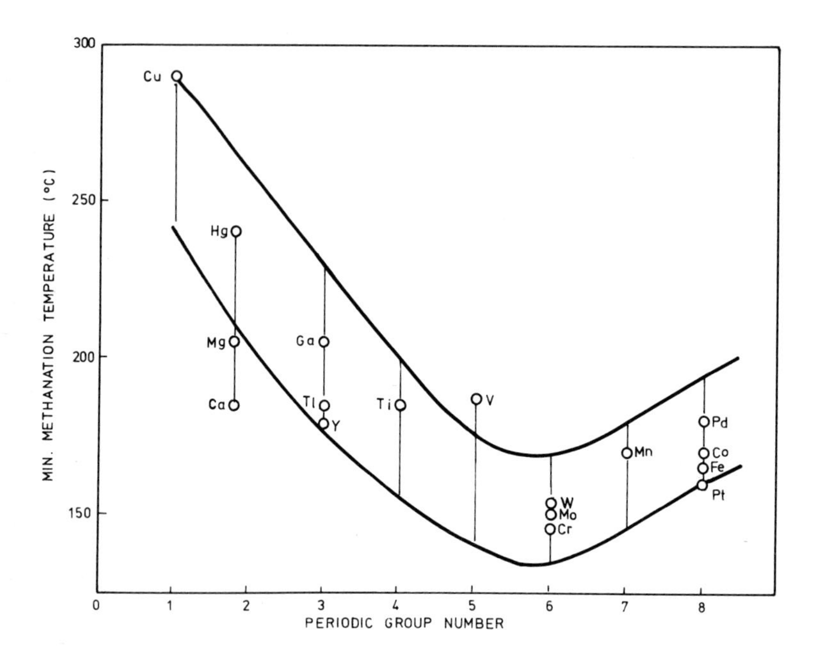

Fig. 15.1. The minimum temperature at which methanation is observed as a function of periodic group number.
The catalysts involve a combination of nickel and the element shown (present as a metal or an oxide) supported on alumina.

In attempting to relate catalytic activity to the ease of production/ease of hydrogenation of M/O, several correlations were tested. The first involved the heat of adsorption of oxygen on various materials as measured by Boreskov (ref. 31). Although the data is limited, a satisfactory correlation was found (Figure 15.2).

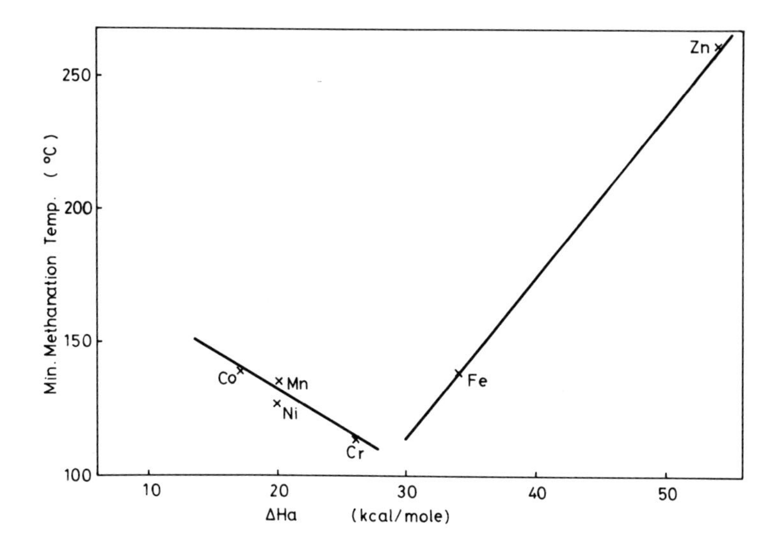

Fig. 15.2. Catalytic activity as a function of heats of adsorption of oxygen (ref. 31).
Catalysts are combinations as described in the legend of Fig. 15.1.

This approach was extended by comparing catalytic activity with the heat of formation per oxygen atom of the additive (ref. 32). Although the data are less accurate than the measurements by Boreskov et al. (ref. 31), a good correlation was observed. (Figure 15.3).

It was suggested that the ease of production/reduction of M'O might also be reflected in the redox potential of the ions involved (ref. 33). Although the systems are very different, this was, indeed, found to be the case (Figure 15.4).

Further to these studies, it was realised that the correlation established between crystal field stabilisation energies and the catalytic oxidation of hydrogen over oxides (ref. 34) could be relevant. As seen from Figure 15.5, a similar correlation could be established for catalytic methanation.

As a result, there seems little doubt that catalytic activity can, indeed, be related to the production/reduction of M'O during methanation.

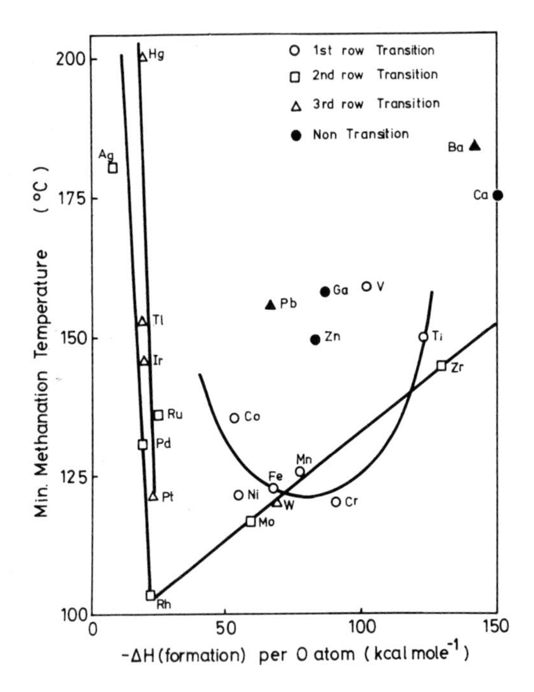

Fig. 15.3. Catalytic activity as a function of the heat of formation per oxygen atom (ref. 32).

Catalysts are combinations as described in the legend of Fig. 15.1. The oxide present was calculated on thermodynamic grounds only.

(d) Removal of products

The third approach to design was based on these results. Since it would appear that the removal of adsorbed species from the surface is rate controlling, is there any way to accelerate this removal? Dealing with the O fragment initially, one possible means could be suggested. Presumably the process is rate determining because O blocks the surface. If there is any way to diffuse O into the bulk and, possibly, to hydrogenate at a site other than that used in methanation, then the methanation site will remain free and methanation will be faster. As a consequence, the literature was searched for materials that would favour bulk diffusion of oxide.

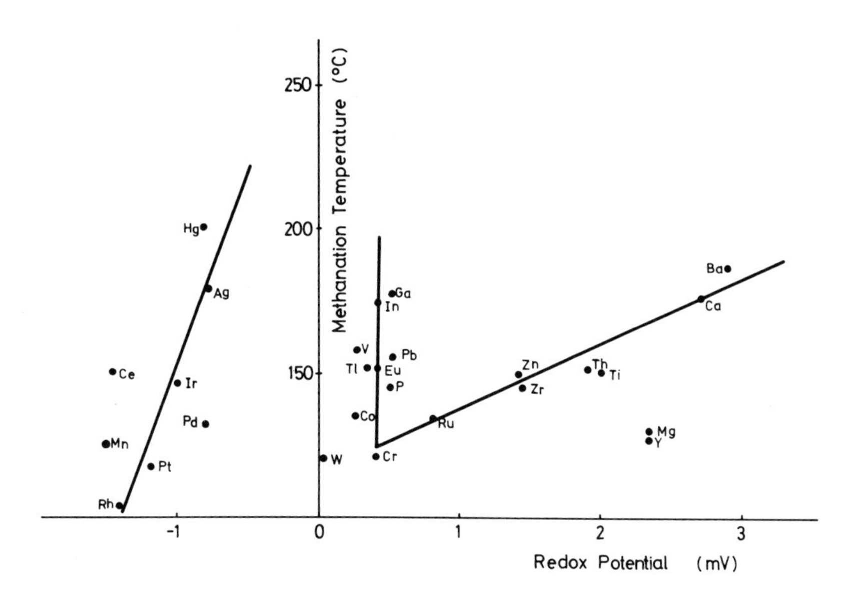

Fig. 15.4. Catalytic activity as a function of redox potential (ref. 33)
Catalysts are combinations as described in the legend of Fig. 15.1. The oxide present was calculated on thermodynamic grounds only.

There is good evidence that traces of oxygen adsorbed on nickel (ref. 35) and on rhodium (ref. 36) do not stay on the surface, but diffuse into the bulk. In addition, bismuth ions have been suggested to favour the migration of oxygen (ref. 37), and zirconia and cobalt oxides have been suggested to "store" oxygen in their bulk. These arguments would suggest, then, that combinations of Ni or Ru (to favour reaction 19) with rhodium, bismuth oxide, cobalt oxide or zirconia could be active catalysts.

One other product might be expected to block the surface, and this is water produced in reaction (26). Anhydrous sodium sulphate or urania (ref. 14) have been suggested to influence the supply of water on the catalyst surface, and combinations of Ni or Ru with either of these compounds should be tested.

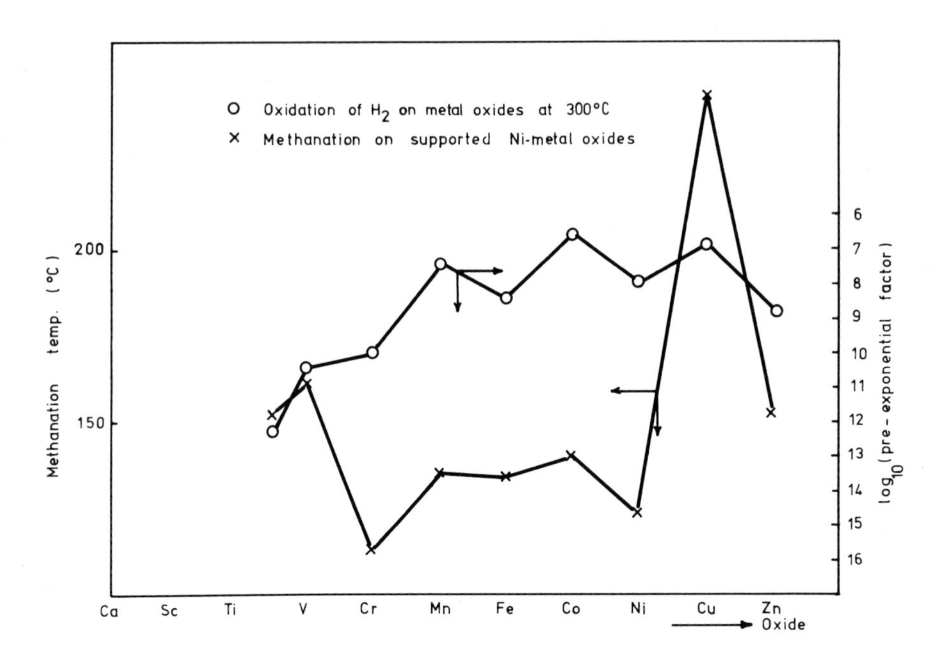

Fig. 15.5. Catalytic activity as a function of crystal field stabilisation energy.

(e) Other approaches to design

In considering the general framework of design described in Chapter 2, several difficulties arise. An approach based on coordination chemistry is possible in connection with the production or reduction of M'O (reactions 19 and 26) and, as seen from Figures 15.1-15.5, this would be successful. For MC, however, little is known about the nature or the coordination of the carbon, and such an approach will fail. Similar arguments can be advanced in considering possible valency changes in the catalyst, where an approach based on the valency change caused by acceptance or loss of O led to successful prediction (Figure 15.4), but no comment could be made concerning the formation or removal of MC.

Arguments based on geometric factors can also be partially successful. It would be expected (and, indeed, it has been shown (ref. 24)) that an ensemble of active centres is necessary on the surface. However, it is also shown that surface migration of species does occur. As a result, it is unlikely that arguments based on bond length vs lattice distance would give an accurate prediction.

More success can be achieved in considering the nature of the support. Methanation is exothermic, and the support must be thermally stable. Methane is, however, a terminal product of high stability, and selectivity problems will not arise. Chemical interactions with the catalyst are undesired (for example, the formation of nickel aluminate), but surface migration of adsorbed species should be favoured. As a result, alumina and carbon (which are known to favour hydrogen spillover (refs. 27,28)) were selected, with the proviso that reaction conditions should be such that attack of the carbon support by hydrogen should be minimal i.e. the temperature should be less than 375°C (ref. 38).

II. SUMMARY OF THE DESIGN

The catalysts suggested on the basis of different arguments are summarised in Table 15.2, together with the results of experimental testing (ref. 29). For this preliminary survey, the minimum temperature at which methanation activity could be observed was used as a test parameter: subsequent experimental testing showed that this was a realistic yardstick for activity. For brevity, only Ni based catalysts are shown in Table 15.2: Ru based catalysts gave the same trends.

TABLE 15.2
Summary of the suggested primary components and of preliminary testing

Basis of approach	Catalysts suggested, together with minimum temperatures of methanation ^{o}C
(a) Large amounts of hydrogen required	Ni-Pd (132), TC (>300), WC (> 300) Ni boride (125)
(b) Enhanced spillover required	Ni-Pd (132), Ni-Ir (145), Ni-Ag (180)
(c) Formation and hydrogenation of M'O	See figures
(d) Removal of O from the surface	Ni-Rh (104), Ni-bismuth oxide (235), Ni-cobalt oxide (135), Ni-sodium sulphate (> 300), Ni-molybdate (125), Ni-molybdate/ urania (116), Ni-zirconia (145)
(e) Support should be Al_2O_3 or C	Ni-Pt/Al_2O_3 (118), Ni-Pt/C (180), Ni/Cr oxide Al_2O_3 (120), Ni/Cr oxide-SiC (126): Ni/Cr oxide-c (125): Ni/Cr oxide-zeolon 200H (125)
(f) Comparison	Ni (120)

Unless otherwise stated, all catalysts were supported on γ-Al_2O_3

The success or otherwise of these predictions can be seen from Table 15.2. Generally, catalysts based on the concept that large amounts of hydrogen were required did not perform particularly well (compare (a), (b) and (f), Table 15.2), but catalysts based on the concept that oxygen removal from the surface was desirable did perform well (see (c) and (d), Table 15.2).

III. DESIGN OF SECONDARY COMPONENTS

A limited study of the design of secondary components was carried out as based on the results summarised in Table 15.2. These show that the optimum methanation activity can be related with the hydrogenation of M'O, and that activity can be improved if O can be removed from the surface by diffusion into the bulk. As a result, multicomponent systems were tested involving (a) Ni, to facilitate dissociative adsorption (b) chromium oxide, platinum or molybdenum oxide, to facilitate hydrogenation (Figure 15.4) (c) Rh, to favour diffusion of O into the bulk and (d) urania, to facilitate displacement of water. Typical results include

Ni molybdate/Al_2O_3 (125^oC)
Ni molybdate-Pt/Al_2O_3 (125^oC) > Ni molybdate-urania/Al_2O_3 (116^oC)
Nickel chromate/Al_2O_3 (125^oC) > Ni/Al_2O_3 (120^oC)
Nickel chromate-Rh-Pt/Al_2O_3 (105^oC)

It is seen that inclusion of several components does, indeed, improve performance of the catalysts.

Other aspects of secondary design that were studied included the improvement of temperature stability of the catalysts and measurements of sulphur resistance. Although important in their own right, the reasons for investigating such factors are exterior to the logical framework of the design and are not considered here.

IV. COMMENTS

It is useful to consider both the 1974 and the 1978 designs together, since they give valuable pointers individually and collectively.

The survey of thermodynamic data carried out in 1974 was valuable, and there was recognition of the role of adsorption in the process. For reasons beyond their control, the designers did not consider the correct mechanism as established in 1978. Their proposed mechanism (mechanism b, above) seemed reasonable, but they failed to recognise the importance of geometric factors in controlling carbon formation. They were affected by the literature in considering catalysts that could accelerate this reaction and, as a result, considered only metallic catalysts. Amongst those suggested were catalysts known to be active (Ni and Ru).

The 1978 design is, in some respects, unconventional, since it rests more on the application of common sense to the proposed surface reaction rather than to the application of the design procedure. It can be seen that the procedure could well lead to the same conclusions (see above), but the fact of the matter is that it was not used. The basis of the design - the reaction on the surface - was certainly established by 1978, and the predictions benefited from this. The design led to the recognition of a new type of methanation catalyst and, as such, can be deemed a success. Further attention should be paid to the design of secondary components of the catalyst.

REFERENCES

1 Gas Making and Natural Gas, B.P. Trading Ltd., London (1972).
2 J.A. Cusumano, R.A. Dalla Betta and R.B. Levy, Catalysis in Coal Conversion, Academic Press, New York, 1978.
3 M. Greyson, in Catalysis, Vol. IV (Ed. P.A. Emmett), Reinhold, Princetown, 1956.
4 G.A. Mills and F.W. Steffgen, Catal. Rev. Sci. Eng., 8, 159 (1973).
5 V.M. Vlasenko and G.E. Yuzefovich, Russ. Chem. Revs., 38, 728 (1969).
6 M.A. Vannice, Catal. Rev. Sci. Eng., 14, 153 (1976).
7 R.B. Anderson, in "Catalysis" Vol. IV (Ed. P.A. Emmett), Reinhold, Princetown 1956.
8 G.C. Bond, Catalysis by Metals, Academic Press, New York, 1962.
9 D.W. McKee, J. Catal., 8, 240 (1967).
10 O.V. Krylov, Catalysis by Non-Metals, Academic Press, New York, 1970.
11 R.E. Cunningham and A.T. Gwathmey, Adv. Catal., 9, 25 (1957).
12 R.T.K. Baker, M.A. Barber, P.S. Harris, F.S. Feates and R.J. Waite, J. Catal. 26, 51 (1972).
13 J. Rostrup-Nielsen, Steam Reforming Catalysts, Teknisk Forslag (Copenhagen) 1975.
14 Catalyst Handbook, Wolfe Scientific Texts, London, 1970.
15 J.M. Thomas and W.J. Thomas, Introduction to the principles of heterogeneous catalysis, Academic Press, New York, 1967.
16 T.D. Ralston, W.P. Haynes, A.J. Forney and R.R. Schehl, U.S. Bureau of Mines Rep. Invest. R1, 7941 (1974).
17 M.A. Vannice, J. Catal., 37, 449, 462 (1975).
18 M.A. Vannice, J. Catal., 50, 228 (1977).
19 M. Araki and V. Ponec, J. Catal., 44, 439 (1976).
20 J.L. Bousquet, P.C. Gravelle and S. Teichner, Bull. Soc. Chim. Fr., 3693 (1972).
21 P.R. Wentrcek, B.J. Wood and H. Wise, J. Catal., 43, 363 (1976).
22 J.A. Rabo, P. Risch and M.C. Poutsma, J. Catal., 53, 295 (1978).
23 P.R. Wentrcek, J.G. McCarty, B.J. Wood and H. Wise, Amer. Chem. Soc. Div. Fuel Preprints, 21, 52 (1976).
24 G.C. Bond and B.D. Turnham, J. Catal., 45, 128 (1976).
25 Chemical Engineering, Sept. 12, 98 (1977).
26 R.W. Mitchell, L.W. Randolphi, P.C. Maybury, J. Chem. Soc.:Chem. Comm. 172 (1976).
27 P.A. Sermon and G.C. Bond, Catal. Revs. Sci. Eng., 8, 211 (1973).
28 M. Boudart, A.W. Aldog and M.A. Vannice, J. Catal., 18, 46 (1970).
29 G. Carlesen, E. Rockova and D.L. Trimm, Unpublished results.
30 R.A. Dalla Betta and M. Shelef, J. Catal., 49, 383 (1977).
31 G.K. Boreskov, Kinet. Catal., 8, 878 (1967).
32 Y. Moro-oka and A. Ozaki, J. Catal., 5, 116 (1966).
33 CRC Handbook of Chemistry and Physics, 58th Edition, Ed.(R.C. Weast), CRC Press West Palm Beach, Florida (1977).
34 D.A. Dowden, Catal. Revs. Sci. Eng., 5, 1 (1972).
35 R.W. Joyner and M.W. Roberts, J. Chem. Soc. Farad. I, 70, 1819 (1974).
36 B.A. Sexton and G.A. Somorjai, J. Catal., 46, 167 (1977).
37 A.W. Sleight and W.J. Linn, Anaals, N.Y. Acad. Sci., 272, 22 (1976).
38 D.L. Trimm, Catal. Revs. Sci. Eng., 16, 155 (1977).

CHAPTER 16

THE DESIGN OF A CATALYST FOR THE REDUCTION OF NITROGEN OXIDES TO NITROGEN

Having discussed various catalyst design exercises that were, in general, successful, it may be useful to consider a design which went badly wrong. It will be seen that the reactions considered were complex, and that the designers failed to follow the design procedure and went badly astray in the course of the exercise.

I. THE IDEA

The production of large quantities of nitrogen oxides is highly desirable in the context of the manufacture of nitric acid. However, the production of small quantities of the oxides (as, for example, in engine exhausts) is highly undesirable, largely because they are a significant source of air pollution. As a result, there has been considerable interest focused on the removal of nitrogen oxides.

The problem is far from simple. It is theoretically possible to decompose nitrogen oxides e.g.

$$N_2O \rightleftarrows N_2 + \tfrac{1}{2}O_2 \qquad (1)$$

but the reaction is slow at lower temperatures. As an alternative, nitrogen oxides can be reduced e.g.

$$N_2O + H_2 \rightleftarrows N_2 + H_2O \qquad (2)$$

$$N_2O + CO \rightleftarrows N_2 + CO_2 \qquad (3)$$

but care must be exercised that reduction to ammonia does not occur e.g.

$$N_2O + 4H_2 = 2NH_3 + H_2O \qquad (4)$$

If this does occur, then reoxidation of ammonia to nitrogen oxides is always possible

$$NH_3 + 2O_2 = N_2O + \tfrac{3}{2}H_2O \qquad (5)$$

In 1976, the development of an exhaust gas clean up catalyst had proceeded to the point where unburnt hydrocarbons and carbon monoxide could be reduced to insignificant levels. This involved catalytic oxidation over precious metals supported on a ceramic monolith. Then, as now, no completely satisfactory catalyst had been developed for the removal of nitrogen oxides, and the design team chose to tackle this problem. In many ways the design produced was of considerable interest, not least - in the present context - that the designers made good use of the literature, but carried out a poor design.

II. DESCRIPTION OF THE IDEA

The project was defined in terms of the production of nitrogen from nitrogen oxides in an exhaust gas environment. As a result, the designers looked just at the possibility of decomposing the oxides. Nitrogen dioxide starts to decompose to nitric oxide and oxygen at 150^oC, and decomposition is complete by 600^oC. The decomposition of nitric oxide is thermodynamically feasible, but the process - whether catalysed or not - is very slow (ref. 1). As a result, the team did not consider decomposition as a probable route, and turned their attention to the general reaction.

$$NO_x + \text{reducing agent} \rightleftarrows N_2 + \text{oxidised products}$$

In an exhaust environment, hydrogen, carbon monoxide or unburnt hydrocarbons are the most probable reducing agents, and the general reactions may be written

$$NO + H_2 = \tfrac{1}{2}N_2 + H_2O \qquad (6)$$

$$NO + CO + H_2 = \tfrac{1}{2}N_2 + CO_2 + H_2O \qquad (7)$$

$$4NO + CH_4 = 2N_2 + 2H_2O + CO_2 \qquad (8)$$

$$NO + CO = \tfrac{1}{2}N_2 + CO_2 \qquad (9)$$

$$NO + \tfrac{5}{2}H_2 = NH_3 + H_2O \qquad (10)$$

It was recognised that the overall reaction was far from simple, with the possibility existing of many side reactions e.g.:

$$\tfrac{1}{2}N_2 + \tfrac{3}{2}H_2 = NH_3 \qquad (11)$$

$$5CH_4 + 8NO + 2H_2O = 5CO_2 + 8NH_3 \qquad (12)$$

$$3H_2 + CO = CH_4 + H_2O \qquad (13)$$

$$NH_3 + CO = HCN + H_2O \qquad (14)$$

$$NO + \tfrac{5}{2}H_2 + CO = HCN + 2H_2O \qquad (15)$$

$$CO + H_2O = CO_2 + H_2 \qquad (16)$$

It was realised that the feasibility of many reactions would depend on the exhaust gas composition and on the thermodynamics of the reactions. As a first step, reactions with hydrogen were examined.

$$NO + H_2 = \tfrac{1}{2}N_2 + HO \qquad (6)$$

$$NO + \tfrac{5}{2}H_2 = NH_3 + H_2O \qquad (10)$$

$$\tfrac{1}{2}N_2 + \tfrac{3}{2}H_2 = NH_3 \qquad (11)$$

TABLE 16.1

Free energy changes for reactions between nitric oxide and hydrogen (kcal $mole^{-1}$) (ref. 2)

Temp. (K)	Reaction (6)	Reaction (10)	Reaction (11)
300	- 75.30	- 79.22	- 39.26
400	- 73.90	- 75.36	- 14.64
500	- 72.45	- 71.33	11.21
600	- 71.02	- 67.23	37.88
700	- 69.40	- 62.87	65.33
800	- 67.82	- 58.53	92.97
900	- 66.25	- 54.15	121.06
1000	- 64.60	- 49.69	149.15

Arguing that the decomposition of ammonia (reverse reaction 11) could be expected to eliminate one potential problem, the next step was to consider thermodynamic data for reaction (11) in detail (refs. 2,3). From known equilibrium constants, the concentration of ammonia expected to be present in a typical exhaust gas was calculated.

TABLE 16.2

The formation of ammonia (ref. 3)

Temp. (K)	Equilibrium Constant, K_p for reaction (11)	Concentration of ammonia (a) p.p.m
300	7.211×10^{2}	22,044
400	6.324	1,933
500	3.258×10^{-1}	99.6
600	4.188×10^{-2}	12.8
700	9.247×10^{-3}	2.8
800	2.994×10^{-3}	0.9
900	1.156×10^{-3}	0.35
1000	5.470×10^{-4}	0.17

(a) Calculated for a typical exhaust gas composition, made up of 75% N_2, 0.5% H_2 balance CO_2, H_2O etc.

Although this table shows that the concentration of ammonia should be negligible above ca. 650 K, experimental data shows that this is not so (ref. 4). As a result, it was realised that a more effective ammonia decomposition catalyst would be advantageous.

The designers then turned their attention to the feasibility of other reactions, and produced the table below (Table 16.3).

TABLE 16.3

Free energy data for typical exhaust gas reactions (kcal $mole^{-1}$) (ref. 2)

Temp. (K)	Free energy changes (kcal $mole^{-1}$) for reactions 8	9	12	13	14	15
300		- 82.18		- 33.89	10.82	- 68.40
400	-272.83	- 79.75	-489.07	- 28.60	10.78	- 64.59
500	-271.61	- 77.35	-479.69	- 23.06	10.70	- 60.62
600	-270.42	- 74.94	-470.60	- 17.15	10.59	- 56.56
700	-269.73	- 72.53	-463.81	- 11.49	10.46	- 52.40
800		- 70.13		- 5.56	10.31	- 48.23
900		- 67.74		0.42	10.14	- 43.99
1000				6.44	9.96	- 39.74

Reactions

$CH_4 + 4NO = 2N_2 + 2H_2O + CO_2$ (8)

$NO + CO = \frac{1}{2}N_2 + CO_2$ (9)

$5CH_2 + 8NO + 2H_2O = 5CO_2 + 8NH_3$ (12)

$3H_2 + CO = CH_4 + H_2O$ (13)

$NH_3 + CO = HCN + H_2O$ (14)

$NO + \frac{5}{2}H_2 + CO = HCN + 2H_2O$ (15)

It is seen that all reactions are feasible, but that reaction (14) is least likely to occur. Hydrogen cyanide can, however, be formed via reaction (15).

The designers then turned their attention to published information on catalytic processes involving nitrogen oxides. In one way, the information they retrieved was not organised well, as it is spread throughout the design. However, in keeping with the philosophy of following actual case studies, the present format follows their design.

The existence of some twenty catalysts for the reduction of nitric oxide was established from the literature, although no logical/universal approach to the mechanism of the reaction was discovered. As a result, initial arguments were based on an analogous reaction - the oxidation of carbon monoxide, hydrogen or hydrocarbons by oxygen. This involves the sequence (ref. 5):

$$O_2 \rightarrow O_2^- \rightarrow O^- \rightarrow O^{2-} \qquad (17)$$

from which it was suggested that a catalyst should be capable of giving an electron to nitric oxide, possibly accepting an electron from the other reactant/products and should be capable of rupturing the N-O bond.

While accepting that most catalysts had been identified on a trial and error basis, Ru was recognised as a catalyst that selectively reduced nitric oxide to nitrogen rather than to ammonia (refs. 1,6). The designers also recognised that ruthenium oxides, if formed, could volatilise (ref. 7), and they found evidence to suggest that a mixed catalytic-homogeneous reaction was possible. This could involve volatilised oxides or could involve atoms/free radicals produced on the surface and reacting in the gas phase (refs. 8,9).

Bearing in mind the volatilisation problems and obvious economic arguments, the team also searched the literature for evidence of catalytic activity among the semi-conducting oxides. These were chosen for their known ability to accept or donate electrons (ref. 10). It was found that the catalytic activity of oxides for the oxidation of hydrogen, carbon monoxide or hydrocarbons by oxygen had been related to the energy of binding of oxygen to the surface (ref. 5), with the activity falling off in the order

$$CO_3O_4 > MnO_2 > NiO > CuO > Cr_2O_3 > Fe_2O_3 > ZnO > V_2O_5 > TiO_2 > Sc_2O_3$$

It was suggested that oxidation by nitric oxide should show a similar activity pattern and, indeed, subsequent retrieval of information on oxidation by nitric oxide showed that this was so (ref. 11). There are, however, significant differences between oxidation by oxygen and oxidation by nitric oxide, which may be seen from consideration of the general reaction in terms of a redox model (ref.5)

Writing the reaction as

(a) Reducing agent + oxidised surface = products adsorbed on reduced surface
(b) Adsorbed products = gas phase products + reduced surface
(c) Reduced surface + oxidising agent = oxidised surface

It is known that the rate-limiting step is usually step (a) (or, at low temperatures, step (b)) when oxygen is the oxidant. For oxidising agents such as NO, the rate limiting step may well be step (c) and, if there is any residual oxygen in the exhaust stream, the selectivity of step (c) may well be so much in favour of reaction with oxygen that nitric oxide could pass over the catalyst unchanged. What is required, then, is a catalyst which will favour the dissociative adsorption of nitric oxides, to give either chemisorbed N atoms or N_2 in the gas phase. At the same time, the catalyst should favour, for example, the reaction between carbon monoxide and nitric oxide at the expense of reaction between carbon monoxide and oxygen.

Taking the last point first, the activity of oxides for the CO-NO reaction has been reported to be (ref. 1):

Fe_2O_3 > $CuCr_2O_4$ > Cu_2O > Cr_2O_3 > NiO > Pt > Co_3O_4 > Al_2O_3 (5% SiO_2) > MnO > V_2O_5

while the activity sequence for the CO-O_2 reaction is reported above.
In terms of the dissociative adsorption requirements the designers argued that the decomposition of nitrous oxide was a suitable comparison, in view of the fact that the initial step in the sequence involved scission of the N-O bond:

$$N_2O + e \longrightarrow N_2 + O^-_{ads} \qquad (18)$$

For the decomposition, the activity of the catalysts has been found to decrease in the order

CoO > Cu_2O > NiO > MnO_2 > CuO > Fe_2O_3 > ZnO > CeO_2 > TiO_2 > Cr_2O_3 > ThO_2 > ZrO_2 > V_2O_5 > HgO > Al_2O_3

Filing this information, the design team then moved on to consider reaction mechanism. They concentrated their attention on three reactions:

$$NO + H_2 = H_2O + \tfrac{1}{2}N_2 \qquad (6)$$

$$NO + CO = CO_2 + \tfrac{1}{2}N_2 \qquad (9)$$

$$NO + \tfrac{5}{2}H_2 = NH_3 + H_2O \qquad (10)$$

and began by considering the nature of the species involved, both in the gas phase and on the surface.

Recognising that the desired reaction involves, at some stage, the scission of a N-O bond, they began by looking at the nature of the NO molecule. The paramagnetic molecule : N$\dot{\overset{\cdot}{=}}$O : can exhibit a variety of possibilities on entering into a complex. Just as a free radical, it can

(a) form a covalent bond;
(b) Use the N lone pair of electrons to enter a co-ordination bond, filling an empty d-orbital;
(c) lose the anti-bonding lone electron, thereby forming a "nitrosium" ion, NO^+
(d) acquire an electron, to form an NO^- anion (ref. 12).

Removal of an electron is reasonably easy, but the resulting NO^+ ion should have a stronger N-O bond than NO itself.

Spectroscopic study of the adsorption of NO led Terenin et al. (ref. 12) to report that the covalent bonding of NO to cations is strong but that adsorption of ionic NO is very weak. Some reference was found to the exact nature of the adsorbed species on tungsten (ref. 13), where decomposition was suggested to involve the following species on the surface:

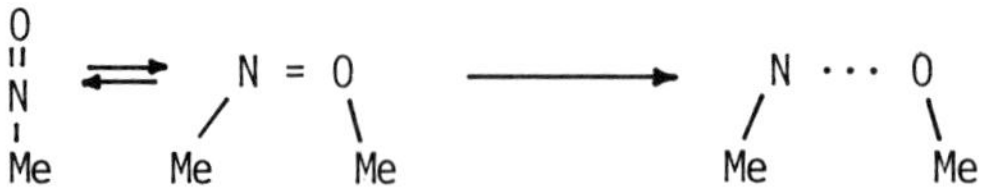

$$\underset{\substack{| \\ Me}}{\overset{O}{\underset{\|}{N}}} \rightleftarrows \underset{Me \quad\quad Me}{N = O} \longrightarrow \underset{Me \quad\quad Me}{N \cdots O} \tag{19}$$

The team then went on to consider the adsorption of carbon monoxide in the context of oxidation by oxygen, for which there is a considerable amount of information (refs. 11,14,15). Since the desorption of carbon dioxide is not rate limiting under conditions pertinent to an exhaust (refs. 1,14), no attention was paid to the adsorption of products. Adsorption and reaction of a carbonate ion or complex was favoured, following the suggestion originally proposed by Garner (ref. 15):

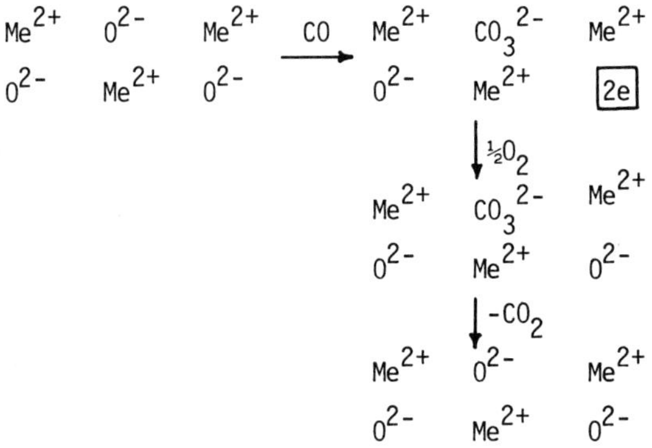

$$\begin{matrix} Me^{2+} & O^{2-} & Me^{2+} \\ O^{2-} & Me^{2+} & O^{2-} \end{matrix} \xrightarrow{CO} \begin{matrix} Me^{2+} & CO_3^{2-} & Me^{2+} \\ O^{2-} & Me^{2+} & \boxed{2e} \end{matrix} \xrightarrow{\frac{1}{2}O_2} \begin{matrix} Me^{2+} & CO_3^{2-} & Me^{2+} \\ O^{2-} & Me^{2+} & O^{2-} \end{matrix} \xrightarrow{-CO_2} \begin{matrix} Me^{2+} & O^{2-} & Me^{2+} \\ O^{2-} & Me^{2+} & O^{2-} \end{matrix} \tag{20}$$

or, in electron terms:

$$CO_g + 2O^{2-}_{ads} \longrightarrow CO^{2-}_{3ads} + 2e \tag{21}$$

$$\tfrac{1}{2}O_2 + 2e \rightleftarrows O^{2-}_{ads} \tag{22}$$

$$CO^{2-}_{3ads} \longrightarrow CO_{2g} + O^{2-}_{ads} \tag{23}$$

or (ref. 11)

$$CO^{2-}_{3ads} + CO_g \longrightarrow 2CO_2 + 2e \tag{24}$$

The structure of the carbonate ion most commonly quoted is shown below (I), although the difficulties in achieving this structure from that formed on original adsorption of CO(presumed to be as II below) was recognised (ref. 14):

The designers did not consider possible forms of adsorption of hydrogen (see, for example, Chapter 11).

What was obvious at this stage in the design was that the design team had not distinguished in their minds the possibility (known from their reading) that the desired reaction could occur on both metals and metal oxides. In fact, to some degree, they had lost the logical framework of the design, by which the distinction and separate development of different possibilities is encouraged. We shall see that they retrieved this position to some extent, but the distinction is never completely clear.

Having considered possible adsorbed species, alternative reaction mechanisms were discussed. Despite the knowledge that the catalytic decomposition of nitrogen oxides is slow (ref. 1), the importance of the N-O bond scission to any overall process led to renewed consideration of the decomposition reaction. The mechanisms (and kinetics) proposed may be considered in turn. Bearing in mind the environment of an exhaust, it is necessary to consider reactions important over a fairly wide temperature range.

a) Decomposition of NO (direct)

The reactions may be considered in chemical or electron terms (ref. 16):

$$NO_{gas} \rightleftarrows NO_{ads} \qquad (25)$$

$$2NO_{ads} \rightarrow N_{2gas} + 2O^{-}_{ads} \qquad (26)$$

$$2O_{ads} \rightleftarrows O_{2gas} \qquad (27)$$

or

$$NO_{gas} + e_1e_2 \rightleftarrows NO^{-}_{(1)} + e_2 \qquad (28)$$

$$NO_{gas} + e_2 \rightleftarrows NO^{-}_{(2)} \qquad (29)$$

$$NO^{-}_{(1)} + NO^{-}_{(2)} \rightarrow N_2 + O^{-}_{(1)}\,O^{-}_{(2)} \qquad (30)$$

$$O^{-}_{(1)}O^{-}_{(2)} = O_{2gas} + e_1e_2 \qquad (31)$$

b) Decomposition of NO via N_2O

In addition to the direct decomposition of nitric oxide, nitrous oxide could also be formed before complete dissociation occurs:

$$2NO_g + e \rightleftarrows N_2O_g + O^-_{ads} \quad (32)$$

$$N_2O_g + e \rightleftarrows N_2O^-_{ads} \quad (33)$$

$$N_2O^-_{ads} \rightleftarrows N_{2g} + O^-_{ads} \quad (34)$$

$$O^-_{ads} \rightleftarrows \tfrac{1}{2}O_2 + e \quad (31)$$

$$O^-_{ads} + N_2O_g \rightarrow N_2 + O_2 + e \quad (35)$$

There is some evidence that this reaction path could be important at low temperatures (refs. 16,17), since nitrous oxide has been identified among the products of nitric oxide decomposition. Once formed, the decomposition of nitrous oxide has been well established (ref. 18), although reaction (35) is unlikely to be important under the conditions pertinent to an exhaust.

c) Decomposition of NO via ionic intermediates

It could also be suggested that the nitrosium ion plays a role in the decomposition

$$NO_g \rightleftarrows NO_{ads} \quad (25)$$

$$NO_{ads} - e \rightleftarrows NO^+_{ads} \quad (36)$$

$$\tfrac{1}{2}O_{2g} + e \rightleftarrows O^-_{ads} \quad (31)$$

$$NO^+_{ads} + O^-_{ads} \rightarrow \tfrac{1}{2}N_2 + O_2 \quad (37)$$

It is difficult to assess the probability of this suggestion. The strength of the N-O bond in NO^+ and the weakness of the NO^+-catalyst bond (ref. 12) argues against this mechanism, but the proposed interaction between oppositely charged adsorbed species is a reaction step which has been well established for many reactions (ref. 19).

d) Reaction between carbon monoxide and nitric oxide

Arguing on the basis of the adsorption of carbon monoxide as a carbonate species (see above), the designers moved to consider a reaction sequence that had been proposed for perovskite catalysts (ref. 20). These are oxides of general formula ABO_3 where A is La, Pb, Sr, K or B and B is, for example, Ru. In essence they have been found to act as if Ru metal is well distributed in a matrix of oxide (see Chapter 4).

The catalysts are of particular interest in that they offer a selective route to the production of nitrogen (rather than ammonia) even in the presence of hydrogen.

At temperatures below $300^{o}C$, the consumption of hydrogen is small, and the predominant reaction involves

$$M\text{-}O\text{-}M + CO = M\text{-}\square\text{-}M + CO_2 \qquad (38)$$

where $\square$ is an oxygen vacancy. Some $50^{o}C$ higher, hydrogen also reduces the lattice.

Nitric oxide can react with the reduced lattice in two ways, either to replace oxygen in the lattice or to adsorb dissociatively on the lattice

$$NO_3 + M\text{-}\square\text{-}M \nearrow M\text{-}O\text{-}M + N_{ads} \qquad (39)$$

$$NO_3 + M\text{-}\square\text{-}M \searrow M\text{-}\square\text{-}M + N_{ads} + O_{ads} \qquad (40)$$

For the second case, NO and CO can compete for the same sites, and the fate of N_{ads} is dependent on the relative importance of the reactions below (ref. 20):

$$N_{ads} + NO_{ads} \rightleftarrows N_2O \qquad (41)$$

$$N_{ads} + CO_{ads} \rightleftarrows NCO_{ads} \qquad (42)$$

$$N_{ads} + 2H_{ads} \rightleftarrows NH_{2ads} \qquad (43)$$

$$2N_{ads} \rightarrow N_2 \qquad (44)$$

$$NO_{ads} + NH_{2ads} \rightarrow N_2 + H_2O \qquad (45)$$

$$NH_{2ads} + H_{ads} \rightleftarrows NH_3 \qquad (46)$$

The formation of an isocyanate intermediate has been established on a variety of noble metal catalysts and on CuO catalysts.

e) <u>The formation of ammonia</u>

In considering the desired reactions, it is also necessary to postulate reaction mechanisms leading to ammonia, the unwanted product. In addition to the direct formation discussed above

$$N_{ads} + 3H_{ads} \rightleftarrows NH_3 \qquad (47)$$

ammonia may also be formed by the hydrolysis of an isocyanate species (ref. 20)

$$NCO_{ads} + H_2O = NH_3 + CO_2 \qquad (48)$$

It has been suggested (ref. 21) that formation of ammonia via reaction (47) is unlikely under exhaust conditions, as a result of the requirement for correct adsorption on four adjacent atoms. This argument should not be taken too seriously, as it ignores the possibility of mobile adsorption and the fact that the synthesis of ammonia is believed to involve reaction (47), albeit when the concentrations of nitrogen and hydrogen are higher than in an exhaust.

At this point the design team stopped their description of the idea although, as we shall see, they return to the concept at a later stage. It is interesting to note their mistakes, some of which were fairly fundamental.

Perhaps most important was the minimal attention paid to the possibility of reaction with hydrogen. This reducing agent is known to be present in an exhaust gas (both as free hydrogen and chemically bound in unburnt hydrocarbons), but the possible reactions of the gas were not dealt with in sufficient detail.

Secondly the designers made no attempt to describe the reaction on the surface. Electronic requirements of the reactions were considered in a few cases, but little attention was paid to the form of possible adsorbed species and to what effects this could have on the proposed mechanism. This was partially corrected in the next stage of the design.

III. CO-ORDINATION AND GEOMETRIC CONSIDERATIONS AND DESIGN

Having described, however inadequately, the reactions which could be expected to occur, the team then returned to the activity pattern reported for the CO-NO reaction (ref. 1):

$$Fe_2O_3 > CuCr_2O_4 > Cu_2O > Cr_2O_3 > N{:}O > Pt > Co_3O_4 > Al_2O_3\ (5\%\ SiO_2) > MnO > V_2O_5$$

They recognised that a catalytic reaction involved adsorption of reactants, which could be likened to formation of a complex. In these terms, adsorption of one or two reactants involved either an increase in the co-ordination number of the metal ion involved in the complex, or displacement of a ligand by the adsorbing species. Of the list above, Cu(I) is two co-ordinate and Cu(II) is four co-ordinate while Ni(II) is four co-ordinate and Ni(III) is six co-ordinate. As a result, these two metal centres _could_ adsorb by increasing their co-ordination number.

On the other hand, NO can form strong bonds with a metal by donating up to three electrons (see above). As a result, it is capable of displacing one or more ligands attached to a metal centre: thus, for example, NO can readily displace a carbonyl and a halide ligand in the presence of a halogen acceptor (ref. 22). In view of the potential ease of this reaction, it is not surprising that complexes of NO with V, Fe, Cr and Co ions have been reported (ref. 23).

Having proceeded to this point, the designers recognised that the geometric requirements of the system demanded, amongst other things, that the structure of the catalyst should allow the facile formation of nitrogen. However, no attempt was made to match bond distances and, indeed, this would have been very difficult, bearing in mind that they had not proposed any surface structures for reacting adsorbed species. A mistake early in the design was leading to promulgation of the error throughout the treatment.

An interesting and instructive feature of the design was now apparent from the report. It is common practice to allocate different tasks to different members of the team and, in this case, it was obvious that half of the team had adopted a second approach which, although related, was distinct from the considerations summarised above. Given the task of investigating the reactions on metallic catalysts, they chose to re-start the design.

IV. CATALYSIS BY METALS

a) Description of the idea.

The overall desired reactions were, once again, summarised as below:

$$2NO + H_2 = N_2 + H_2O \qquad (6)$$

$$2NO + 2CO = N_2 + 2CO_2 \qquad (9)$$

$$N_2O + H_2 = H_2O + N_2 \qquad (49)$$

$$N_2O + CO = CO_2 + N_2 \qquad (50)$$

$$NO_2 + H_2 = H_2O + NO \qquad (51)$$

$$NO_2 + CO = CO_2 + NO \qquad (52)$$

The possibility of producing ammonia was noted but not expressed as an equation.

b) Modes of adsorption.

From the above reactions, the adsorption of hydrogen, carbon monoxide and nitrogen oxides were seen to be important. On metals, it was suggested that dissociative adsorption of hydrogen was the only significant mode of adsorption, although several possible adsorbed species were suggested for carbon monoxide:

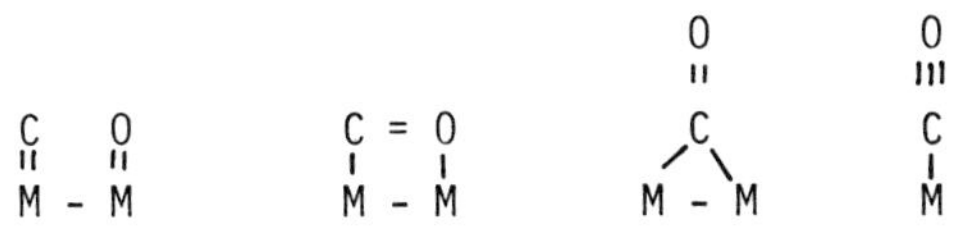

```
                                 O        O
                                 ||       |||
C    O         C = O             C        C
||   ||        |   |            / \       |
M  - M         M - M           M - M      M
```

Although the various possibilities of NO adsorption discussed above were recognised, the predominant form was suggested to involve pairing of the odd electron with unpaired d-electrons from a metal: this was expected to lead to weakening of the N-O bond.

c) Reaction mechanisms.

Several mechanisms for the reduction of nitrogen oxides have been proposed (ref. 24).

(i) Reduction of NO_2 by hydrogen

Reduction was suggested to occur either via molecular hydrogen or atomic hydrogen

$$H_2 + NO_2 = H + HNO_2 \quad (53)$$

$$H + NO_2 = NO + OH \quad (54)$$

$$OH + H_2 = H_2O + H \quad (55)$$

$$OH + NO_2 + M = HNO_3 + M \quad (56)$$

$$OH + NO + M = HNO_2 + M \quad (57)$$

$$H + NO_2 = OH + NO \quad (58)$$

$$2OH = H_2O + O \quad (59)$$

$$OH + O = O_2 + H \quad (60)$$

It is interesting to note that no attempt was made to relate these reactions to a catalyst surface and, indeed, they were based on similar reactions that have been suggested to occur in the gas phase.

(ii) Reduction of NO by hydrogen

In contrast, this mechanism was related to the surface:

$$H_2 = 2H_{ads} \quad (61)$$

$$NO = NO_{ads} \quad (25)$$

$$NO_{ads} + H_{ads} = HNO \quad (62)$$

$$HNO_{ads} + H_{ads} = H_2O + N_{ads} \quad (63)$$

$$2N_{ads} = N_2 \quad (44)$$

$$N_{ads} + 3H_{ads} = NH_3 \quad (47)$$

(iii) Reduction of NO by CO

This mechanism was considered in the absence and presence of possible side reactions involving oxygen

$$CO = CO_{ads} \tag{64}$$

$$NO = NO_{ads} \tag{25}$$

$$CO_{ads} + NO_{ads} = CO_2 + N_{ads} \tag{65}$$

$$2N_{ads} = N_2 \tag{44}$$

$$O_2 = 2O_{ads} \tag{27}$$

$$O_{ads} + CO = CO_2 \tag{66}$$

The designers recognised that CO, NO and O_2 could compete for the same sites on the surface.

(iv) Reduction of N_2O by hydrogen

Reaction was envisaged between atomic hydrogen and nitrous oxide in the gas phase or adsorbed:

$$H_2 = 2H_{ads} \tag{61}$$

$$H_{ads} + N_2O = N_2 + OH_{ads} \tag{67}$$

$$H_{ads} + N_2O_{ads} = N_2 + OH_{ads} \tag{68}$$

$$OH_{ads} + H_{ads} = H_2O \tag{69}$$

Again it is interesting to note that a detailed description of the reaction on the surface was not presented, nor was it made clear exactly what were the requirements of the proposed reactions. Some attempt was made to retrieve this position at a later stage in the design.

d) Design of primary components

The initial approach was based on the electronic requirements of reaction, as loosely defined above. It was recognised that an active metal should have vacancies in the d shell and/or unpaired d electrons in order for bonding with NO to occur. This allowed the elimination of Cu, Ag, Au, Zn, Cd and Hg: the fact that complexes of Cu and NO have been reported (ref. 22) was ignored.

It was then argued that a very strong heat of adsorption of a reactant on a metal would lead to poisoning and, as a result, La, Ti, Zr, Hf, Mo and W (strong adsorption of carbon monoxide) as well as V, Nb, Ta and Cr (strong adsorption of hydrogen) were excluded from consideration. The products of the desired reactions (N_2, CO_2 and H_2O) were considered to be adsorbed only weakly, although NH_3 may be more strongly bonded to the surface.

Of the remaining metals, Th, Pa, U and Tc were eliminated since they were radioactive, and Fe and Re were eliminated because they favoured the production of ammonia. This left the following metals as possible catalysts

Mn, Ru, Os, Co, Rh, Ir, Pd and Pt

The designers moved on to consider geometric factors in the desired reactions, based on a N-O bond length of 1.15 Å and a N-N bond length of 1.097 Å. Ru and Os adopt a close packed hexagonal crystal lattice, while Co, Rh, Ir, Pd and Pt favour face centred cubic geometry. As a result of the short bond lengths of the reactant and product, Ru or Os were considered as desirable catalysts.

This conclusion was supported by a literature report that Pt and Pd catalysts may be poisoned for the CO-NO reaction as a result of strong adsorption of CO, but Ru and Os catalysts were not (ref. 4). It was recognised that Ru could form volatile oxides in the presence of oxygen (ref. 7).

V. CONCLUSIONS

Before attempting to tie together the two parts of the design, the team did attempt to consider mass and heat transfer in the systems. The treatment was rudimentary and no conclusions emerged.

In presenting the conclusions, further mistakes were made. In particular, new information was introduced which was pertinent to the design but which had not been previously considered. Following the report, the points raised can be listed:

(a) Treating the CO-NO reaction as a specific example of the oxidation of carbon monoxide, metal oxides with small metal-oxygen bond distances are known to be active catalysts (ref. 10). This information was available during the design and had not been considered. MnO_2, Ni_2O_3 and Co_3O_4 were suggested as active catalysts.

(b) The strength of oxygen bonding to the surfaces of oxides is lowest for Co_3O_4, followed by MnO_2, NiO and CuO (ref. 5). This is new information, and no attempt was made to relate it to the design. It was suggested (but no reason was given) that Co_3O_4 should be an active catalyst for the oxidation of carbon monoxide and, by analogy, for the CO-NO reaction.

(c) Work on the catalytic reduction of NO was discovered and reported (ref. 10). The most active catalyst was reported to be Co_3O_4, although there were indications that CoO was more stable with time.

(d) Ru or Os were expected to be the most active metallic catalysts, although Co, Rh, Ir, Pd, Pt and Mn may show some activity.

VI. COMMENTS

Even though this design is a superb example of how not to tackle a problem, it does have some good points. In particular, the use of published literature was much better than in the average design.

However, in the approach to the problem, the exercise was a disaster. It has been clearly stated that the design procedure provides a convenient and logical framework of thought. This is particularly important where the overall reaction network is complex. The team showed no evidence of a logical approach to the problem, they arranged and used the literature in a piecemeal fashion, they lacked co-ordination through the design and they produced, in essence, a badly organised series of comments on the literature. As an example of how not to tackle a design, it is hard to beat.

In fact, as it turns out, cobalt oxides are fairly good additives for an exhaust catalyst. The design failed to identify vanadia based catalysts, which can be used to promote the reduction of nitrogen oxides (ref. 25).

REFERENCES

1 M. Shelef, K. Otto and H.S. Gandhi, J. Catal. 12, 361 (1968).
2 F.D. Rossini, K.S. Pitzel, R.J. Arnett, R.M. Brann and G.C. Pimenkel, Selected Values of Physical and Thermodynamic Properties, Carnegie Press, 1953.
3 A. Nielsen, An investigation on promoted iron catalysts for the synthesis of ammonia, ul Gjellerups Forlag, Copenhagen, 1968.
4 M. Shelef and H.S. Gandhi, Ind. Eng. Chem. Prod. Res. Dev., 11, 2 (1972).
5 G.K. Boreskov, Kin. i Kat., 14, 1 (1973).
6 M. Shelef, Catal. Rev. Sci. Eng., 11, 1 (1975).
7 A.S. Darling, Internat. Metall. Rev., 91 (1973).
8 M.A. Bogoyartenskaya and A.A. Kovalskii, Zh. fiz. Khim., 20, 1325 (1948)
9 R.D. Holton and D.L. Trimm, Proc. V Internat. Congr. on Catal. (London), The Chemical Society, London, p.79 (1976).
10 O.V. Krylov, Catalysis by Non-metals, Academic Press, London, 1970.
11 J.M. Thomas and W.J. Thomas, Introduction to the Principles of Heterogeneous Catalysis, Academic Press, London, 1967.
12 A. Terenin and L. Roev, Actes due deuxieme Congress, International de Catalyse Paris. Technip. Paris, p.111 (1960).
13 J.T. Yates, Jr. and T.G. Madey, J. Chem. Phys., 45, 1623 (1966).
14 F.S. Stone, Adv. in Catal., 13, 5 (1962).
15 B.J. Garner, (Ed.), Chemistry of the Solid State, Butterworths Scientific Publications, London, 1955.
16 E.R.S. Winter, J. Catal., 22, 158 (1971).
17 E.R.S. Winter, J. Catal., 34, 431 (1974).
18 R.A. Baker and R.C. Doerr, Ind. Eng. Chem. Proc. Des. Dev., 4, 188 (1965).
19 J.E. Germain, The Catalytic Conversion of Hydrocarbons, Academic Press, New York, 1969.
20 D.W. Johnson, P.K. Gallagher, G.K. Wertheim and E.M. Vogel, J. Catal., 48, 87 (1977).
21 E.R.S. Winter, J. Chem. Soc. (A) (London), 2889 (1968).
22 J.P. Candlin, E.A. Taylor and F.N. Thompson, Reactions of Transition Metal Complexes, Elsevier, Amsterdam, 1968.
23 F.A. Cotton and G. Wilkinson, Advanced Inorganic Chemistry, Interscience, New York, 1972.
24 C.H. Bamford and C.F.H. Tipper, Comprehensive Chemical Kinetics, Vol. 6, Elsevier, Amsterdam, 1972.
25 M. Tokagi, T. Kawai, M. Soma, T. Onishi and K. Tamaru, J. Phys. Chim., 80, 430 (1976).

AUTHOR INDEX

SUBJECT INDEX